Adsorption Processes for Water Treatment and Purification

AF752592

Adrián Bonilla-Petriciolet
Didilia Ileana Mendoza-Castillo
Hilda Elizabeth Reynel-Ávila
Editors

Adsorption Processes for Water Treatment and Purification

Springer

Editors
Adrián Bonilla-Petriciolet
Instituto Tecnológico de Aguascalientes
Aguascalientes, México

Didilia Ileana Mendoza-Castillo
Instituto Tecnológico de Aguascalientes
Aguascalientes, México

Hilda Elizabeth Reynel-Ávila
Instituto Tecnológico de Aguascalientes
Aguascalientes, México

ISBN 978-3-319-58135-4 ISBN 978-3-319-58136-1 (eBook)
DOI 10.1007/978-3-319-58136-1

Library of Congress Control Number: 2017943969

© Springer International Publishing AG 2017
This work is subject to copyright. All rights are reserved by the Publisher, whether the whole or part of the material is concerned, specifically the rights of translation, reprinting, reuse of illustrations, recitation, broadcasting, reproduction on microfilms or in any other physical way, and transmission or information storage and retrieval, electronic adaptation, computer software, or by similar or dissimilar methodology now known or hereafter developed.
The use of general descriptive names, registered names, trademarks, service marks, etc. in this publication does not imply, even in the absence of a specific statement, that such names are exempt from the relevant protective laws and regulations and therefore free for general use.
The publisher, the authors and the editors are safe to assume that the advice and information in this book are believed to be true and accurate at the date of publication. Neither the publisher nor the authors or the editors give a warranty, express or implied, with respect to the material contained herein or for any errors or omissions that may have been made. The publisher remains neutral with regard to jurisdictional claims in published maps and institutional affiliations.

Printed on acid-free paper

This Springer imprint is published by Springer Nature
The registered company is Springer International Publishing AG
The registered company address is: Gewerbestrasse 11, 6330 Cham, Switzerland

Preface

Adsorption is a competitive technology for the treatment and purification of wastewater, groundwater, and industrial effluents. This separation process may offer a satisfactory cost-effectiveness tradeoff for water pollution control. Overall, the adsorption process engineering implies several factors that should be analyzed, studied, and optimized to develop economic and technically feasible water treatment strategies. The complex physicochemical nature of real-life multicomponent systems imposes new challenges for water pollution control using adsorption systems. Therefore, the adsorption process intensification is fundamental to face water pollution caused by anthropogenic and geogenic compounds.

This book contains eight chapters that cover the state of the art of relevant topics of adsorption process and its application for water sanitation. In particular, this book describes and discusses relevant themes involved in the modeling and design of adsorption systems and the synthesis and application of alternative adsorbents for water pollution control caused by different aquatic pollutants. These key aspects are important to develop effective and low-cost water treatment technologies. This book was motivated by the desire that we and our colleagues have had to present the developments and applications in this area.

Editors are grateful to all authors and reviewers for their contribution to this book. We also acknowledge the support of Susan Westendorf and the staff of Springer Nature Editorial for their assistance during the preparation of this book.

Aguascalientes, México

Adrián Bonilla-Petriciolet
Didilia Ileana Mendoza-Castillo
Hilda Elizabeth Reynel-Ávila

Contents

Contributors

Pietro Altimari Department of Chemistry, Sapienza University of Rome, Rome, Italy

Abolfazl Bayrami Department of Biology, Faculty of Basic Sciences, University of Mohaghegh Ardabili, Ardabil, Iran

Adrián Bonilla-Petriciolet Instituto Tecnológico de Aguascalientes, Aguascalientes, México

Almudena B. Botet-Jiménez Departamento de Química Orgánica e Inorgánica, Universidad de Extremadura, Badajoz, Spain

Tito Roberto Sant'Anna Cadaval Jr. Industrial Technology Laboratory, School of Chemistry and Food, Federal University of Rio Grande, Rio Grande, RS, Brazil

Fabrizio Di Caprio Department of Chemistry, Sapienza University of Rome, Rome, Italy

Wan Mohd Ashri Wan Daud Department of Chemical Engineering, Faculty of Engineering, University of Malaya, Kuala Lumpur, Malaysia

Guilherme Luiz Dotto Chemical Engineering Department, Federal University of Santa Maria, Santa Maria, RS, Brazil

Carlos Javier Durán-Valle Departamento de Química Orgánica e Inorgánica, Universidad de Extremadura, Badajoz, Spain

Instituto Universitario de Investigación del Agua, Cambio Climático y Sostenibilidad, Universidad de Extremadura, Badajoz, Spain

Gordon McKay Division of Sustainability, College of Science and Engineering, Hamad Bin Khalifa University, Doha, Qatar

Didilia Ileana Mendoza-Castillo Instituto Tecnológico de Aguascalientes, Aguascalientes, México

Raúl Ocampo-Pérez Center of Research and Postgraduate Studies, Faculty of Chemical Science, Autonomous University of San Luis Potosí, San Luis Potosí, SLP, Mexico

Delia Omenat-Morán Departamento de Química Orgánica e Inorgánica, Universidad de Extremadura, Badajoz, Spain

Francesca Pagnanelli Department of Chemistry, Sapienza University of Rome, Rome, Italy

Jeferson Steffanello Piccin Food Engineering Department, Passo Fundo University, UPF, Passo Fundo, RS, Brazil

Luiz Antonio Almeida de Pinto Industrial Technology Laboratory, School of Chemistry and Food, Federal University of Rio Grande, Rio Grande, RS, Brazil

Shima Rahim Pouran Department of Chemical Engineering, Faculty of Engineering, University of Malaya, Kuala Lumpur, Malaysia

Research Laboratory of Advanced Water and Wastewater Treatment Processes, Department of Applied Chemistry, Faculty of Chemistry, University of Tabriz, Tabriz, Iran

Abdul Aziz Abdul Raman Department of Chemical Engineering, Faculty of Engineering, University of Malaya, Kuala Lumpur, Malaysia

Hilda Elizabeth Reynel-Ávila Instituto Tecnológico de Aguascalientes, Aguascalientes, México

José Rivera-Utrilla Inorganic Chemistry Department, University of Granada, Granada, Spain

Manuel Sánchez-Polo Inorganic Chemistry Department, University of Granada, Granada, Spain

Nina Paula Gonçalves Salau Chemical Engineering Department, Federal University of Santa Maria, Santa Maria, RS, Brazil

Mohammad Saleh Shafeeyan School of Chemical Engineering, College of Engineering, University of Tehran, Tehran, Iran

Meng Xu Department of Civil and Environmental Engineering, Hong Kong University of Science and Technology, Kowloon, Hong Kong

Chapter 1
Introduction

Adrián Bonilla-Petriciolet, Didilia Ileana Mendoza-Castillo, and Hilda Elizabeth Reynel-Ávila

Abstract This chapter covers fundamental aspects of adsorption process engineering. In particular, the importance of adsorption processes for water treatment is discussed and analyzed. Environmental impact of some key aquatic pollutants is also reviewed. Opportunity areas for adsorption process intensification are also highlighted in this chapter including a brief overview of the content of this book.

Keywords Adsorption • Water sanitation • Process intensification • Aquatic pollutants

Contents

A. Bonilla-Petriciolet (✉)
Instituto Tecnológico de Aguascalientes, Av. Lopez Mateos 1801, C.P. 20256, Aguascalientes, Mexico
e-mail: petriciolet@hotmail.com

D.I. Mendoza-Castillo • H.E. Reynel-Ávila
Instituto Tecnológico de Aguascalientes, Aguascalientes, Mexico
e-mail: didi_men@hotmail.com; helizabeth_00@hotmail.com

© Springer International Publishing AG 2017

A. Bonilla-Petriciolet et al. (eds.), *Adsorption Processes for Water Treatment and Purification*, DOI 10.1007/978-3-319-58136-1_1

1.1 Adsorption: A Cost-Effective Technology for Water Treatment

Water pollution control is a major environmental concern worldwide. For example, it has been estimated that one third of world population can be affected by the lack of safe drinking water (Schwarzenbach et al. 2010). Pollution of water resources is caused by synthetic and natural chemicals that are released from a variety of anthropogenic and natural sources including the geological composition of aquifers. Different technologies for water treatment and purification have been extensively discussed in the literature where the design and operation of affordable methods can be still considered as a challenge. Herein, it is convenient to highlight that all treatment methods have their own technical and economic boundaries for real-life applications.

In particular, the adsorption process has been recognized as a viable technology for water sanitation. Adsorption processes are widely used for the treatment of wastewaters, groundwater, and industrial effluents including the production of drinking water. This treatment method may offer several advantages for water purification because it can be operated at different scenarios besides its easy use, flexibility, versatile design, low-energy requirements, and cost-effectiveness trade-off. Overall, the economic and technical feasibility of adsorption processes depends on several factors including the adsorbent type, fluid properties and pollutants to be removed, operating conditions, process configuration, regeneration, and waste disposal.

Batch and continuous adsorption systems can be employed for water treatment, and they offer different capabilities. Batch reactors are useful to determine adsorption rates, maximum adsorption capacities, and thermodynamic parameters including the analysis of the adsorbate(s)-adsorbent interactions. On the other hand, packed-bed columns are suitable for water purification in large-scale applications where the treatment of significant volumes of fluids can be performed in short operating times. Adsorption tests in packed columns are required to calculate relevant parameters for scale-up such as the breakthrough and saturation times, the bed adsorption capacity, and mass transfer parameters. This process configuration also allows to determine the maximum performance of the adsorbent and to identify the best dynamic operating scenario. Note that the operational conditions of dynamic adsorption systems imply residence times lower than the equilibrium time, and, consequently, the mass transfer resistances play a major role in the removal of pollutants. Therefore, the removal effectiveness of continuous adsorption systems is usually lower than that obtained for batch processes.

Activated carbon is the universal adsorbent for liquid phase and still prevails as the main commercial product for water pollution control (Rivera-Utrilla et al. 2011; Bhatnagar et al. 2013). However, a variety of alternative adsorbents has been synthesized and proposed for the adsorption of aquatic pollutants. For illustration, Table 1.1 shows a survey of different materials that can be employed in the removal of anthropogenic and geogenic pollutants in the context of environmental water

Table 1.1 Survey of materials used as adsorbents for water treatment and purification

Adsorbent	Heavy metals	Metalloids	Halogens	Dyes	Pharmaceuticals	Toxins	Radioisotopes	Organic pollutants
Alumina	√	√	√	√	√	√		
Activated carbon	√	√	√	√	√	√	√	√
Biological materials	√	√	√	√				√
Bone char	√	√	√	√	√	√	√	√
Clays	√	√	√	√	√	√		√
Composites	√	√	√	√	√		√	√
Graphene-based adsorbents	√	√	√	√	√	√	√	√
Low cost adsorbents[a]	√	√	√	√	√		√	√
Nano adsorbents	√	√	√	√	√	√	√	√
Polymeric materials	√	√	√	√	√	√	√	√

[a]Biomasses, agricultural, and industrial wastes and by-products

protection. They include aluminas (Kasprzyk-Hordern 2004), zeolites (Koshy and Singh 2016), clays (Vinati et al. 2015), and novel adsorbents such as nanomaterials (Santhosh et al. 2016), graphene-based adsorbents (Peng et al. 2017), magnetic materials (Mehta et al. 2015), metal-organic frameworks (Kumar et al. 2017), and others. Additionally, some studies have suggested the application of low-cost adsorbents that comprise agricultural and industrial wastes and by-products (Bhatnagar and Sillanpaa 2010; Adegoke and Bello 2015; De Gisi et al. 2016; Ahmed and Ahmaruzzaman 2016) and biomasses (Bhatnagar et al. 2015).

To date, the vast amount of studies performed on the adsorption of aquatic pollutants has mainly focused on the analysis and understanding of single solutions (i.e., one adsorbate in solution). However, the multicomponent adsorption is relevant for the design, optimization, and operation of real-life systems for purification of wastewaters, industrial effluents, and groundwater. The simultaneous adsorption of several adsorbates may imply synergic, antagonistic, or non-interaction effects depending on the adsorbent, the number and type of adsorbates (i.e., pollutants), and their concentrations, besides the fluid properties such as temperature and pH. The presence of several pollutants in the same solution may significantly affect the adsorbent performance. For instance, the adsorption of heavy metal ions in multimetallic systems is affected by the properties and concentrations of the co-ions (Choy and McKay 2005), while some dyes may favor the adsorption of metallic species in mixtures metal + dye (Hernandez-Eudave et al. 2016). The complex physicochemical nature of real-life multicomponent systems imposes new challenges for water treatment technologies including adsorption processes.

Although adsorption is a consolidated technology, it is clear that there are opportunity areas to improve its performance for facing current pollution problems of groundwater and wastewater including the treatment of industrial effluents. With this in mind, this book covers scientific advances related to the intensification of adsorption processes for water treatment and purification. Chapters contained in this book focus on the discussion and analysis of current topics from adsorption research and gaps for further study that can contribute to the consolidation of this separation process as a robust, economically feasible and environmentally friendly method for water sanitation.

1.2 Priority Pollutants in Water Purification

There is a great variety of water pollutants with a wide range of physicochemical properties and toxicological profiles. They include anthropogenic chemicals, geogenic pollutants, and persistent compounds that imply different challenges for adsorption processes. The most relevant aquatic pollutants include heavy metal ions, metalloids, pesticides, biocides, pharmaceuticals, and other emerging inorganic and organic compounds (Schwarzenbach et al. 2010). These pollutants can be toxic for human beings and biota at the nanogram to milligram per liter level. In some cases, they are considered priority in the context of water pollution control due to their persistence in the environment, bioaccumulation, and toxicological profile. With illustrative purposes, key aquatic pollutants are discussed in the next subsections.

1.2.1 Heavy Metals

Metallic species represent a significant proportion of the pollutants contained in the effluents generated by a wide variety of industries. Heavy metal ions imply an environmental risk due to their cumulative, toxic, and non-biodegradability characteristics (Mouni et al. 2009). At certain concentration threshold, heavy metals are toxic for human beings causing damages in the gastrointestinal, cardio-vascular, renal, and peripheral central nervous systems (Ibrahim et al. 2006). The toxicity of heavy metals is due to their ability to form compounds with cellular components that contain sulfur, oxygen, or nitrogen producing the inhibition of enzymes or the modification of protein structures leading to cellular dysfunctions in the organism (Ibrahim et al. 2006). Mercury, cadmium, copper, zinc, nickel, lead, chromium, aluminum, and cobalt are considered as priority aquatic pollutants due to their toxicity (Dias et al. 2007; Fu and Wang 2011). According to literature, several studies have focused on the adsorption of heavy metal ions (e.g., Rao et al. 2010; Barakat 2011; Gupta and Bhattacharyya 2012; Malik et al. 2016; Shah et al. 2016; Yu et al. 2016; Uddin 2017).

1.2.2 Dyes

Pigments and dyes are pollutants generated by food, rubber, paper, cosmetic, pharmaceutical, automotive, and textile industries (Gong et al. 2008; Wong et al. 2009; Ghodbane and Hamdaoui 2010). It has been estimated that $\sim 7 \times 10^5$ tons of dyes per year are generated worldwide due to inefficient dyeing techniques (Wong et al. 2009). Therefore, residual industrial effluents often contain a diversity of dyes. These compounds can be classed in acid, basic, disperse, reactive, and direct dyes (Ghodbane and Hamdaoui 2010). They have high molecular weights, complex structures, and, consequently, a persistent and recalcitrant nature (Qui and Zheng 2009; Hernández-Montoya et al. 2013). Dyes and their metabolites can be toxic, mutagenic, and carcinogenic to a wide variety of living organisms (Al-Ghouti et al. 2003). Adsorption process has proved to be an effective option to treat dye polluted fluids. Several studies have been performed to evaluate the dye uptakes of a vast number of adsorbents (e.g., Allen and Koumanova 2005; Demirbas 2009; Chincholi et al. 2014; Yagub et al. 2014; Adeyemo et al. 2015; Seow and Lim 2016; Vital et al. 2016).

1.2.3 Pharmaceuticals

Pharmaceuticals are considered as one of the emerging hazardous water pollutants (Sarici-Ozdemir and Önal 2010; Bagheri et al. 2014). These pollutants have been reported in trace levels (typically in ng/L and μg/L) in surface waters, wastewaters, groundwaters and even drinking water (World Health Organization 2011). Pharmaceuticals are generally better chemically characterized than other environmental pollutants (World Health Organization 2011). The most common pharmaceuticals are antibiotics, analgesics, anti-inflammatories, painkillers, and hormones. They can be absorbed by the tissues of animals and humans (especially liver and kidney) obstructing the metabolic processes and inhibiting the hormone receptors (World Health Organization 2011). They can enter into the environment via the human or animal excreta, and, consequently, these pharmaceuticals can reach the wastewater treatment plants (Baccar et al. 2012; Bagheri et al. 2014). Recent literature has reported that the conventional sewage treatment plants are not effective to remove/degrade these species because they were not specifically designed for handling trace pollutants (Baccar et al. 2012; Bagheri et al. 2014). Therefore, the pharmaceuticals can be continually introduced into water resources representing an environmental risk for both ecosystems and human beings (Önal et al. 2007; Yu et al. 2008; Sarici-Ozdemir and Önal 2010; Baccar et al. 2012). Adsorption has been proposed to remove these pollutants with satisfactory results. Adsorbents used for pharmaceutical removal include activated carbons, silica, and zeolites (Fukahori et al. 2011; Baccar et al. 2012; Martucci et al. 2012; Kim et al. 2014; Kyzas and Deliyanni 2015). On the other hand, the most studied pharmaceuticals in adsorption tests are naproxen, ibuprofen, diclofenac, erythromycin, atenolol, and carbamazepine (Kim et al. 2014).

1.2.4 Fluoride

Fluoride is recognized by the World Health Organization as a toxic geogenic pollutant that is present in groundwater (Jadhav et al. 2016). Fluoride pollution of water resources is considered as a relevant environmental issue that is associated to public health problems in several developing countries. It has been estimated that over 260 million people worldwide consume fluoride-polluted groundwater (Sani et al. 2016). Water consumption with fluoride concentrations higher than 1.5 mg/L is associated to dental and skeletal fluorosis in human beings. Other health effects of a chronic fluoride ingestion may include endocrine gland injury, thyroid disorder, renal dysfunction, gastrointestinal disorders, gastric irritation, arthritis, cancer, infertility, Alzheimer's syndrome, and osteoporosis (Das et al. 2005; Gupta et al. 2007; Ozvath 2009; Tomar and Kumar 2013; Jadhav et al. 2016). Fluoride may interfere with proteins, carbohydrates, vitamins, and enzymes that mediate coagulation, glycolysis, oxidative phosphorylation, and neurotransmission causing serious disruptions to these important biochemical processes (Tomar and Kumar 2013; Kut et al. 2016). Water defluoridation can be performed via adsorption (e.g., Mohapatra et al. 2009; Bhatnagar et al. 2011; Miretzky and Cirelli 2011; Jagtap et al. 2012; Loganathan et al. 2013; Velazquez-Jimenez et al. 2015; Vinati et al. 2015). However, activated carbons and other conventional adsorbents may show a poor performance for fluoride removal.

1.2.5 Arsenic

Arsenic is also a geogenic pollutant that is widely distributed in the earth crust (Mandal and Suzuki 2002). A chronic exposure to arsenic via mainly drinking water may cause melanosis, keratosis, cancer (in skin, bladder, kidney, and lung), arterial hypertension, and reproductive disorders (Shankar et al. 2014). The international concentration and threshold of arsenic for drinking water is 10 μg/L. As stated by Schwarzenbach et al. (2010), this pollutant exemplifies the dilemma between public health concerns and economic feasibility of water sanitation technologies to reach concentration levels of arsenic lower than the safety concentration threshold. The arsenic adsorption from water implies both economic and technical challenges. Several authors have analyzed the advantages and limitations of adsorption processes for the treatment of water polluted by arsenic (Verma et al. 2014; Baig et al. 2015) and various adsorbents have been reported for the arsenic removal from water (e.g., Mohan and Pittman 2007; Gallegos-Garcia et al. 2011; Verma et al. 2014; Baig et al. 2015; Habuda-Stanić and Nujić 2015; Jadhav et al. 2015).

1.2.6 Emerging Pollutants

In recent years, the advances in analytical instrumentation and techniques have facilitated to detect very low concentrations (i.e., ng/L to μg/L) of several chemicals present in groundwater and wastewaters (Nghiem and Fujioka 2016). This is 2the case for the emerging chemicals, which have been recently recognized as relevant aquatic pollutants despite they had occurred in the environment for a long time (Jeirani et al. 2016; Nghiem and Fujioka 2016). These emerging pollutants are generally defined as any synthetic or naturally occurring chemicals that are not routinely monitored but come mainly from wastewaters of municipal, agricultural, and industrial sources (Nghiem and Fujioka 2016). They can be classified depending on their origin, use, or properties (Nghiem and Fujioka 2016). These compounds include steroid hormones, phytoestrogens, personal care products, pesticides, hydrocarbons, and disinfection by-products (Nghiem and Fujioka 2016). They have the potential to cause adverse effects to biota and long-term human health effects such as cancer. Note that several of these pollutants can be classified as endocrine disrupting chemicals (Shi et al. 2012; Nghiem and Fujioka 2016). Conventional treatment processes are not effective to control the water pollution problem caused by these micro-pollutants. Adsorption has been explored for this purpose where activated carbon (Zhang and Hofmann 2013; Jeirani et al. 2016), zeolites (Lule and Atalay 2014), clays (Bojemueller et al. 2001), and some polymers have been employed as adsorbents of these emerging aquatic pollutants.

1.3 Adsorption Process Intensification

Process intensification is an engineering strategy to improve the performance, metrics, and attributes of a system under investigation and, overall, it implies the development of smaller, cleaner, and more energy-efficient technologies (Stankiewicz and Moulijn 2000). In particular, adsorption process involves several technical aspects that can be analyzed from a process intensification perspective; see Fig. 1.1. It is clear that research efforts have been directed toward to improve the different stages of adsorption process engineering for facing water pollution control. This section provides an overview of the opportunity areas for adsorption process intensification in water treatment and purification.

1.3.1 Synthesis of Tailored Adsorbents

The synthesis of novel adsorbents with outstanding properties for the adsorption of aquatic pollutants is a permanent research area. Several studies have shown that the effectiveness of adsorbents for the removal of organic pollutants is usually higher

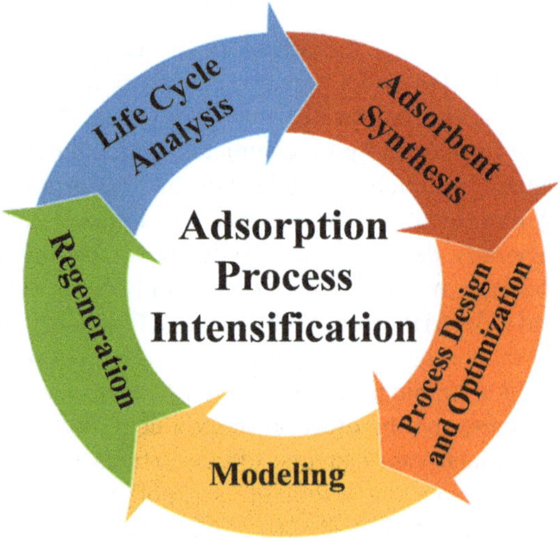

Fig. 1.1 Engineering aspects involved in the adsorption process intensification for water sanitation

than that obtained for inorganic compounds including geogenic chemicals. It is important to remark that textural properties and surface chemistry are fundamental to establish the performance of adsorbents in wastewater treatment and water purification. Physicochemical properties of adsorbents can be tailored for a specific application using proper synthesis protocols. The synthesis route and feedstock (i.e., precursor) determine both the surface chemistry and textural parameters of the adsorbent. The surface chemistry of adsorbents can be modified using thermal and chemical treatments with acids, bases, ozone, reactive gases, and surfactants. Also, impregnation procedures with organic and inorganic species (e.g., rare earth elements), biological modification, plasma, and microwave treatments can be used to modify the adsorbent surface properties (Bhatnagar et al. 2013). Novel synthesis technologies based on microwave and ultrasound offer several possibilities for obtaining new adsorbents for water pollution control (Jamshidi et al. 2016). The selection of a proper surface modification protocol is paramount to impact the cost-efficacy trade-off of adsorption processes. It is convenient to remark that a significant number of variables can be involved in the synthesis of an adsorbent, which should be manipulated using proper experimental designs to identify their effect on adsorbent properties and the best conditions for the adsorbent preparation. This type of studies also allows to understand the solid-liquid interfaces involved in the adsorption of water pollutants.

During the last years, several studies have given an emphasis of tailoring the adsorption properties of activated carbons and other adsorbents for the removal of water inorganic and organic pollutants. However, the synthesis of effective adsorbents for the simultaneous removal of several pollutants in the same solution is still a challenge. It is expected that the synthesis conditions that improve the adsorbent performance for one pollutant could be contradictory with those conditions required to maximize the adsorption of other pollutant. Therefore, there is a lack of studies related to the multiobjective optimization of adsorbent synthesis for an effective simultaneous removal of different aquatic pollutants. Green and energy-saving technologies are also required for the industrial production of outstanding adsorbents for water treatment.

1.3.2 Optimization and Design of Adsorption Systems

Adsorbent performance in water treatment is a nonlinear function of several operating parameters including fluid properties (e.g., pH, temperature), pollutant characteristics and concentration, and the presence of other adsorbates. It is well known that both adsorbent and adsorbate characteristics determine the adsorption mechanism, which may include ion exchange, electrostatic interactions, surface precipitation, chemical reactions, or a combination of them. The operating conditions of adsorption processes can be optimized via experimental designs and statistical analysis to maximize the adsorbent performance for the removal of a specific compound or set of pollutants. For instance, the adsorption effectiveness in packed-bed columns is a function of both the operating conditions and bed geometry where adsorbent size and distribution, fluid flow rate, adsorbate concentration, pH, temperature, and adsorbent mass are the main design parameters. These design variables can be manipulated to identify those conditions that provide the maximum bed adsorption capacity.

In terms of adsorption design, the reduction of mass transfer resistances is paramount to improve the separation effectiveness and to reduce the operational costs especially for continuous adsorption systems. In this direction, some studies have evaluated the use of alternative bed geometries/configurations to enhance the removal performance of packed-bed columns. For example, the adsorption of water pollutants with tapered convergent and stratified beds (Pota and Mathews 2000; Sze and McKay 2012) and helical coil-packed columns (Moreno-Pérez et al. 2016) have been analyzed. Other authors have reported the application of external forces to improve the adsorption of aquatic pollutants. They include the application of ultrasound (Dashamiri et al. 2017), electrical, and magnetic fields (Lounici et al. 2004; González Vázquez et al. 2016) with promising results. These emerging technologies require further investigation to determine their capabilities and limitations for scale-up and industrial applications.

1.3.3 Modeling of Adsorption Processes

The modeling of adsorption processes is relevant for the design, operation, optimization, and control of water purification technologies for real-life applications. Adsorption models can be classed as theoretical, semiempirical, and empirical. They may involve simple analytical expressions or complex systems of algebraic and differential equations when mass transfer phenomena are considered. Adsorption data can be also modeled using computationally intelligent data processing algorithms such as artificial neural networks (Tovar-Gómez et al. 2013) or fuzzy logic (Asl et al. 2017).

Adsorption models are indeed nonlinear and multivariable especially when they are applied in the analysis of multicomponent systems (i.e., groundwater, industrial effluents, and wastewater). Therefore, the modeling of adsorption data is considered as a challenging mathematical problem that should imply a nonlinear data regression procedure for the determination of model parameters with a proper error function and a detailed statistical analysis for establishing the model performance (Dotto et al. 2013). Note that, depending on the model complexity, the application of other computational strategies is often necessary in adsorption data correlation, e.g., the use of numerical methods for solving partial differential equations when mass transfer models are applied. It is desirable that an adsorption model can provide a reliable estimation of the process performance without requiring the use of extensive experimental data and it should have a suitable mathematical complexity, a capability for providing an acceptable accuracy in the correlation and estimation of adsorption performance at different operating conditions, and the model should be useful for assessing the effect of operating variables on the adsorption effectiveness. In summary, the selected model should reflect a compromise between its mathematical complexity and the accuracy obtained for process description.

It is convenient to remark that the numerical performance of adsorption models cannot be determined a priori and they may show limitations for data analysis depending on the adsorption system under study and its operating conditions. In fact, there is no general model applicable to all adsorbate(s)/adsorbent systems, and different adsorption models should be tested for identifying the best option for the case of study at hand. To date, there is a lack of studies to identify and to compare the relative advantages and limitations of available models in the analysis of adsorption processes for the removal of priority water pollutants especially in multicomponent systems. As stated, multicomponent adsorption systems impose challenges for available models due to the simultaneous presence of antagonistic, synergic, and non-interaction effects. For example, the simultaneous removal of heavy metals and acid dyes is complex due to the synergistic adsorption promoted by the dye molecule and the competitive adsorption between metallic species (Tovar-Gómez et al. 2015). For this type of mixtures, traditional models may fail to reproduce and predict the adsorbent performance at different operating scenarios. It is clear that reliable adsorption models are required for the analysis of multicomponent systems containing different organic and inorganic pollutants, which should be representative of groundwater, wastewater, and industrial effluents.

1.3.4 Regeneration and Final Disposal of Exhausted Adsorbents

Regeneration of exhausted adsorbents is a fundamental stage to establish the economic feasibility of adsorption as a water purification technology. The adsorbent performance is gradually reduced due to the progressive accumulation of adsorbates on adsorbent surface causing its exhausting. The strength of the interactions adsorbate(s)-adsorbate and the adsorbate concentration loaded on the adsorbent play a major role in the regeneration process. Therefore, the regeneration is used to remove the adsorbates accumulated on the surface and to recover the adsorbent capacity. Note that there is a decrement of the adsorption capacity of the adsorbent in each regeneration cycle.

Regeneration strategies can be classified in thermal, chemical, microbiological, and vacuum regeneration (Salvador et al. 2015a, b). The mechanisms used in adsorbent regeneration may imply heating, pH variation, changes in the medium for adsorbate extraction, chemical reactions, and degradation (Salvador et al. 2015a, b). These regeneration methods can use different reagents such as solvents, organic and inorganic chemical compounds, electrical current, physical waves, and microorganisms (Salvador et al. 2015a, b). The recovery of adsorption capacity is determined by the regeneration agent, the physicochemical properties of adsorbent and loaded pollutants, and the operating conditions. Desorption and regeneration efficiencies of these methods may vary substantially, and the best option for the adsorbent and pollutants at hand should be identified. Regeneration conditions should be optimized to improve the adsorbent lifetime favoring its uses in several adsorption-desorption cycles and to reduce the damage in the adsorbent structure avoiding a significant loss of mass and active sites.

For the case of adsorbents used in water treatment, the chemical regeneration is the common approach to perform the adsorption-desorption cycles. Recently, novel regeneration methods have been proposed, and they include the use of microwaves or ultrasound (Salvador et al. 2015a, b). Electromagnetic and mechanical waves are applied in microwaves and ultrasound regeneration processes, which have been recognized as energy-saving methods. However, the industrial use of these technologies requires further research to solve their disadvantages. Some studies have also analyzed the potential of microbiological regeneration for adsorbents employed in water treatment (e.g., Aguayo-Villarreal et al. 2016). In conclusion, low-cost and eco-friendly regeneration technologies are required to enhance the cost-efficacy trade-off of adsorption processes for water treatment and purification.

Finally, the disposal of adsorbents at their end of operational life after several adsorption-desorption cycles is also a key aspect to reduce the environmental impact caused by solid waste generation. The traditional approach for the disposal of exhausted adsorbents includes their incineration and use in landfills. However, the improper disposal of exhausted adsorbents may increase the costs of water treatment. The use of spent adsorbents in other industrial applications is an attractive approach for waste minimization. For example, Rao et al. (2009) have analyzed the disposal of fluoride-loaded bone char as a fine aggregate of concrete. Further studies are required in this direction to determine alternatives for the reusing and disposing of spent adsorbents.

1.3.5 Life Cycle Analysis

Life cycle analysis is useful to assess and quantify all the environmental interactions over the stages of the life cycle of an adsorbent manufacturing chain, its use, and disposal in terms of inputs of energy and natural resources and of outputs of emissions to the different environmental compartments (Arena et al. 2016). This analysis should include the effects on natural resources, environment, and human health.

The capabilities and limitations of different adsorbents can be identified via the life cycle assessment providing a detailed sustainability analysis of the water purification process. For example, Alhashimi and Aktas (2017) have analyzed and compared the environmental and economic performance of biochar and activated carbon for the adsorption of heavy metals. Results showed that biochar is a better adsorbent with lower energy demand and global warming potential impact than those of activated carbon for heavy metal removal. Authors of this chapter consider that this type of comparisons should be extended for other adsorbents.

Although several promising adsorbents have been recently proposed for water treatment, the assessment of environmental impacts involved in the manufacturing chain of these materials is still lacking. Therefore, the industrial production and commercialization of these adsorbents can be a matter of debate. Further studies in this direction are required to identify and develop sustainable synthesis routes for the industrial production of adsorbents with outstanding properties for water sanitation. Therefore, it is mandatory to analyze the advantages and limitations, from environmental impact and economical perspectives, of novel adsorbents in comparison to activated carbon using the life cycle analysis.

1.4 Scope and Outline of Chapters

This book covers the state of the art of relevant topics of adsorption processes for water pollution control. Chapters contain the analysis and discussion of data modeling, synthesis of new adsorbents, and the application of adsorbents for the removal of aquatic pollutants.

In particular, Chap. 2 covers different fundaments related to adsorption equilibrium in liquid phase including its experimental determination, analysis, and modeling in batch reactors. Guidelines and experimental procedures to obtain adsorption isotherms are discussed. The Giles classification of adsorption isotherms in liquid phase is introduced and analyzed. Different models commonly used for data fitting of adsorption isotherms in liquid phase are discussed. In particular, this chapter introduces the basis of statistical physical models for representing and understanding the liquid-phase adsorption. Advantages and limitations of linear and nonlinear regression analysis are also discussed including statistical criteria for determining the model accuracy. Finally, the thermodynamics of adsorption process is described given a special emphasis in the calculation of equilibrium constant and its implication in the estimation of thermodynamic parameters.

Kinetics approaches used for adsorption modeling are covered and reviewed in Chap. 3. The description and fundaments of diffusional mass transfer and reaction models are provided. Additionally, the characteristics of the breakthrough curves obtained for dynamic adsorption systems (e.g., packed-bed columns) are analyzed including mass transfer and empirical models for their correlation and representation. This chapter also contains the design of fixed-bed columns, numerical methods used for solving mass transfer-based models, parameter estimation procedures, and statistical criterions for data analysis.

Hydrothermal carbonization is examined in detail in Chap. 4. Specifically, this chapter discusses the basis, synthesis variables, operating conditions, and advantages of using this thermochemical route for the preparation of adsorbents for water treatment. Authors have reviewed different raw materials and synthesis conditions for the preparation of hydrochars. Physicochemical properties of hydrochars are also compared with those adsorbents obtained with pyrolysis. A detailed review of the application of hydrochars in the adsorption of dyes, pesticides, drugs, endocrine disruptive chemicals, phosphorus, phenols, and heavy metals is reported. Some key points related to the reusability of hydrochars are also presented in this chapter.

The application of low-cost adsorbents for the removal of lead, cadmium, and zinc from aqueous solution is analyzed and discussed in Chap. 5. This discussion is focused on agricultural wastes, industrial by-products and wastes, marine materials, zeolites, and clays. The chapter provides the characteristics of these low-cost adsorbents, their sources, and metal adsorption capacities at different operating conditions. This analysis also includes the best kinetic and isotherm models used in adsorption data fitting of heavy metal ions.

The adsorption of tetracyclines and nitroimidazole drugs on different adsorbents is reported in Chap. 6. Commercial activated carbons, sludge-derived materials, and activated carbons obtained from petroleum coke with different chemical and textural natures have been studied. This chapter describes the results of adsorption studies in static and dynamic regime using ultrapure water, surface water, groundwater, and urban wastewater. Authors have highlighted the influence of solution chemical nature (pH and ionic strength) on the adsorption of these compounds analyzing the adsorbent-adsorbate interactions. Also, the impact of microorganisms on the performance of activated carbon for the removal of these pollutants has been analyzed.

In Chap. 7, authors have studied the biosorption of copper using *Saccharomyces cerevisiae*. This chapter illustrates the importance of developing a model capable of describing the effect of operating conditions on biosorption performance. Batch and packed-bed biosorption studies are reported given an especial emphasis on the modeling of experimental data.

Finally, Chap. 8 introduces the application of transition metal-substituted magnetite as a material with a dual function: adsorbent and heterogeneous catalyst for water sanitation. Authors have covered the synthesis of this type of adsorbents, their physicochemical properties, and its performance as both adsorbent and Fenton catalyst.

Editors and authors of this book consider that topics covered in these chapters are relevant to contribute for the development of adsorption processes for water treatment and sanitation.

References

Adegoke KA, Bello OS (2015) Dye sequestration using agricultural wastes as adsorbents. Water Resour Ind 12:8–24

Adeyemo AA, Adeoye IO, Bello OS (2015) Adsorption of dyes using different types of clay: a review. Appl Water Sci. doi:10.1007/s13201-015-0322-y

Aguayo-Villarreal IA, Hernández-Montoya V, Ramírez-López EM, Bonilla-Petriciolet A, Montes-Morán MA, (2016) Effect of surface chemistry of carbons from pine sawdust for the adsorption of acid, basic and reactive dyes and their bioregeneration using Pseudomona putida. Ecol Eng 95:112–118

Ahmed MJK, Ahmaruzzaman M (2016) A review on potential usage of industrial waste materials for binding heavy metal ions from aqueous solutions. J Water Process Eng 10:39–47

Al-Ghouti MA, Khraisheh MAM, Allen SJ, Ahmad MN (2003) The removal of dyes from textile wastewater: a study of the physical characteristics and adsorption mechanisms of diatomaceous earth. J Environ Manag 69:229–238

Alhashimi HA, Aktas CB (2017) Life cycle environmental and economic performance of biochar compared with activated carbon: a meta-analysis. Resour Conserv Recycl 118:13–26

Allen SJ, Koumanova B (2005) Decolourisation of water/wastewater using adsorption (review). J Univ Chem Technol Metallurgy 40:175–192

Arena N, Lee J, Clift R (2016) Life cycle assessment of activated carbon production from coconut shells. J Clean Prod 125:68–77

Asl SMH, Masomi M, Hosseini M, Javadian H, Ruiz M, Sastre AM (2017) Synthesis of hydrous iron oxide/aluminum hydroxide composite loaded on coal fly ash as an effective mesoporous and low-cost sorbent for Cr(VI) sorption: fuzzy logic modeling. Process Saf Environ 107:153–167

Baccar R, Sarra M, Bouzid J, Feki M, Blánquez P (2012) Removal of pharmaceutical compounds by activated carbon prepared from agricultural by-product. Chem Eng J 211–212:310–317

Bagheri H, Roostaie A, Baktash MY (2014) A chitosan-polypyrrole magnetic nanocomposite as μ-sorbent for isolation of naproxen. Anal Chim Acta 816:1–7

Baig SA, Sheng T, Hu Y, Xu J, Xu X (2015) Arsenic removal from natural water using low cost granulated adsorbents: a review. Clean Soil Air Water 43:13–26

Barakat MA (2011) New trends in removing heavy metals from industrial wastewater. Arab J Chem 4:361–377

Bhatnagar A, Sillanpaa M (2010) Utilization of agro-industrial and municipal waste materials as potential adsorbents for water treatment – a review. Chem Eng J 157:277–296

Bhatnagar A, Kumar E, Sillanpaa M (2011) Fluoride removal from water by adsorption – a review. Chem Eng J 171:811–840

Bhatnagar A, Hogland W, Marques M, Sillanpaa M (2013) An overview of the modification methods of activated carbon for its water treatment applications. Chem Eng J 219:499–511

Bhatnagar A, Sillanpaa M, Witek-Krowiak A (2015) Agricultural waste peels as versatile biomass for water purification – a review. Chem Eng J 270:244–271

Bojemueller E, Nennemann A, Lagaly G (2001) Enhanced pesticide adsorption by thermally modified bentonites. Appl Clay Sci 18:277–284

Chincholi M, Sagwekar P, Nagaria C, Kulkarni S, Dhokpande S (2014) Removal of dye by adsorption on various adsorbents: a review. Int J Eng Sci Technol 3:835–840

Choy KKH, McKay G (2005) Sorption of metal ions from aqueous solution using bone char. Environ Int 31:845–854

Das N, Pattanaik P, Das R (2005) Defluoridation of drinking water using activated titanium rich bauxite. J Colloid Interf Sci 292:1–10

Dashamiri S, Ghaedi M, Asfaram A, Zare F, Wang S (2017) Multi-response optimization of ultrasound assisted competitive adsorption of dyes onto $Cu(OH)_2$-nanoparticle loaded activated carbon: central composite design. Ultrason Sonochem 34:343–353

De Gisi S, Lofrano G, Grassi M, Notarnicola M (2016) Characteristics and adsorption capacities of low-cost sorbents for wastewater treatment: a review. Sustain Mater Technol 9:10–40

Demirbas A (2009) Agricultural based activated carbons for the removal of dyes from aqueous solutions: a review. J Hazard Mater 167:1–9

Dias JM, Alvim-Ferraz MCM, Almeida MF, Rivera-Utrilla J, Sánchez-Polo M (2007) Waste materials for activated carbon preparation and its use in aqueous-phase treatment: a review. J Environ Manag 85:833–846

Dotto GL, Costa JAC, Pinto LAA (2013) Kinetic studies on the biosorption of phenol by nanoparticles from Spirulina s.p. LEB 18. J Environ Chem Eng 1:1137–1143

Fu F, Wang Q (2011) Removal of heavy metal ions from wastewaters: a review. J Environ Manag 92:407–418

Fukahori S, Fujiwara T, Ito R, Funamizu N (2011) pH-dependent adsorption of sulfa drugs on high silica zeolite: modeling and kinetic study. Desalination 275:237–242

Gallegos-Garcia M, Ramírez-Muñiz K, Song S (2011) Arsenic removal from water by adsorption using iron oxide minerals as adsorbents: a review. Min Proc Ext Met Rev 33:301–315

Ghodbane H, Hamdaoui (2010) Decolorization of antraquinonic dye, C.I. Acid Blue 25, in aqueous solution by direct UV irradiation, UV/H_2O_2 and UV/Fe(II) processes. Chem Eng J 160:226–231

Gong R, Zhu S, Zhang D, Chen J, Ni S, Guan R (2008) Adsorption behavior of cationic dyes on citric acid esterifying wheat straw: kinetic and thermodynamic profile. Desalination 230:220–228

González Vázquez OF, Moreno Virgen MR, Hernández Montoya V, Tovar Gómez R, Alcántara Flores JL, Pérez Cruz MA, Montes Morán MA (2016) Adsorption of heavy metals in the presence of a magnetic fiel don adsorbents with different magnetic properties. Ind Eng Chem Res 55:9323–9331

Gupta SS, Bhattacharyya KG (2012) Adsorption of heavy metals on kaolinite and montmorillonite: a review. Phys Chem Chem Phys 14:6698–6723

Gupta VK, Ali I, Saini VK (2007) Defluoridation of wastewaters using waste carbon slurry. Water Res 41:3307–3316

Habuda-Stanić M, Nujić M (2015) Arsenic removal by nanoparticles: a review. Environ Sci Pollut Res Int 22:8094–8123

Hernandez-Eudave MT, Bonilla-Petriciolet A, Moreno-Virgen MR, Rojas-Mayorga CK, Tovar-Gomez R (2016) Design analysis of fixed-bed synergic adsorption of heavy metals and acid blue 25 on activated carbon. Desalin Water Treat 57:9824–9836

Hernández-Montoya V, Pérez-Cruz MA, Mendoza-Castillo DI, Moreno-Virgen MR, Bonilla-Petriciolet A (2013) Competitive adsorption of dyes and heavy metals on zeolitic structures. J Environ Manag 116:213–221

Ibrahim D, Froberg B, Wolf A, Rusyniak DE (2006) Heavy metal poisoning: clinical presentations and pathophysiology. Clin Lab Med 26:67–97

Jadhav SV, Bringas E, Yadav GD, Rathod VK, Ortiz I, Marathe KV (2015) Review arsenic and fluoride contaminated groundwaters: a review of current technologies for contaminants removal. J Environ Manag 162:306–325

Jadhav SV, Marathe KV, Rathod VK (2016) A pilot scale concurrent removal of fluoride, arsenic, sulfate and nitrate by using nanofiltration: competing ion interaction and modelling approach. J Water Process Eng 13:153–167

Jagtap S, Yenkie MK, Labhsetwar N, Rayalu S (2012) Fluoride in drinking water and defluoridation of water. Chem Rev 112:2454–2466

Jamshidi M, Ghaedi M, Dashtian K, Hajati S, Bazrafshan AA (2016) Sonochemical assisted hydothermal synthesis of ZnO: Cr nanoparticles loaded activated carbon for simultaneous ultrasound-assisted adsorption of ternary toxic organic dye: derivative spectrophotometric, optimization, kinetic and isotherm study. Ultrason Sonochem 32:119–131

Jeirani Z, Niu CH, Soltan J (2016) Adsorption of emerging pollutants on activated carbon. Rev Chem Eng 27:2–32

Kasprzyk-Hordern B (2004) Chemistry of alumina, reactions in aqueous solution and its application in water treatment. Adv Colloid Interface Sci 110:19–48

Kim Y, Bae J, Park J, Suh J, Lee S, Park H, Choi H (2014) Removal of 12 selected pharmaceuticals by granular mesoporous silica SBA-15 in aqueous phase. Chem Eng J 256:475–485

Koshy N, Singh DN (2016) Fly ash zeolites for water treatment applications. J Environ Chem Eng 4:1460–1472

Kumar P, Pournara A, Kim KH, Bansal V, Rapti S, Manos MJ (2017) Metal-organic frameworks: challenges and opportunities for ion-exchange/sorption applications. Prog Mater Sci 86:25–74

Kut KMK, Sarswat A, Srivastava A, Pittman CU Jr, Mohan D (2016) A review of fluoride in African groundwater and local remediation methods. Groundwater Sust Develop 2:190–212

Kyzas GZ, Deliyanni EA (2015) Modified activated carbons from potato peels as green environmental-friendly adsorbents for the treatment of pharmaceutical effluents. J Chem Eng Res Des 97:135–144

Loganathan P, Vigneswaran S, Kandasamy J, Naidu R (2013) Defluoridation of drinking water using adsorption processes. J Hazard Mater 248:1–19

Lounici H, Belhocine D, Grib H, Drouiche M, Pauss A, Mameri N (2004) Fluoride removal with electro-activated alumina. Desalination 161:287–293

Lule GM, Atalay MU (2014) Comparison of fenitrothion and trifluralin adsorption on organo-zeolites and activated carbon. Part II: thermodynamic parameters and the suitability of the kinetic models of pesticide adsorption. Particul Sci Technol 32:426–430

Malik DS, Jain CK, Yadav AK (2016) Removal of heavy metals from emerging cellulosic low-cost adsorbents: a review. Appl Water Sci. doi:10.1007/s13201-016-0401-8

Mandal BK, Suzuki KT (2002) Review arsenic round the world: a review. Talanta 58:201–235

Martucci A, Pasti L, Marchetti N, Cavazzini A, Dondi F, Alberti A (2012) Adsorption of pharmaceuticals from aqueous solutions on synthetic zeolites. Micropor Mesopor Mat 148:174–183

Mehta D, Mazumdar S, Singh SK (2015) Magnetic adsorbents for the treatment of water/waste-water – a review. J Water Process Eng 7:244–265

Miretzky P, Fernandez Cirelli A (2011) Fluoride removal from water by chitosan derivatives and composites: a review. J Fluor Chem 132:31–240

Mohan D, Pittman CU Jr (2007) Arsenic removal from water/wastewater using adsorbents: a critical review. J Hazard Mater 142:1–53

Mohapatra M, Anand S, Mishra BK, Giles DE, Singh P (2009) Review of fluoride removal from drinking water. J Environ Manag 91:67–77

Moreno-Pérez J, Bonilla-Petriciolet A, Rojas-Mayorga CK, Mendoza-Castillo DI, Mascia M, Errico M (2016) Adsorption of zinc ions on bone char using helical coil-packed bed columns and its mass transfer modeling. Desalin Water Treat 57:24200–24209

Mouni L, Merabet D, Robert D, Bouzaza A (2009) Batch studies for the investigation of the sorption of the heavy metals Pb^{2+} and Zn^{2+} onto Amizour soil (Algeria). Geoderma 154:30–35

Nghiem LD, Fujioka T (2016) Removal of emerging contaminants for water reuse by membrane technology. In: Hankins NP, Singh R (eds) Emerging membrane technology for sustainable water treatment. Elsevier, Amsterdam, pp 217–247

Önal Y, Akmil-Basar C, Sarici-Ozdemir C (2007) Elucidation of the naproxen sodium adsorption onto activated carbon prepared from waste apricot: kinetic, equilibrium and thermodynamic characterization. J Hazard Mater 148:727–734

Ozvath DL (2009) Fluoride and environmental health: a review. Rev Environ Sci Biotechnol 8:59–79

Peng W, Li H, Liu Y, Song S (2017) A review on heavy metal ions adsorption from water by graphene oxide and its composites. J Mol Liq 230:496–504

Pota AA, Mathews AP (2000) Adsorption dynamics in a stratified convergent tapered bed. Chem Eng Sci 55:1399–1409

Qui W, Zheng Y (2009) Removal of lead, copper, nickel, cobalt, and zinc from water by a cancrinite-type zeolite synthesized from fly ash. Chem Eng J 145:483–488

Rao SM, Reddy BVV, Lakshmikanth S, Ambika NS (2009) Re-use of fluoride contaminated bone char sludge in concrete. J Hazard Mater 166:751–756

Rao KS, Mohapatra M, Anand S, Venkateswarlu P (2010) Review on cadmium removal from aqueous solutions. Int J Eng Sci Technol 2:81–103

Rivera-Utrilla J, Sánchez-Polo M, Gómez-Serrano V, Álvarez PM, Alvim-Ferraz MCM, Dias JM (2011) Activated carbon modifications to enhance its water treatment applications. J Hazard Mater 187:1–23

Salvador F, Martin-Sanchez N, Sanchez-Hernandez R, Sanchez-Montero MJ, Izquierdo C (2015a) Regeneration of carbonaceous adsorbents. Part I: thermal regeneration. Microporous Mater 202:259–276

Salvador F, Martin-Sanchez N, Sanchez-Hernandez S-MMJ, Izquierdo C (2015b) Regeneration of cabonaceous adsorbents. Part II: chemical, microbiological and vacuum regeneration. Microporous Mater 202:277–296

Sani T, Gómez-Hortiguela L, Pérez-Pariente J, Chebude Y, Díaz I (2016) Defluoridation performance of nano-hydroxyapatite/stilbite composite compared with bone char. Sep Purif Technol 157:241–248

Santhosh C, Velmurugan V, Jacob G, Jeong SK, Grace AN, Bhatnagar A (2016) Role of nanomaterials in water treatment applications: a review. Chem Eng J 306:1116–1137

Sarici-Ozdemir C, Önal Y (2010) Study to investigate the importance of mass transfer of naproxen sodium onto activated carbon. Chem Eng Process 49:1058–1065

Schwarzenbach RP, Egli T, Hofstetter TB, von Gunten U, Wehrli B (2010) Global water pollution and human health. Annu Rev Environ Resour 35:109–136

Seow TW, Lim CK (2016) Removal of dye by adsorption: a review. Int J Appl Eng Res 11:2675–2679

Shah J, Kumar S, Sharma S, Sharma R, Sharma R (2016) Removal of nickel from aqueous solution by using low cost adsorbents: a review. Int J Sci Eng Appl Sci 2:48–73

Shankar S, Shanker U, Shikha (2014) Arsenic contamination of groundwater: a review of sources, prevalence, health risks, and strategies for mitigation. Sci World J. doi:10.1155/2014/304524

Shi H, Cheng X, Wu Q, Mu R, Ma Y (2012) Assessment and removal of emerging water contaminants. Environ Anal Toxicol 2:1–14

Stankiewicz AI, Moulijn JA (2000) Process intensification: transforming chemical engineering. Chem Eng Prog 96:22–34

Sze MFF, McKay G (2012) Enhanced mitigation of para-chlorophenol using stratified activated carbon adsorption columns. Water Res 46:700–710

Tomar V, Kumar D (2013) A critical study on efficiency of different materials for fluoride removal from aqueous media. Chem Cent J 7:51

Tovar-Gómez R, Moreno-Virgen MR, Dena-Aguilar JA, Hernández-Montoya V, Bonilla-Petriciolet A, Montes-Morán MA (2013) Modeling of fixed-bed adsorption of fluoride on bone char using a hybrid neural network approach. Chem Eng J 228:1098–1109

Tovar-Gómez R, Moreno-Virgen MR, Moreno-Pérez J, Bonilla-Petriciolet A, Hernández-Montoya V, Durán-Valle CJ (2015) Analysis of synergistic and antagonistic adsorption of heavy metals and acid blue 25 on activated carbon from ternary systems. Chem Eng Res Des 93:755–772

Uddin MK (2017) A review on the adsorption of heavy metals by clay minerals, with special focus on the past decade. Chem Eng J 308:438–462

Velazquez-Jimenez LH, Vences-Alvarez E, Flores-Arciniega JL, Flores-Zuñiga H, Rangel-Mendez JR (2015) Water defluoridation with special emphasis on adsorbents-containing metal oxides and/or hydroxides: a review. Sep Purif Technol 150:292–307

Verma P, Agarwal A, Singh VK (2014) Arsenic removal from water through adsorption-a review. Recent Res Sci Technol 6:219–226

Vinati A, Mahanty B, Behera SK (2015) Clay and clay minerals for fluoride removal from water: a state-of-the-art review. Appl Clay Sci 114:340–348

Vital RK, Saibaba KVN, Shaik KB (2016) Dye removal by adsorption: a review. J Bioremed Biodegr 7:1–4

Wong SY, Tan YP, Abdullah AH, Ong ST (2009) Removal of basic blue 3 and reactive orange 16 by adsorption onto quartenized sugar cane bagasse. Malays J Anal Sci 13:185–193

World Health Organization (2011) Pharmaceuticals in drinking water. WHO Press, Geneva

Yagub MT, Sen TK, Afroze S, Ang HM (2014) Dye and its removal from aqueous solution by adsorption: a review. Adv Colloid Interfac 209:172–184

Yu Z, Peldszus S, Huck PM (2008) Adsorption characteristics of selected pharmaceuticals and endocrine disrupting compound-naproxen, carbamazepine and nonylphenol-on activated carbon. Water Res 42:2873–2882

Yu JG, Yue BY, Wu XW, Liu Q, Jiao FP, Jiang XY, Chen XQ (2016) Removal of mercury by adsorption: a review. Environ Sci Pollut Res Int 23:5056–5076

Zhang J, Hofmann R (2013) Modeling the adsorption of emerging contaminants on activated carbon: classical and quantum QSAR approaches. Water Sci Technol 13:1543–1552

Chapter 2
Adsorption Isotherms in Liquid Phase: Experimental, Modeling, and Interpretations

Jeferson Steffanello Piccin, Tito Roberto Sant'Anna Cadaval Jr., Luiz Antonio Almeida de Pinto, and Guilherme Luiz Dotto

Abstract Adsorption is a fundamental unit operation used for several purposes in the academy and industry. Particularly, adsorption in liquid phase is used to remove recalcitrant compounds from effluents (dyes, heavy metals, phenols, pharmaceuticals, and others), to recover valuable metals from leachates (gold, silver, cobalt, and others), and to purify products during the industrial processing (fuels, juices, liquors, wines, and others). For all these applications, the obtainment, modeling, and interpretation of the equilibrium isotherms are a key and fundamental study. Based on the abovementioned, this chapter presents the particularities of adsorption equilibrium isotherms in liquid phase from scientific and technological viewpoints. From the scientific viewpoint, the importance of adsorption isotherms will be addressed. For example, the equilibrium isotherms provide parameters for decision-making of the researcher in relation to the adsorption capacity of a particular adsorbent, give an idea how the interaction of adsorbent–adsorbate occurs, and provide means to find thermodynamic parameters, among others. From technological viewpoint, the adsorption capacity of the material is a basic parameter for the project. Thus, in this chapter the following points are highlighted: experimental procedures to obtain equilibrium curves, isotherm analysis, models used to correlate the equilibrium data and interpretation of its parameters, regression methods (comparison between linear and nonlinear regression methods), error analysis, adsorption thermodynamics, and the use of these data for equipment design.

J.S. Piccin
Food Engineering Department, Passo Fundo University, UPF, Br. 285, Km 171, 99052-900, Passo Fundo, RS, Brazil
e-mail: jefersonpiccin@gmail.com

T.R.S. Cadaval Jr. • L.A.A. de Pinto
Industrial Technology Laboratory, School of Chemistry and Food, Federal University of Rio Grande, km 08 Itália Avenue, 96203-900, Rio Grande, RS, Brazil
e-mail: titoeq@gmail.com; dqmpinto@furg.br

G.L. Dotto (✉)
Chemical Engineering Department, Federal University of Santa Maria, 1000 Roraima Avenue, 97105-900, Santa Maria, RS, Brazil
e-mail: guilherme_dotto@yahoo.com.br

© Springer International Publishing AG 2017
A. Bonilla-Petriciolet et al. (eds.), *Adsorption Processes for Water Treatment and Purification*, DOI 10.1007/978-3-319-58136-1_2

Keywords Adsorption • Equilibrium • Isotherms • Liquid phase • Modeling

Contents

2.1 Introduction

Adsorption is a unit operation that involves the contact of a solid phase with a fluid phase (liquid or gas) (Ruthven 1984). In this work, only the solid–liquid adsorption is addressed. The solid phase is known as adsorbent and the liquid phase (the solvent, normally water) contains one or more compounds to be adsorbed (the adsorbates). Due to unbalanced forces, the adsorbate is attracted to the adsorbent surface, and consequently, the degrees of freedom and the surface free energy are reduced (Suzuki 1993). The transference of the adsorbate from the liquid phase to the solid phase continues until the equilibrium to be reached between the amount of adsorbate linked in the adsorbent and the amount of adsorbate remaining in the solution. The affinity degree between the adsorbent and adsorbate determines this distribution in liquid and solid phases (Rouquerol et al. 2014).

In general, the adsorption can be classified according to the type of interaction that occurs between the adsorbent and adsorbate. If there is an electron transfer between the adsorbent and adsorbate, then, it is a chemical adsorption or chemisorption. In this case, the adsorption involves electron transfer, and it is of high energy, ranging from 40 to 800 kJ/mol and, consequently, desorption is difficult, and thus the process is irreversible and only a monolayer is observed. In chemisorption, the interactions can occur mainly by ionic or covalent bonds (Crini and Badot 2008). Otherwise, if no electron exchange is observed, a physical adsorption or physisorption occurs. In this case, the adsorption energies are low, ranging from 5 to 40 kJ/mol and, consequently, desorption is possible and the process can be reversible and multilayer adsorption is possible. In physisorption, the interactions can be electrostatic, hydrogen bonds, van der Waals, or dipole–dipole (Bergmann and Machado 2015). It should be highlighted that this classification between physical and chemical adsorption is a general behavior, but is not a dogma. Each case should be examined separately.

Several advantages have been cited regarding the adsorption operation. For example, in comparison with other unit operations, the adsorption in liquid medium has a low energetic requirement and its implementation and operation are easy. Many materials can be used as adsorbents, which can be regenerated and reused several times. This practice becomes the adsorption in a low-cost operation. Also, adsorption is efficient since it can remove or recover all the adsorbate from the solution, providing a perfect separation. After, the adsorbent and adsorbate could be reused. In some cases, adsorption is also selective (Do 1998). On the other hand, after the adsorption, an additional operation can be necessary to provide a good solid–liquid separation, for example, filtration, sedimentation, or centrifugation. Another drawback is high cost of activated carbon, the adsorbent most commonly used, mainly due its high surface area. In some cases, a secondary problem can occur with the disposal of the generated sludge (Bansal and Goyal 2005).

To develop an adequate adsorption system, the adsorbent choice is the first and fundamental aspect. A good adsorbent should have the following characteristics: low cost; availability; efficiency; high surface area and pore volume; mechanical, chemical, and thermal stability; ease of desorption and reuse; and able to provide a fast kinetics and, mainly, present a high adsorption capacity (Rodrigues 2015). For this proposal, the activated carbon is the adsorbent most utilized (Bansal and Goyal 2005). Other common adsorbents are clays, silica, and zeolites (Rouquerol et al. 2014). However, in the last years, several researches have been focused on the preparation, characterization, and application of nonconventional adsorbents. Some examples are powdered agro-wastes, powdered industrial wastes, chitosan, chitin, fungi, bacteria, and algae (Dotto et al. 2015a). Surely, these studies are relevant, but, to choose an adequate adsorbent, among other information, the following question is fundamental: in which process the adsorbent will be used?

In operational terms, some configurations are possible for an adsorption process, for example, discontinuous batch adsorption, continuous stirred-tank reactor (CSTR), fixed bed adsorption, expanded bed adsorption, fluidized bed adsorption, simulated moving bed adsorption, and others (Rouquerol et al. 2014; Dotto et al. 2015a; Rodrigues 2015). In this chapter, the discontinuous batch adsorption and the fixed bed adsorption will be addressed, since they account for about 90% of the scientific literature.

Figure 2.1 shows a schematic representation of a discontinuous batch adsorption operation. In this case, a certain amount m of a pure adsorbent is put in contact with a solution with an initial volume V_0 and an initial concentration of adsorbate C_0. The solution is stirred at temperature constant until the end of operation (e.g., the equilibrium). During the operation period, the adsorbate is transferred to the adsorbent surface, decreasing its concentration in the solution until C_e and increasing its quantity in the solid phase until q_e (Crini and Badot 2008).

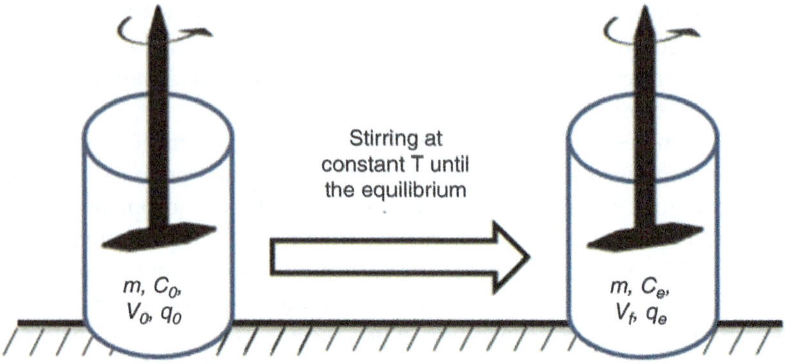

Fig. 2.1 Schematic representation of a discontinuous batch adsorption operation

In discontinuous batch operations, a simple mathematic treatment is normally performed to obtain the amount of adsorbate adsorbed into the adsorbent at equilibrium (q_e). A global mass balance of the adsorbate is given by Eq. (2.1) (Geankoplis 1998):

$$C_0V_0 + q_0m = C_eV_f + q_em \tag{2.1}$$

For a virgin adsorbent, the amount of adsorbate at the beginning is equal to 0, leading to Eq. (2.2):

$$C_0V_0 = C_eV_f + q_em \tag{2.2}$$

or

$$q_e = \frac{C_0V_0 - C_eV_f}{m} \tag{2.3}$$

In the majority of the experimental cases, the aliquot removed for quantification of the adsorbate is negligible regarding the total volume of the solution, leading to $V_0 = V_f = V$. So, the amount of adsorbate adsorbed into the adsorbent at equilibrium is given by Eq. (2.4):

$$q_e = \frac{(C_0 - C_e)V}{m} \tag{2.4}$$

The discontinuous batch adsorption systems are useful and fundamental to verify the quality of an adsorbent and define some operational parameters, such as pH, temperature, amount of adsorbent, and operation time, in laboratory scale. Also it is used for industrial applications for small volumes (Piccin et al. 2009, 2011).

Figure 2.2 shows a schematic representation of a fixed bed adsorption operation. In fixed bed adsorption systems, a solution with initial adsorbate concentration C_0 (normally named influent) is pumped at a flow rate Q, through a column with

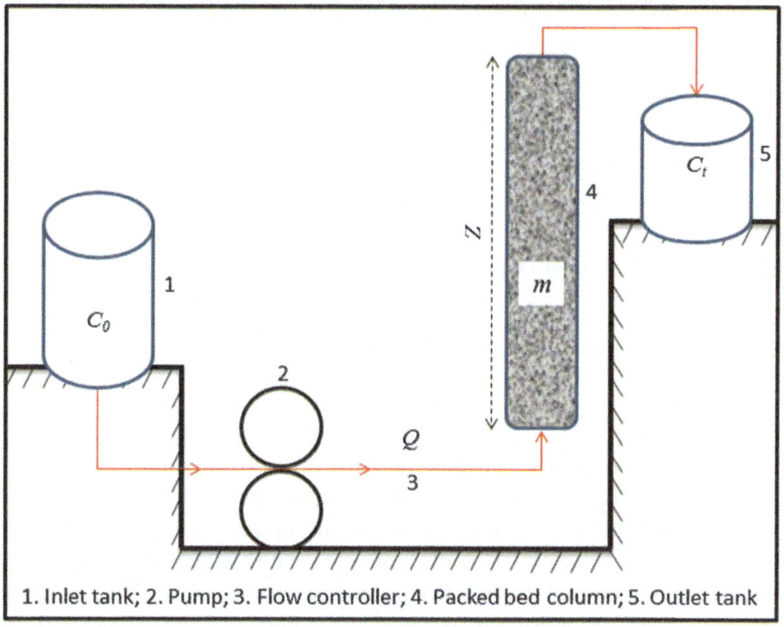

Fig. 2.2 Schematic representation of a fixed adsorption operation

Z height, which is packed with a certain amount m of adsorbent. During the operation, the adsorbate is transferred from the solution to the adsorbate surface. As a consequence, the solution is clarified, achieving an output concentration of C_t. The output solution is normally named effluent. The equilibrium is attained when the bed saturation occurs, i.e., $C_t = C_0$ (Shafeeyan et al. 2014).

In the case of fixed bed operations, the data analysis is performed aiming to obtain the breakthrough time (t_b), exhaustion time (t_e), length of mass transfer zone (Z_m), effluent volume (V_{eff}), maximum capacity of the column (q_{eq}), and removal percentage (R). The breakthrough time (t_b) is considered when the outlet adsorbate concentration attains low levels (in general less than 5%) in relation to the initial concentration, and the exhaustion time (t_e) is considered when the outlet concentration attains 95% of the initial concentration. The Z_m reflects the shortest possible adsorbent bed length needed to obtain the breakthrough time t_b at $t = 0$. The metric length of this zone is calculated by Eq. (2.5) (Worch 2008):

$$Z_m = Z\left(1 - \frac{t_b}{t_e}\right) \tag{2.5}$$

The volume of the effluent, V_{eff}, is given by Eq. (2.6):

$$V_{\text{eff}} = Qt_{\text{total}} \tag{2.6}$$

where t_{total} is the total operation time.

The maximum capacity of the column (q_{eq}) is given by Eq. (2.7):

$$q_{\text{eq}} = \frac{QC_0 \int_0^{t_{\text{total}}} \left(1 - \frac{C_t}{C_o}\right) dt}{m} \tag{2.7}$$

The integral in Eq. (2.7) is the area above the breakthrough curve from $C_t/C_0 = 0$ to $C_t/C_0 = 1$.

The removal percentage (R) is given by Eq. (2.8) (Dotto et al. 2015b):

$$R = \frac{\int_0^{t_{\text{total}}} \left(1 - \frac{C_t}{C_o}\right) dt}{t_{\text{total}}} 100 \tag{2.8}$$

The fixed bed systems are useful and fundamental in order to scale up the adsorption operations. The real operational conditions, such as flow rate and bed height, can be simulated, and parameters for scale-up can be obtained. For example, from laboratory fixed bed experiments, it is possible to estimate the column height necessary to obtain a good quality effluent in a determined time. This height can be easily transferred for pilot or industrial scale (Vieira et al. 2014; Dotto et al. 2015b).

To develop an adsorption operation, in discontinuous batch or in fixed bed systems, the first step is the adsorbent choice and the second is the obtainment of the adsorption isotherms. Adsorption isotherms are a relation between the amount of adsorbate adsorbed in the adsorbent (q_e) and the amount of adsorbate remaining in the liquid phase (C_e), when the two phases are in dynamic equilibrium at a determined temperature. In liquid phase adsorption systems, the isotherm curves are important due to the following aspects:

- From the isotherm parameters, it is possible to obtain the maximum adsorption capacity of a determined adsorbent under different experimental conditions. The maximum adsorption capacity is an indicative of the adsorbent quality.
- Also from the isotherm parameters, it is possible to obtain information about the energetic, steric, and affinity viewpoints.
- The isotherm shape can provide information about the interaction mechanism that occurs between the adsorbent and the adsorbate.
- In terms of the adsorption rate modeling, a local equilibrium is generally considered, in order to solve the partial differential equations. This local equilibrium is mathematically described by the adsorption isotherms.
- Thermodynamic adsorption parameters, such as standard Gibbs free energy change (ΔG^0), standard enthalpy change (ΔH^0), and standard entropy change (ΔS^0), can be found from the isotherms. These parameters are fundamental to verify the spontaneity and nature of the adsorption operation.

In the light of this knowledge, the equilibrium isotherms should be examined carefully. Firstly, the equilibrium experiments should be performed with several experimental points. After, the isotherm curves should be correctly classified. Suitable models should be fitted to the curves, in order to find an adequate representation and consistent parameters. For this, a correct statistic treatment is necessary. Finally, the isotherms can be used to find information about the adsorption operation.

2.2 Experimental Procedures to Obtain Equilibrium Curves

For the construction of adsorption isotherm, it is necessary that a series of equilibrium concentration data of the liquid phase with its respective adsorption capacity is obtained. These data not only should be in temperature equilibrium but also in all other system conditions, such as adsorbent characteristics, agitation, solution volume, and especially pH, in the case of adsorption in liquid phase. Then, in the batch adsorption tests, an adsorbent dosage (m) is mixed with a certain volume (V_0) of a solution at an initial solute concentration (C_0).

In this context, a number of methods have been standardized to obtain the adsorption isotherms and this section will handle some of these methods. The American Society for Testing and Materials (ASTM) reports two methodologies for the determination of adsorption isotherms in liquid phase. The ASTM D-3860 is the standard practice for the determination of adsorptive capacity of activated carbon by aqueous phase isotherm technique. This practice covers the determination of the adsorptive capacity of activated carbon to remove undesirable constituents from water and wastewater. The method suggested by the ASTM, in this case, reports that different adsorbent dosages are placed into contact with a solution containing known solute concentration. Thus, when equilibrium is reached, different values of equilibrium concentrations (C_e) are obtained and the adsorption capacities (q_e) are calculated by Eq. (2.4). According to the method, for the activated carbon, usually after few hours equilibrium is reached. However, the time required for equilibrium to be reached will be treated later.

Already the ASTM D-4706 test method covers the determination of the relative activation level of unused or reactivated carbons by adsorption of iodine from aqueous solution. This test method is based upon a three-point adsorption isotherm, and the standard iodine solution is treated with three different weights of activated carbon under specified conditions, according to the ASTM D-3860. The equilibrium data are processed using the Langmuir model (which will be discussed later) and the iodine number is the maximum adsorption capacity of the monolayer (q_m).

Van Den Hul and Lyklema (1968) and Hang and Brindley (1970) proposed a method for determining the available surface area for adsorption in an aqueous medium based on adsorption isotherm Methylene Blue dye (CAS No. 61-73-4). This method is based on the adsorption of a layer of Methylene Blue on the surface (internal and external) material. Based on the maximum monolayer adsorption capacity (q_m) obtained by Langmuir model (Eq. (2.18)) that is treated subsequently, it is possible to obtain the number of molecules adsorbed per unit area (the projected area of the Methylene Blue molecule is 1.08×10^{-18} m^2/molecule) and the total area (a_p, in m^2/g) of the adsorbent according to Eq. (2.9):

$$a_p = 1.7388 q_m \tag{2.9}$$

Although the methods of ASTM suggest that the adsorption isotherms are performed by different dosages of adsorbent, more recent works by adsorption, especially those that use nonconventional adsorbents, have chosen to use fixed adsorbent dosages, and different point isotherms are obtained varying the solute initial concentration of the solution. By this method, variations in the adsorption system conditions are less susceptible. In this case, for example, stock solutions containing 300 or 400 mg/L (or more) are diluted in the ratio 1:1 (solution/solvent) obtaining different initial conditions. Both methods (different dosages of adsorbent and different initial concentrations of the solute solution) lead to the same result if the technique is properly developed.

However, we want to draw attention to the considerations regarding the adsorption equilibrium. Several authors have presented data kinetic adsorption capacity justifying that equilibrium is achieved within hours after consecutive measurements performed in relatively short times (few minutes or a few hours) showing similar results. However, we consider that for determining the correct balance, this should be measured in longer periods of time, i.e., from 8 to 12 h, being performed until there are no observed changes in the equilibrium concentration. The evaluation of these changes can be detected by lower coefficients of variation of 5% in the equilibrium concentration in three consecutive measurements, as suggested by some authors.

2.3 Classification of the Equilibrium Isotherms

As described above, the equilibrium isotherms show the amount of adsorbate that can be adsorbed by the adsorbent (q_e) in relation to the equilibrium concentration of the adsorbate in fluid phase (C_e). These are critical parameters in the adsorption system design. Furthermore, the shape of the equilibrium curve helps to explain certain phenomena associated with the interaction between the adsorbate and adsorbent. Therefore, the isotherm shape not only provides information on the affinity between the molecules but also reflects the possible mode of interaction between adsorbate and adsorbent (Wong et al. 2004).

The classification of liquid–solid adsorption isotherms describes a system (Giles et al. 1960) and suggests how their form can be used to diagnose the adsorption mechanism, in order to obtain information regarding the physical nature of the adsorbate and the adsorbent surface and also to measure the specific surface area of the adsorbent. In this classification, the equilibrium curves are identified according

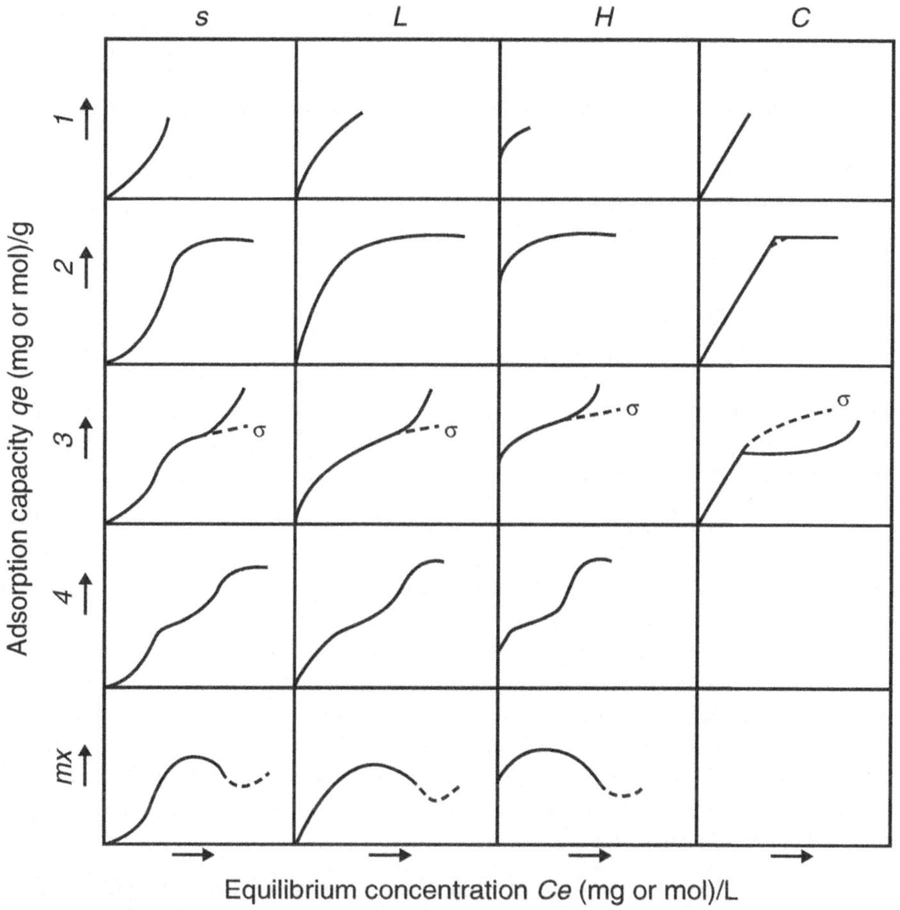

Fig. 2.3 Adsorption isotherm classification (Giles et al. 1960)

to the initial slope into four main classes, and subgroups are described for each class, based on the shapes of the upper parts and slope changes. Figure 2.3 shows the classification proposed by Giles et al. (1960).

The main classes are (i) S curves or vertical orientation isotherm, (ii) L curves or normal or "Langmuir" isotherms, (iii) H curves or high affinity isotherms, and (iv) C curves or constant partition isotherm.

S Curves As can be seen in Fig. 2.3, the S type isotherm has an inclined slope of the curve followed by a vertical orientation. Initially, when the adsorbate concentration increases, there is a chance of the adsorbate to find an available site so that it can occupy, due to competition between solute molecules. Thus, the adsorption capacity is "limited," reaching a plateau. However, this behavior in type S isotherms is opposed, causing the increase of curve slope. This is due to a vertical orientation tendency of the solute molecules in a higher concentration, and then more sites are

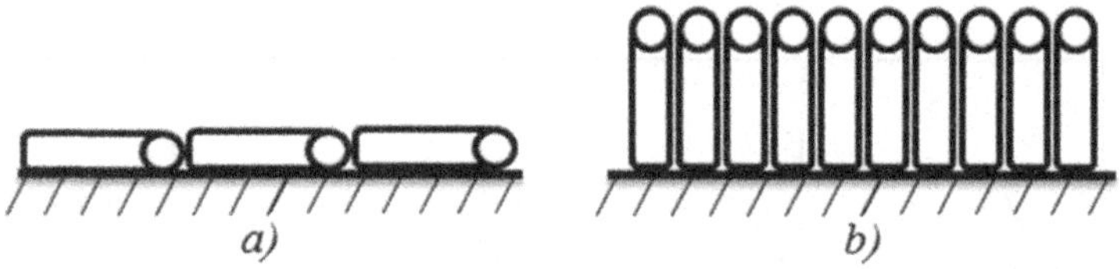

Fig. 2.4 Schematic representation of the molecule orientation in S type isotherms: (**a**) at low concentration; (**b**) at high concentration (Giles et al. 1960)

available for adsorption. In practice, the S curve usually appears when three conditions are fulfilled: (a) the solute molecule is monofunctional; (b) there is a moderate intermolecular attraction, leading to pack vertically in regular array in the adsorbed layer; and (c) there is strong competition, for substrate sites, from molecules of the solvent or another adsorbed species. Figure 2.4 shows a schematic representation of the orientation of the molecules in S type isotherms.

L Curves The normal or Langmuir isotherms are most commonly found in solute adsorption in aqueous solution. The initial shape of the equilibrium curve follows the basic premise that the higher the solute concentration, the greater the adsorption capacity until the number of adsorption site clearance is limited, occurring competition between solute molecules for the available sites. Usually, it is an indicative that the molecules are adsorbed flat on the surface or, sometimes, of vertically oriented adsorbed ions with particularly strong intermolecular attraction. Thus they have one of the following characteristics: (i) the adsorbed molecules are more likely to be adsorbed flat or (ii) are systems with high polar solute and substrate. This isotherm type indicated that the adsorption occurs due to relatively weak forces, such as van der Waals forces.

H Curves The basic difference between the normal isotherms or L type with the high affinity isotherm relates to the beginning of the equilibrium curve. While L type isotherm has its beginning in the origin, H type isotherm shows an initial portion with a vertical orientation, and q_e values are higher than zero, even when the concentration of solute tends to values close to zero. The adsorbed species are often large units, for example, ionic micelles or polymeric molecules However, sometimes, they can be simple ions, which exchange with others of much lower affinity with adsorbent surface, for example, sulfonated dye. This isotherm type indicates chemisorption and adsorption by electrostatic forces. Other classifications commonly used for H type isotherms are like an irreversible isotherm, because when an adsorption occurs at a high concentration, a concentration reduction does not change the adsorption capacity.

C Curves The isotherms with partition constant are characterized by a linear behavior of the equilibrium data at low concentrations of solute. This behavior follows Henry's law for ideal gas equilibrium phases, which translated to adsorption processes, and suggests that the adsorption capacity is proportional to the solute concentration, up until the maximum possible adsorption, where an abrupt change to a horizontal plateau occurs. This is the type of curve obtained for the partition of a solute between two practically immiscible solvents. In such cases, the affinity of the solute by the solid is greater than the affinity for the solvent, or when the adsorption sites are available in quantities sufficient for the adsorption of all solute, but the bonding forces between the solute and the solvent are weak and depend on the liquid phase concentration.

2.3.1 Subclasses

The subclass 1 of S type isotherm indicates a complete vertical behavior of the adsorption capacity, possibly caused by surface precipitation of solute on the surface of the adsorbent. In the case of classes L, H, and C, they occur when the adsorption sites were not fully occupied, or there was not a complete vertical orientation of the molecules of the solvent. This isotherm type is usually described by the Freundlich model (for the case of L and H type) or Henry's law (for the C type).

The subclass 2 indicates that there is no intermolecular interaction between the solute, forming a long plateau, indicating a saturation of the adsorbent monolayer. In this case, a high energy barrier should be overcome before the additional adsorption can occur on new sites, after the surface has been saturated to the first degree. Therefore, the solute has high affinity for the solvent, but low affinity for the layer of solute molecules already adsorbed. In this case, equilibrium data can be represented by Langmuir model and the plateau is represented by the maximum adsorption capacity (q_m) (for the case of L and H isotherm type).

In subclass 3, a short plateau must mean that the adsorbed solute molecules expose a surface, which has nearly the same affinity for more solute as the original surface possessed. This indicates that the solute in the solution has some intermolecular interaction with the solute in the adsorbent surface, leading to the formation of multilayers.

The subclasses 4 are attributed to the development of a fresh surface in which adsorption can occur. The second plateau represents the complete saturation of the new surface. This additional layer may occur when (i) a proportion of the original surface may be uncovered by reorientation of the molecules already adsorbed, due to intermolecular interactions, (ii) formation of new surfaces in crystalline solids, generating new adsorption sites, or (iii) already exposed parts that allow the formation of two layers, for example, due to formation of micelles.

Finally, the *mx* subclass occurs occasionally when a fall in slope occurs after the first inflection. This is probably due to the association of the solutes in solution; with increase in concentration, the solute–solute attraction begins to increase more rapidly than the adsorbent–solute attraction.

2.4 Adsorption Isotherm Models

2.4.1 Henry's Law

Henry's law can be applied for the adsorption on a uniform surface at sufficiently low concentrations, in which all molecules are isolated from their nearest neighbors. The relationship between the fluid phase concentration and the adsorbed phase equilibrium concentration is linear, with a constant of proportionality, which is equal to the adsorption equilibrium constant, known as the Henry constant (K_H). This linear relationship is commonly referred to as Henry's law by analogy with the limiting behavior of dissolution of gases in liquids. The constant of proportionality, which is simply the adsorption equilibrium constant, is referred to as the Henry constant (K_H) and may be expressed in terms of concentration:

$$q_e = K_H C_e \tag{2.10}$$

For physical adsorption, there is no change in molecular state of adsorption, i.e., for adsorption on a uniform surface at sufficiently low concentration, all molecules are isolated from their nearest neighbors. The equilibrium relationship between fluid phase and adsorbed phase concentration will be linear, and the relation to the surface concentration (n_s) can be presented in Eq. (2.11):

$$n_s = \frac{K_H}{a} C_e \tag{2.11}$$

where a is the specific surface area per unit volume of the adsorbate (Ruthven 1984).

2.4.2 Monolayer Adsorption and the Langmuir Isotherm

The adsorbent and the adsorbate are in dynamic equilibrium, and the fractional coverage of the surface depends on the concentration of the adsorbate. The extent of surface coverage is normally expressed as the fractional coverage, θ (Langmuir 1918):

$$\theta = \frac{\text{Number of adsorption sites occupied}}{\text{Number of adsorption sites available}} \tag{2.12}$$

The physical simplicity of the isotherm is based on four assumptions (Atkins and De Paula 2006):

- Adsorption cannot occur beyond monolayer coverage.
- Each site can hold only one adsorbate molecule.

- All sites are energetically equivalent and the surface is uniform.
- The ability of a molecule to adsorb in a given site is independent of the occupation of neighboring.

In the dynamic adsorption equilibrium, the adsorption and desorption rates are the same, so

$$A(\mathrm{aq}) + M(\mathrm{surface}) \Leftrightarrow AM(\mathrm{surface}) \tag{2.13}$$

The rate of surface coverage due to adsorption is proportional to the solution concentration C_{A} of A and the number of vacant sites $N(1-\theta)$, where N is the total number of sites and can be expressed as

$$\frac{d\theta}{dt} = k_a C_{\mathrm{A}} N(1-\theta) \tag{2.14}$$

The change of θ due to desorption is proportional to the number of adsorbed species $N\theta$, so

$$\frac{d\theta}{dt} = -k_d N\theta \tag{2.15}$$

where the kinetic constants are k_a for adsorption and k_d for desorption.

At equilibrium, there is no change in the composition in both phases (the sum of these two rates is equal to zero), and solving for θ results in the Langmuir isotherm:

$$\theta = \frac{K_{\mathrm{L}} C_e}{1 + K_{\mathrm{L}} C_e} \tag{2.16}$$

where the Langmuir constant (K_{L}) is

$$K_{\mathrm{L}} = \frac{k_a}{k_d} \tag{2.17}$$

Considering the exchange of molecules between adsorbed and liquid phase, the fraction covered can be considered the relation between the adsorption capacity at equilibrium (q_e) and the maximum adsorption capacity, which occur when all sites of the monolayer are occupied (q_m).

$$\theta = \frac{q_e}{q_m} \tag{2.18}$$

Replacing Eq. (2.18) in Eq. (2.16), the Langmuir equation becomes

$$q_e = \frac{q_m k_{\mathrm{L}} C_e}{1 + k_{\mathrm{L}} C_e} \tag{2.19}$$

2.4.3 *Multilayer Adsorption and the BET Isotherm*

When the initial adsorbed layer becomes a surface for further adsorption, instead of the isotherm stabilized in a saturated monolayer, the formation of multilayers can be expected. The most widely used isotherm for the multilayer adsorption was derived by Brunauer et al. (1938) and is called the BET isotherm. In liquid systems, the BET isotherm is (Ebadi et al. 2009)

$$q_e = \frac{q_{\mathrm{BET}} k_1 C_e}{(1 - k_2 C_e)(1 - k_2 C_e + k_1 C_e)} \tag{2.20}$$

where q_{BET} is the monolayer adsorption capacity (mg/g), k_1 and k_2 are the BET constants (L/mg).

2.4.4 *Other Isotherm Models*

In the Langmuir isotherm model, the independence and energetic equivalence of the adsorption sites are attributed. Deviations of this supposition can be identified using other models. Various attempts have been made to take these variations into account.

2.4.4.1 Temkin Isotherm

The Temkin isotherm equation assumes that the heat of adsorption of all the molecules in the layer decreases linearly with coverage due to adsorbent–adsorbate interactions and that the adsorption is characterized by a uniform distribution of the binding energies, up to some maximum binding energy. The Temkin model is given by

$$\theta = \frac{RT}{\Delta Q} \ln (K_{\mathrm{T}} C_e) \tag{2.21}$$

where θ is the fractional coverage (defined in Eq. 2.17), R is the universal gas constant (kJ/mol K), T is the temperature (K), $\Delta Q = -\Delta H$ is the variation of adsorption energy (kJ/mol), and K_{T} is the Temkin equilibrium constant (L/mg).

2.4.4.2 Freundlich Isotherm

The Freundlich isotherm assumes that the adsorption occurs on a heterogeneous surface, and the amount that is adsorbed increases infinitely with an increase in concentration (Freundlich 1906). In liquid phase, this isotherm is given by

$$q_e = k_F C_e^{1/n_F} \tag{2.22}$$

where k_F is the Freundlich constant (a common error is noted in k_F unit, i.e., mg^{1-c} L^c/g, where $c = 1/n_F$) and $1/n_F$ is the heterogeneity factor. This isotherm attempts to incorporate the role of adsorbate–adsorbate interactions on the surface.

2.4.4.3 Dubinin–Radushkevich (D-R) Isotherm

The Dubinin–Radushkevich (D-R) isotherm model considers that adsorbent size is comparable to the micropore size, and the adsorption equilibrium relation for a given adsorbate–adsorbent combination can be expressed independently of temperature by using the adsorption potential (ε), according to Eq. (2.23):

$$\varepsilon = RT \ln\left(1 + \frac{1}{C_e}\right) \tag{2.23}$$

The D-R isotherm assumes a Gaussian-type distribution for the characteristic curve and the model can be described by Eq. (2.24):

$$q_e = q_{\max} \exp\left(-\beta \varepsilon^2\right) \tag{2.24}$$

where $q_{\max}$ is the D-R constant (mg/g) and β gives the mean sorption free energy E (kJ/mol) at the moment of its transfer to the solid surface from the bulk solution and can be computed using Eq. (2.25):

$$E = \frac{1}{(2\beta)^{1/2}} \tag{2.25}$$

2.4.4.4 Redlich–Peterson (R-P) Model

The Redlich and Peterson (1959) developed an empirical isotherm model at three parameters used to represent the adsorption equilibrium over a wide concentration range and can be applied in either homogeneous or heterogeneous systems due to its versatility. The R-P model combines elements of Langmuir and Freundlich models and is shown in Eq. (2.26):

$$q_e = \frac{k_R C_e}{1 + a_R C_e^{\beta}} \tag{2.26}$$

where k_R and a_R are the R-P constant (L/g and L^{β}/mg^{β}, respectively) and β is the exponent, which can vary between 1 and 0.

Besides the R-P model, a number of other isotherm models of three and four parameters were developed empirically. However, most are simple modifications of Langmuir and Freundlich models, without a significant relevance in adsorption studies.

2.4.5 *Statistical Physics Models*

Some models based on statistical physics are used to fit and interpret the adsorption isotherms in liquid phase. The hypotheses of the statistical physics models are more complicated and are developed by using the canonical ensemble in statistical physics. Consequently, the interpretations of the adsorption process using the statistical physics models are more useful. The statistical physics models have physicochemical parameters, which are able to explain the adsorption from the macroscopic and microscopic viewpoints.

The statistical physics models suppose that a variable number of ions/molecules are adsorbed on N_M receptor sites per unit surface (identical receptor sites) and independent receptor sites (N_{M1} and N_{M2}) of the adsorbent surface. To establish the statistical physics models, it is necessary to write the expression of the partition function of one receptor site. The general expression is given by

$$Z_{gc} = \sum_{N_i=0,1\ldots} e^{-\beta(-\varepsilon_i-\mu)N_i} \tag{2.27}$$

where $(-\varepsilon_i)$ is the adsorption energy of receptor site, μ is the chemical potential of receptor site, N_i is the occupation state of receptor site, and β is defined as $1/k_BT$ (where k_B is the Boltzmann constant and T the absolute temperature).

If the receptor sites are identically related to N_M receptor sites, the total grand canonical partition function is written as

$$Z_{gc} = \left(z_{gc}\right)^{N_M} \tag{2.28}$$

But, if the receptor sites are independent (two types of receptor sites), the total grand canonical partition function is given by

$$Z_{gc} = \left(z_{gc_1}\right)^{N_{M1}} \left(z_{gc_2}\right)^{N_{M2}} \tag{2.29}$$

According to the literature, the average site occupation number N_o can be written as

$$N_o = k_BT\frac{\partial \ln Z_{gc}}{\partial \mu} \tag{2.30}$$

When the thermodynamic equilibrium is reached, the equality between the chemical potentials can be written as $\mu_m = \mu/n$ where μ is the chemical potential of the adsorbed ions/molecules, n is the number or fraction of ions/molecules per site, and μ_m is the chemical potential of dissolved ions/molecules as

$$\mu_m = k_B T \ln \left(\frac{N}{z_{tr}} \right) \tag{2.31}$$

where z_{tr} is the translation partition function, according to

$$z_{tr} = V \left(\frac{2\pi m k_B T}{h^2} \right)^{3/2} \tag{2.32}$$

where m is the adsorbed mass, h is the Planck constant, and V is the volume of studied system. Finally, the adsorbed quantity as function of concentration describing the expression of the statistical physics model is given by

$$Q = nN_0 \tag{2.33}$$

By the application of the general methodology describing the statistical physics model development, the expressions of the discussed models are presented in Table 2.1.

2.4.6 *Typical Values of Isotherm Parameters for Different Adsorbate–Adsorbent Systems*

Table 2.2 presents typical values of the parameters of the isotherm models for different adsorption systems in liquid phase. Henry's law is applied to represent C1 isotherm type, according to Giles et al. (1960) classification, when the number of

Table 2.1 Statistical physics models based on the general partition function

Model	Equation	
Monolayer model with one energy	$Q = \frac{Q_0}{1 + \left(\frac{c_{1/2}}{c}\right)^n}$	(2.34)
Monolayer model with two energies	$Q = \frac{n_1 N_{1M}}{1 + \left(\frac{c_1}{c}\right)^{n_1}} + \frac{n_2 N_{2M}}{1 + \left(\frac{c_2}{c}\right)^{n_2}}$	(2.35)
Double layer model with two energies	$Q = nN_M \frac{\left(\frac{c}{c_1}\right)^n + 2\left(\frac{c}{c_2}\right)^{2n}}{1 + \left(\frac{c}{c_1}\right)^n + \left(\frac{c}{c_2}\right)^{2n}}$	(2.36)
Multilayer model with saturation	$Q = nN_M \frac{[F_1(c) + F_2(c) + F_3(c) + F_4(c)]}{[G(c)]}$	(2.37)

Table 2.2 Isotherm parameters for several adsorbent–adsorbate systems

Adsorbent	Adsorbate	Parameters		References
Henry law		k_H (L/g)		
Fungal biomass	Acid yellow 49	0.848		Russo et al. (2010)
Chromium tanned leather waste	Acid yellow 194	1.909		Piccin et al. (2013)
Calcined diatomite	Methylene Blue	0.26		Khraisheh et al. (2005)
	Reactive black	0.36		
	Reactive yellow	0.17		
Diatomaceous earth	Reactive black	0.42		Al-Ghouti et al. (2003)
	Reactive yellow	0.18		
Langmuir model		k_L (L/mg)	q_m (mg/g)	
Sugarcane bagasse	Ni(II)	0.173	64.1	Dotto et al. (2016b)
Chitosan/bentonite composite	Methylene Blue	2.055	496.5	Dotto et al. (2016c)
Chitosan	FD&C red 40	0.395	372.9	Piccin et al. (2009)
Papaya seeds	Tartrazine	0.373	51.0	Weber et al. (2014)
Fungal biomass	Acid blue 62	0.024	300	Russo et al. (2010)
Natural palygorskite	Methylene Blue	0.010	187.3	Zhang et al. (2015)
S. melongena leaf powder	Pb(II)	0.274	71.4	Yuvaraja et al. (2014)
$CoFe_2O_4$ graphene oxide	Pb(II)	0.095	299.4	Zhang et al. (2014)
	Hg(II)	0.237	157.9	
Activated carbon	Lanaset gray G	0.208	108.7	Baccar et al. (2010)
Activated carbon	Anthracene	0.260	8.35	Saad et al. (2014)
Activated carbon	Fluoxetine	0.375	1112.0	Nabais et al. (2008)
Freundlich model		k_F ($mg^{1-c}L^c/g$)	n_F	
Ca-bentonite	Congo red	26.91	3.23	Lian et al. (2009)
Dried soya bean meal	Reactive red 2	5.268	1.079	Sahadevan et al. (2009)
NKA-2 resin	Flavones	23.11	1.789	Chen and Zhang (2014)
Fe_3O_4 nanospheres	Neural red	10.62	2.17	Iram et al. (2010)

Activated carbon	Polyphenols	6787.97		0.112	Marsal et al. (2009)
Acid-treated oil shale ash	Deltamethrin	2.406		2.01	Al-Qodah et al. (2007)
Cationic polymer-loaded bentonite	Acid scarlet GR	15.35		4.484	Li et al. (2010)
	Acid dark blue 2G	20.58		5.319	
Cashew nut shell	Congo red	1.357		2.279	Kumar et al. (2010)
Cationic polymer/bentonite	Acid dark blue 2G	12.834		4.47	Qian Li et al. (2010)
Fungal biomass	Acid red 266	114		3.7	Russo et al. (2010)
Temkin model		k_T (L/g)		ΔQ (J/mol)	
Activated carbon from waste biomass	Methylene Blue	33.8		1307	Nunes et al. (2009)
Zeolites	Phenol	16.7		537.6	Yousef et al. (2011)
Surface modified tannery waste	Rhodamine B	1.04		69.9	Anandkumar and Mandal (2011)
	Cr (VI)	7.06		117.3	
BET model		k_1 (L/mg)	k_2 (L/mg)	q_{BET} (mg/g)	
Perfluorooctyl alumina	MTBE	1.2251×10^{-2}	4.6965×10^{-4}	3.5	Ebadi et al. (2007)
Chitosan films	Vanadium	1.3×10^{-2}	2.97×10^{-3}	90.13	Cadaval et al. (2016)
Formaldehyde-pretreated *Pinus pinaster* bark	Phenol	37.84	6.19×10^{-2}	0.086	Vázquez et al. (2007)
Chromium-tanned leather waste	Acid black 210	0.522	1.7×10^{-3}	108.8	Piccin et al. (2013)
Carbonized bark	Pentachlorophenol	16.224	0.184	0.841	Ebadi et al. (2009)[a]
RP model		k_R (L/mg)	α_R (L^β/mg^β)	β	
Cashew nut shell	Congo red	5.548	3.186	0.638	Kumar et al. (2010)
Zeolites	Phenol	51.5	6.28	0.72	Yousef et al. (2011)

[a]Using data of Edgehill (1998)

adsorption sites is very superior to the number of adsorbate molecules, due to a hydrophobic interaction between the adsorbent and adsorbate. The increase of k_H constant represents an increase of adsorption capacity at low concentrations.

The Langmuir model is satisfactory to represent H2 or L2 isotherm type, according to Giles et al. (1960) classification. In practice, the constant k_L is associated with increased affinity of the adsorbate by adsorbent, since k_L represents the inverse of the equilibrium concentration in the liquid phase when the adsorption capacity reaches 50% of the monolayer adsorption capacity (or $f(1/k_L) = 0.5q_m$, where f is the function of the Langmuir isotherm). Therefore, k_L increase leads to a higher initial slope of the adsorption isotherm. On the other hand, q_m is associated with the curve plateau formation and complete saturation of the monolayer adsorbate. q_m is in the order from unity to tens of milligrams per gram in the case of monatomic ion adsorption and in the order from hundreds to thousands of milligrams per gram for dyes and larger molecule adsorption. However, Table 2.2 shows that the maximum monolayer adsorption capacity can vary due to many factors, such as chemical structure of the adsorbate and adsorbent, molecular size, and nature of the adsorbent.

The Freundlich model is satisfactory to describe the adsorption isotherm data of types S, L, and C (subclass 1). The $0 < n_F < 1$ when the isotherm is of class S (or unfavorable), $n_F > 1$ when the isotherm is of class L (or favorable), and $n_F = 1$ when the isotherm is of class C. In the latter case, usually when the number of adsorption sites is greater than the number of molecules to be adsorbed, the Freundlich model is simplified to Henry model. Already, the k_F values are associated with the initial slope of the isotherm curve.

The BET isotherm is an extension of the Langmuir theory for monolayer adsorption to multilayer adsorption, and it is satisfactory to represent the H3 or L3 isotherm type, according to the Giles et al. (1960) classification. q_{BET} and k_1 have the same physical significance as to that of q_m and k_L Langmuir constant, respectively.

Then, an increase in the equilibrium concentration leads to an increase of adsorption capacity. This behavior is due to secondary adsorption at a given site, forming a multilayer and providing a suitable adjustment to the BET model. The multilayer formation may occur due to a change in organizational form of dye molecules arranged on the surface of the adsorbent, in horizontal to vertical alignment, or due to solubility reduction caused by superficial hydrophobic interactions between the adsorbate and the adsorbent (Piccin et al. 2013). The k_2 constant represents the inverse of the concentration value when the isotherm becomes a vertical line and is associated with superficial solubility of adsorbate ($C_s = 1/k_2$). The k_2 value increment is due to a vertical orientation of the isotherm at lower equilibrium concentration in the liquid phase. In this case, Ebadi et al. (2009) demonstrated that the use of adsorbent solubility concentration leads to serious errors in the interpretation of the adsorption data. According to the authors, k_2 or C_s must be obtained by adjusting the model to experimental data. When k_2 tend to

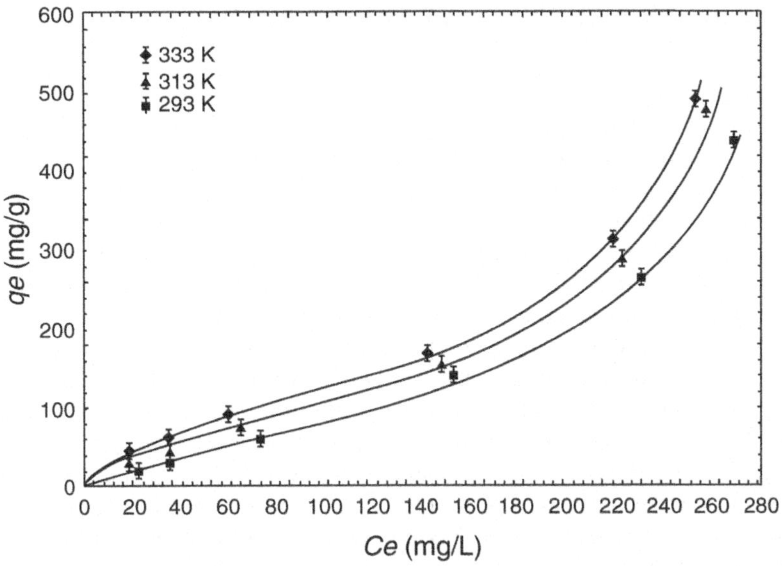

Fig. 2.5 Equilibrium isotherms following the BET multilayer model for vanadium adsorption onto chitosan films (Cadaval et al. 2016)

zero, the BET model can be mathematically simplified to the Langmuir model, releasing degrees of freedom for the model and simplifying the use of the data (Piccin et al. 2013).

Figure 2.5 shows the equilibrium curves of vanadium adsorption onto chitosan films at different temperatures (Cadaval et al. 2016). In this example, the temperature increase led to an increase in the q_{BET} value from 90.9 to 102.3 mg/g. This behavior suggests an endothermic phenomenon. The k_1 and k_2 values increased with increasing temperature. The k_1 increase indicates that lower equilibrium concentrations are necessary to saturate the monolayer, and the k_2 increase indicates that smaller equilibrium concentrations are necessary to the isotherm that becomes a vertical line. However, Piccin et al. (2013) observed contrary behavior toward the concentration of surface saturation (C_s). In this case, a temperature reduction led to a reduction of C_s values, indicating that the highest adsorption capacities by multilayer formation were obtained with lower equilibrium concentrations. This reduction is associated with the solubility of the dye in water, generally lower at high temperatures, and the increased adsorbate–absorbent forces at low temperatures.

Regarding the Redlich–Peterson model, and other models of three and four parameters reported in the literature (Wong et al. 2004; Piccin et al. 2009; Kumar et al. 2010; Yousef et al. 2011), in most cases these do not provide suitable adjustments to the experimental data to the point of its use to be justified. In this case, simpler models such as Langmuir or Freundlich are satisfactory to represent the experimental data.

2.5 Regression Methods and Error Analysis

Several models are able to describe the experimental results of adsorption and are used in equilibrium, kinetics, and mechanisms studies. In the process of statistically analyzing empiric data, errors that lead to the unreliable results can occur. The misuse of linearization is a common error in data analysis. When the data are transformed in order to achieve a linear equation, it is required to know how the error-structure of the data is affected by this tranformation. When the errors are additive on the dependent variable and the usual assumptions of normality and equal variance throughout the range of the data are checked, then the transformation of the dependent variable with a nonlinear function can eliminate the distributional properties. This can occur in the linearization of adsorption isotherms; for example, the Langmuir model (Eq. 2.19) is nonlinear; thus the dependent variable does not depend linearly on the independent variable (El-Khaiary and Malash 2011).

These nonlinear forms can be mathematically manipulated and linearized at different linear forms. Moreover, the statistical tests used to check the fit will often not detect that the parameters are biased. Table 2.3 presents the different linearizations to the Langmuir and Freundlich models.

The statistical regression methods consist in minimizing objective functions through the variation of the model parameters. Table 2.4 summarizes some of these functions and its main characteristics.

From Eqs. (2.42), (2.43), and (2.44), *SSE*, R^2, and *ABS* provide a better fit for higher y_{exp} values, because errors are proportional to their magnitudes. *SSE* is the most common error function in use. In the R^2 function, the objective is to maximize the results, and the *ABS* is similar to the *SSE* to some extent. In relation to Eqs. (2.45), (2.46), (2.47), and (2.48), the values of (χ^2), *ARE*, *HYBRID*, and *MPSD* improve the fit at low concentrations by dividing by the experimental value *HYBRID*, and *MPSD* also includes the number of degrees of freedom of the system, which is important in the analysis of models with different parameter numbers.

The main objective function used for both linear and nonlinear regressions is the sum of the squares of the errors (SSE). In the case of linearized forms of the Langmuir model, the function $y = \beta_0 + \beta_1 \cdot x$ represents the experimental data.

Table 2.3 Linearized form of adsorption isotherm models

Model	Linearized form	Plot	
Langmuir I	$\frac{C_e}{q_e} = \frac{1}{k_L q_m} + \frac{1}{q_m} C_e$	$\frac{C_e}{q_e}$ vs. C_e	(2.38)
Langmuir II	$\frac{1}{q_e} = \frac{1}{q_m} + \frac{1}{k_L q_m C_e}$	$\frac{1}{q_e}$ vs. $\frac{1}{C_e}$	(2.39)
Langmuir III	$\frac{q_e}{C_e} = k_L q_m + k_L q_e$	$\frac{q_e}{C_e}$ vs. q_e	(2.40)
Freundlich	$\log(q_e) = \log(k_F) + \frac{1}{n_F} \log(C_e)$	log (q_e) vs. log (C_e)	(2.41)

Table 2.4 Error function used for isotherm model regression

Function name	Error function	
Sum of square error (SSE)	$\mathrm{SSE}=\sum_{i=1}^{n}\left(y_{i,\exp}-y_{i,\mathrm{mod}}\right)^2$	(2.42)
Coefficient of determination (R^2)	$R^2=1-\dfrac{\sum_{i=1}^{n}\left(y_{i,\exp}-y_{i,\mathrm{mod}}\right)^2}{\sum_{i=1}^{n}\left(y_{i,\exp}-\overline{y_{i,\exp}}\right)^2}=1-\dfrac{\mathrm{SSE}}{\mathrm{SST}}$	(2.43)
Sum of the absolute errors (ABS)	$\mathrm{ABS}=\sum_{i=1}^{n}\left\lvert y_{i,\exp}-y_{i,\mathrm{mod}}\right\rvert$	(2.44)
Chi-square (χ^2)	$\chi^2=\sum_{i=1}^{n}\dfrac{\left(y_{i,\exp}-y_{i,\mathrm{mod}}\right)^2}{y_{i,\mathrm{mod}}}$	(2.45)
Average relative error (ARE)	$\mathrm{ARE}=\dfrac{100}{n}\sum_{i=1}^{n}\left\lvert\dfrac{y_{i,\exp}-y_{i,\mathrm{mod}}}{y_{i,\mathrm{mod}}}\right\rvert$	(2.46)
Hybrid fractional error function (HYBRID)	$\mathrm{HYBRID}=\dfrac{100}{n-n_p}\sum_{i=1}^{n}\dfrac{\left(y_{i,\exp}-y_{i,\mathrm{mod}}\right)^2}{y_{i,\mathrm{mod}}}$	(2.47)
Marquardt's percent standard deviation (MPSD)	$\mathrm{MPSD}=100\sqrt{\dfrac{1}{n-n_p}\sum_{i=1}^{n}\left[\dfrac{\left(y_{i,\exp}-y_{i,\mathrm{mod}}\right)}{y_{i,\mathrm{mod}}}\right]^2}$	(2.48)

Where $y_{i,\,\exp.}$ is the experimental value of independent variable, $y_{i,\,\mathrm{mod}}$ is the modeled value, $\overline{y_{\exp}}$ is the mean of observed values, *SST* is the sum of squares of total deviations, n is the number total of informations, and n_p is the number of parameters of the model

Thus, β_0 and β_1 values can be obtained from the minimization of the objective function (SSE) using the method of linear least squares estimation. However, for nonlinear form of the isotherm models, there is no closed method for obtaining parameters. Instead, in this case, numerical algorithms are used to minimize the objective function and obtain the values of the model parameters. Most algorithms involve choosing initial values for the parameters. Then, the parameters are refined iteratively, that is, the values are obtained by successive approximation. The most commonly used algorithms, in this case, are the Gauss–Newton, Levenberg–Marquardt and the Generalized Reduced Gradient.

2.5.1 Model Accuracy

The coefficient of determination (R^2), defined above, to obtain the more suitable model to represent equilibrium and kinetic and thermodynamic parameters is another common practice in adsorption experiments.

The R^2 value is very sensitive to extreme data points, resulting in misleading indication of the fit. The R^2 is also influenced by the range of the independent variable and increases as the range of independent variable increases and decreases as the range decreases. These issues can be avoided by fitting the data to the model without any transformations and by examination of extreme points.

Another relevant fact is that the R^2 can be made manipulated using more parameters in the model, since the increase in the number of regression parameters leads to decreases in *SSE* value. Therefore, the good fit cannot be based only on *SSE* (and R^2). This is especially common in adsorption studies when it comes to estimating equilibrium parameters. For this fact, the analysis of the adjusted determination coefficient (${R_{\text{adj}}}^2$), which takes into account the experimental degrees of freedom ($n-1$) and the model degrees of freedom ($n-(n_p+1)$), can be a good tool in selecting models. ${R_{\text{adj}}}^2$ is described in Eq. (2.49):

$$ {R_{\text{adj}}}^2 = 1 - \left(\frac{n-1}{1-(n_p+1)}\right)(1+R^2) \tag{2.49} $$

where n is the number of information and n_p is the number of model parameters.

Moreover, Akaike's information criterion (AIC) (Anderson and Burnham 2002) is a well-established statistical method that can be used to compare the models with different numbers or parameters. For a small sample size, AIC is calculated for each model from Eq. (2.50):

$$ \text{AIC} = n\ln\left(\frac{\text{SSE}}{n}\right) + 2n_p + \frac{2n_p(n_p+1)}{n-(n_p+1)} \tag{2.50} $$

A smaller AIC value suggests that the model has more likely to show a better fit. The AIC values can be compared using the evidence ratio (Er), which is defined by

$$ \text{Er} = \frac{1}{e^{-0.5\Delta}} \tag{2.51} $$

where Δ is the absolute value of the difference in AIC between the two models. The evidence ratio means how many times one model is more likely than the other one in relation to the experimental data.

2.5.2 Comparison Between Linear and Nonlinear Regression Methods

In this section, we want to show the effect of the regression method used on the parameters of the model and the adjustment to the experimental data. To exemplify this, we use common adsorption data of a FD&C Red No. 40 dye onto chitosan with deacetylation degree of 84%, particle size of 99 μm, and pH 7.0 at 25 °C (Piccin et al. 2009). The model parameters were obtained by the different linearized forms of models (Table 2.5), using the linear least squares estimation method, and by the nonlinear models (Eqs. (2.19), (2.20), (2.21), and (2.22)), using GRG nonlinear algorithm of add-in solver function of MS Excel (Microsoft, USA). Figure 2.6 shows the experimental data. Figure 2.7 shows the linearized data according to Langmuir I, II, and III and Freundlich forms, respectively.

According to Fig. 2.7d, the linearized form of Freundlich model shows a best fit ($R^2 = 0.9797$). However, the forms I and III of Langmuir model have poor adjustment ($R^2 < 0.900$) to the experimental data. However, in Table 2.5, it can be observed that the R^2 and SSE values for Langmuir, independent of the way of obtaining the model parameters, provide a better fit to the experimental data. These data make it clear that the different forms of linearization can lead to serious errors

Table 2.5 Model parameters and respective determination coefficient (R^2) and sum of square error (SSE) from adsorption of acid red No. 40 onto chitosan obtained by different regression methods

Model	Parameters			
Langmuir	$k_L \times 10^3$ (L/mg)	q_m (mg/g)	R^2	SSE
Nonlinearized	4.656	181.6	0.9918	174.41
Langmuir I	4.121	193.6	0.9911	188.63
Langmuir II	2.659	269.4	0.9684	675.36
Langmuir III	3.739	207.4	0.9899	218.65
Freundlich	k_F (mg^{1-c} L^c/g)	n_F	R^2	SSE
Nonlinearized	2.896	1.559	0.9821	383.44
Linearized	1.659	1.328	0.9690	662.51

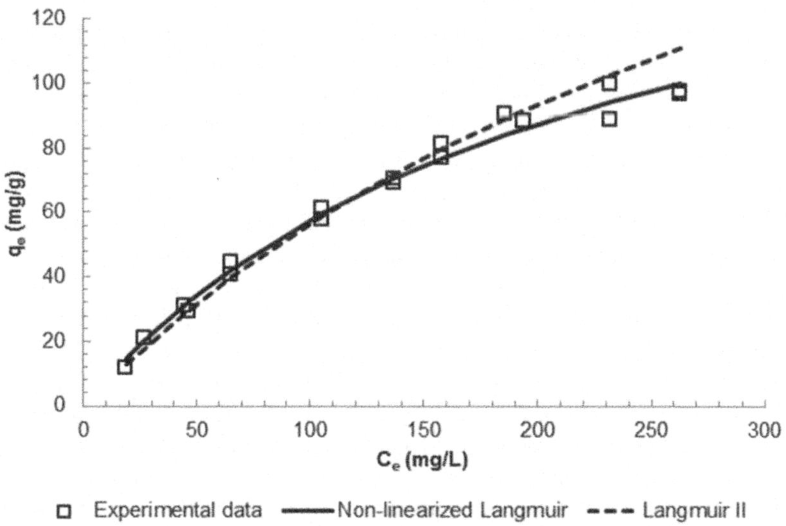

Fig. 2.6 Adsorption equilibrium data of acid red No. 40 dye by chitosan (Piccin et al. 2009)

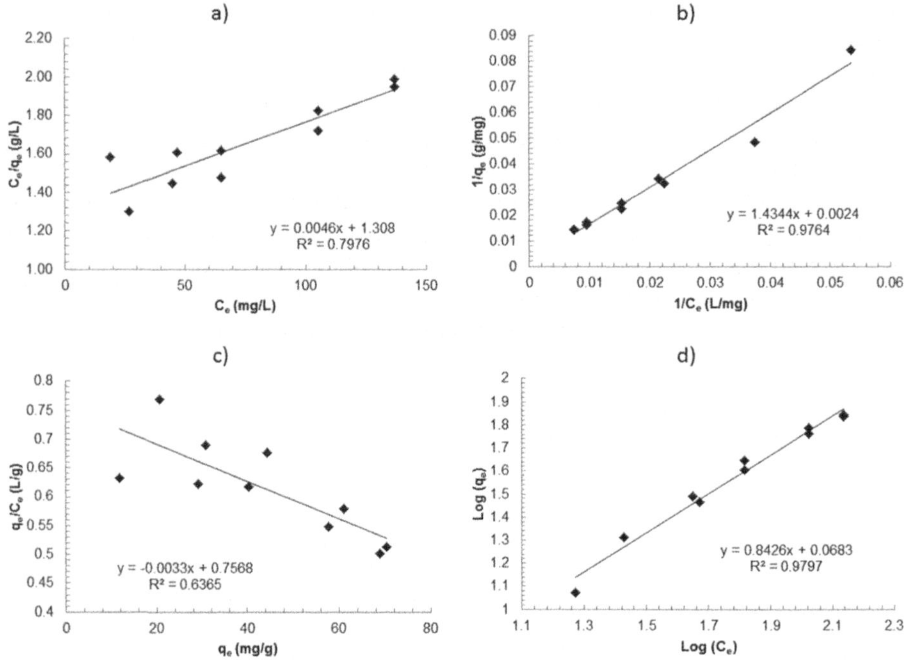

Fig. 2.7 Adjustment of the linearized forms of models to equilibrium experimental data of FD&C red No. 40 dye onto chitosan: (**a**) Langmuir I; (**b**) Langmuir II; (**c**) Langmuir III; and (**d**) Freundlich

in the conclusion of which model is most suitable to represent the experimental data. The use of one variable both as dependent and independent variable, as in the case of Langmuir I (C_e) and Langmuir III (q_e), leads to an inadequate correlation. The error distribution of the dependent variable (C_e/q_e or q_e/C_e) is different from both the error distributions of C_e and q_e. In the case of Langmuir II, the reversal of relative weights of data points because of $1/q_e$ and $1/C_e$ in dependent and independent variables, respectively, leads to distortion of error distribution. This distortion can be observed in Fig. 2.7b, where most of the data are grouped together in a small space of the Cartesian plane, near to the origin. For the Freundlich linearized model, an alteration of relative weights of the data because of $\log(q_e)$ in the dependent variable and a distortion of relative weights of data because of $\log(q_e)$ and $\log(C_e)$ in the dependent and independent variables, leading to distortion of error distribution, occur.

Furthermore, it is observed that Langmuir II form, which has best linearized fit, overestimates the q_m values, when compared to the other forms of the model. In this case, the model fit curve introduces distortions of the experimental data for high adsorption capacity, as shown in Fig. 2.7 (solid lines), which can lead to errors in the analysis and design of the equipment. For this reason, the correct form of the analysis and modeling equilibrium curves is through normalized form (or not linearized) of the equilibrium data. The linearization can be used as a method of visual analysis of the data range to which a model will fit properly to the experimental data (e.g., equilibrium curves with multilayer adsorption), but not as the conclusion about the adjustment of or obtaining the model parameters.

2.6 Adsorption Thermodynamics

In solid–liquid adsorption systems, the estimation of the thermodynamic parameters is essential. In general, the adsorption thermodynamics is studied by the estimation of standard Gibbs free energy change (ΔG^0), standard enthalpy change (ΔH^0), and standard entropy change (ΔS^0) (Crini and Badot 2008). From these parameters, it is possible to verify if the adsorption is favorable, spontaneous, endothermic, or exothermic. It is possible to obtain information regarding the disorder in the solid–liquid interface during the adsorption. Also, it is possible to infer about the adsorption nature, i.e., physisorption or chemisorption, and verify if the operation is controlled by enthalpy or entropy (Ruthven 1984; Dotto et al. 2016a). Indeed, the correct calculation of the adsorption thermodynamic parameters is fundamental.

In the thermodynamic sense, the majority of the studies has considered the adsorption as a reaction (Liu 2009):

$$A + B \longleftrightarrow AB \tag{2.52}$$

The adsorbent (A) interacts with the adsorbate (B), forming AB. When this reaction attains the thermodynamic equilibrium, the chemical potentials in the liquid phase (μ_l) and in the solid–liquid interface (μ_{s-l}) are equal, and the Gibbs free energy change (ΔG) tends to zero, leading to Eq. (2.53) (Chen and Zhang 2014):

$$\Delta G = \mu_{s-l} - \mu_l = \Delta G^0 + RT \ln(K_e) = 0 \tag{2.53}$$

then,

$$\Delta G^0 = -RT \ln(K_e) \tag{2.54}$$

where R is the universal gas constant, T is the temperature, and K_e is the equilibrium thermodynamic constant.

The relationship of ΔG^0 with ΔH^0 and ΔS^0 can be expressed as (Liu 2009)

$$\Delta G^0 = \Delta H^0 - T\Delta S^0 \tag{2.55}$$

Substituting Eq. (2.54) in Eq. (2.55), the following relation is obtained:

$$\ln(K_e) = -\frac{\Delta H^0}{RT} + \frac{\Delta S^0}{R} \tag{2.56}$$

Then, by the plot of ln (K_e) versus (1/*T*), the values of ΔH^0 and ΔS^0 can be found. The graph is known as the Van't Hoff plot. This methodology is used in several works in order to estimate the adsorption thermodynamic parameters (Crini and Badot 2008; Dotto et al. 2015a).

The use of Van't Hoff plot is relatively simple but is dependent of the correct calculation of the equilibrium thermodynamic constant K_e. Indeed, the K_e values are calculated by different manners in the literature, and some of these manners are unreasonable (Milonjic 2007). For example, in most cases, K_e is used with units. However, from the mathematical viewpoint, a parameter that has a dimension cannot be computed logarithmically. The parameter in transcendental functions must be dimensionless; otherwise, the computation for this parameter does not make sense (Zhou et al. 2012). In other cases, K_e is obtained from the distribution constant ($K_e = C_{\mathrm{ad}}/C_e$) (being C_e and C_{ad}, the adsorbate concentrations in solution and in solid phase at equilibrium, respectively). This is valid only at very low adsorbate concentrations. Another way is the use of the isotherm parameters (e.g., Langmuir, $K_e = \rho q_m K_{\mathrm{L}}$). This is not completely correct but is reasonable, since the initial slope of the isotherm can be compared with the Henry constant K_{H} (Dotto et al. 2013).

As presented above, the correct calculation of K_e is a discussed topic, without common sense. Here, a reasonable mean to find K_e is presented. In a solid–liquid adsorption system, the equilibrium thermodynamic constant is given by Eq. (2.57) (Liu 2009):

$$K_e = \frac{\text{activity of occupied sites}}{(\text{activity of vacant sites})(\text{activity of adsorbate in solution})} \tag{2.57}$$

Assuming that the activity of the occupied and unoccupied sites is the same, Eq. (2.57) becomes (Zhou et al. 2012)

$$K_e = \frac{\theta}{(1-\theta)\alpha_e} \tag{2.58}$$

where α_e is the activity of the adsorbate in solution at equilibrium and θ is the fraction of the surface covered at equilibrium. For the Langmuir model, θ is given by Eq. (2.18) (for other isotherms, q_m can be replaced by the parameter relative to the maximum adsorption capacity in mol/g). The activity of a substance can be related to its molar concentration (C_e) according to Eq. (2.59) (Smith et al. 2005):

$$\alpha_e = \frac{\gamma_e C_e}{\gamma^0 C^0} \tag{2.59}$$

where γ_e is the activity coefficient at the adsorption equilibrium, γ^0 is the activity coefficient at the standard state, and C^0 is the molar concentration of the standard reference solution (1 mol/L).

Substituting Eqs. (2.18) and (2.59) in Eq. (2.58), for very dilute solutions ($\gamma_e = \gamma^0$), Eq. (2.60) is obtained (Zhou et al. 2012):

$$K_e = \frac{\frac{q_e}{q_m}}{\left(1 - \frac{q_e}{q_m}\right)\frac{C_e}{C^0}} \tag{2.60}$$

or

$$q_e = \frac{q_m K_e\left(\frac{C_e}{C^0}\right)}{1 + K_e\left(\frac{C_e}{C^0}\right)} \tag{2.61}$$

In this way, the dimensionless K_e can be found by fitting of the Eq. (2.61) with the experimental data of q_e (mol/g) versus C_e (mol/L), considering $C^0 = 1$ mol/L. The regression and the parameter estimation should have good statistical indicators, as presented in Sect. 2.5.

After the correct estimation of the thermodynamic parameters, some important information about the adsorption can be obtained. For example, the negative values of ΔG^0 show a spontaneous and favorable process. The higher the ΔG^0 magnitude, the more favorable and spontaneous the adsorption. Negative values of ΔH^0 indicate an exothermic process, while positive values of ΔH^0 show an endothermic process. The magnitude of ΔH^0 can give an idea about the interactions that occur between the adsorbent and adsorbate. Physisorption, such as van der Waals interactions, is usually lower than 20 kJ/mol, and electrostatic interaction ranges from 20 to 80 kJ/mol. Chemisorption bond strengths can be from 80 to 450 kJ/mol. In relation to the ΔS^0, negative values show that the randomness decreases at the solid solution interface during the adsorption, and positive values suggest the possibility of some structural changes or readjustments in the adsorbate–adsorbent complex. Finally, if ΔH^0 contributes more than the $T\Delta S^0$ to find negative values of ΔG^0, the adsorption is an enthalpy controlled process; otherwise, if $T\Delta S^0$ contributes more than ΔH^0, the adsorption is an entropy controlled process (Crini and Badot 2008; Bergmann and Machado 2015).

2.7 Concluding Remarks

This chapter presented some fundamental aspects about the equilibrium isotherms in liquid phase adsorption, taking into account the academic and industrial viewpoints. In order to obtain accurate and correct interpretations about the adsorption operation from the equilibrium isotherms, some paramount aspects should be remarked:

- The correct determination of the adsorption equilibrium requires at least 8 h of experiment. The experiments should be performed until the liquid phase concentration remains constant (coefficients of variation lower than 5%) after three consecutive measurements.
- The choice of the equilibrium models to be fitted with the experimental data cannot be performed indiscriminately. The experimental equilibrium curves should be classified according to the shape, and then only the adequate models should be used to fit the experimental data.
- For a correct parameter estimation from the isotherm models, the use of nonlinear estimation method is strongly suggested. To ensure the fit quality, the coefficient of determination (R^2) and at least one error analysis should be used. In some cases, the adjusted determination coefficient (R_{adj}^2) and AIC are also necessary.
- The Van't Hoff plot (ln (K_e) versus ($1/T$)) is a simple and reasonable way to find ΔH^0 and ΔS^0 values. However the equilibrium thermodynamic constant (K_e) should be used without units. Otherwise, the thermodynamic parameters have no sense.

References

Al-Ghouti MA, Allen SJ, Ahmad MN (2003) The removal of dyes from textile wastewater: a study of the physical characteristics and adsorption mechanisms of diatomaceous earth. J Environ Manag 69:229–238

Al-Qodah Z, Shawaqfeh AT, Lafi WK (2007) Two-resistance mass transfer model for the adsorption of the pesticide deltamethrin using acid treated oil shale ash. Adsorption 13:73–82

Anandkumar J, Mandal B (2011) Adsorption of chromium (VI) and rhodamine B by surface modified tannery waste: kinetic, mechanistic and thermodynamic studies. J Hazard Mat 186:1088–1096

Anderson DR, Burnham KP (2002) Avoiding pitfalls when using information-theoretic methods. J Wildl Manag 66:912–918

ASTM D3860-98 (2008) Standard practice for determination of adsorptive capacity of activated carbon by aqueous phase isotherm technique. ASTM International, West Conshohocken, PA, USA

ASTM D4607-94 (2006) Standard test method for determination of iodine number of activated carbon. ASTM International, West Conshohocken, PA, USA

Atkins P, De Paula J (2006) Atkins' physical chemistry. Oxford University Press, New York

Baccar R, Blánque P, Bouzid J, Feki M, Sarrá M (2010) Equilibrium, thermodynamic and kinetic studies on adsorption of commercial dye by activated carbon derived from olive-waste cakes. Chem Eng J 165:457–464

Bansal RC, Goyal M (2005) Activated carbon adsorption. CRC Press Taylor & Francis Group, New York

Bergmann CP, Machado FM (eds) (2015) Carbon nanomaterials as adsorbents for environmental and biological applications. Springer, Cham

Brunauer S, Emmett PH, Teller E (1938) Adsorption of gases in multimolecular layers. J Am Chem Soc 60:309–319

Cadaval TRS Jr, Dotto GL, Seus ER, Mirlean N, Pinto LAA (2016) Vanadium removal from aqueous solutions by adsorption onto chitosan films. Desalin Water Treat 57:16583–16591

Chen Y, Zhang D (2014) Adsorption kinetics, isotherm and thermodynamics studies of flavones from *Vaccinium bracteatum* Thunb leaves on NKA-2 resin. Chem Eng J 254:579–585

Crini G, Badot PM (2008) Application of chitosan, a natural aminopolysaccharide, for dye removal from aqueous solutions by adsorption processes using batch studies: a review of recent literature. Prog Polym Sci 33:399–447

Do DD (1998) Adsorption analysis: equilibria and kinetics. Imperial College Press, London

Dotto GL, Moura JM, Cadaval TRS Jr, Pinto LAA (2013) Application of chitosan films for the removal of food dyes from aqueous solutions by adsorption. Chem Eng J 214:8–16

Dotto GL, Nascimento dos Santos JM, Rosa R, Pinto LAA, Pavan FA, Lima EC (2015a) Fixed bed adsorption of methylene blue by ultrasonic surface modified chitin supported on sand. Chem Eng Res Des 100:302–310

Dotto GL, Sharma SK, Pinto LAA (2015b) Biosorption of organic dyes: research opportunities and challenges. In: Sharma SK (ed) Green chemistry for dyes removal from waste water: research trends and applications. Wiley, Hoboken, pp 295–329

Dotto GL, Sellaoui L, Lima EC, Ben Lamine A (2016a) Physicochemical and thermodynamic investigation of Ni(II) biosorption on various materials using the statistical physics modeling. J Mol Liq 220:129–135

Dotto GL, Meili L, de Souza Abud AK, Tanabe EH, Bertuol DA, Foletto EL (2016b) Comparison between Brazilian agro-wastes and activated carbon as adsorbents to remove Ni(II) from aqueous solutions. Water Sci Technol 73:2713–2721

Dotto GL, Rodrigues FK, Tanabe EH, Fröhlich R, Bertuol DA, Martins TR, Foletto EL (2016c) Development of chitosan/bentonite hybrid composite to remove hazardous anionic and cationic dyes from colored effluents. J Environ Chem Eng 4:3230–3239

Ebadi A, Soltan Mohammadzadeh JS, Khudiev A (2007) Adsorption of methyl tert-butyl ether on perfluorooctyl alumina adsorbents-high concentration range. Chem Eng Technol 30:1666–1673

Ebadi A, Mohammadzadeh JSS, Khudiev A (2009) What is the correct form of BET isotherm for modeling liquid phase adsorption? Adsorption 15:65–73

Edgehill RU (1998) Adsorption characteristics of carbonized bark for phenol and pentachlorophenol. J Chem Technol Biot 71:27–34

El-Khaiary MI, Malash GF (2011) Common data analysis errors in batch adsorption studies. Hydrometallurgy 105:314–320

Freundlich HMF (1906) Over the adsorption in solution. J Phys Chem 57:1100–1107

Geankoplis CJ (1998) Procesos de transporte y operaciones unitarias. Compañía Editorial Continental, Ciudad de México

Giles CH, MacEwan TH, Nakhwa SN, Smith D (1960) Studies in adsorption. Part XI. A system of classification of solution adsorption isotherms, and its use in diagnosis of adsorption mechanisms and in measurement of specific surface areas of solids. J Chem Soc 3973–3993

Hang TP, Brindley GW (1970) Methylene blue absorption by clay minerals. Determination of surface areas and cation exchange capacities. Clay Miner 18:203–212

Iram M, Guo C, Guan Y, Ishfaq A, Liu H (2010) Adsorption and magnetic removal of neutral red dye from aqueous solution using Fe_3O_4 hollow nanospheres. J Hazard Mat 181:1039–1050

Khraisheh MAM, Al-Ghouti MA, Allen SJ, Ahmad MN (2005) Effect of OH and silanol groups in the removal of dyes from aqueous solution using diatomite. Water Res 39:922–932

Kumar PS, Ramalingam S, Senthamarai C, Niranjanaa M, Vijayalakshmi P, Sivanesan S (2010) Adsorption of dye from aqueous solution by cashew nut shell: studies on equilibrium isotherm, kinetics and thermodynamics of interactions. Desalination 261:52–60

Langmuir I (1918) The adsorption of gases on plane surfaces of glass, mica and platinum. J Am Chem Soc 40:1361–1403

Li Q, Yue QY, Su Y, Gao BY, Sun HJ (2010) Equilibrium, thermodynamics and process design to minimize adsorbent amount for the adsorption of acid dyes onto cationic polymer-loaded bentonite. Chem Eng J 158:489–497

Lian L, Guo L, Guo C (2009) Adsorption of Congo red from aqueous solutions onto Ca-bentonite. J Hazard Mat 161:126–131

Liu Y (2009) Is the free energy change of adsorption correctly calculated? J Chem Eng Data 54:1981–1985

Marsal A, Bautista E, Ribosa I, Pons R, García MT (2009) Adsorption of polyphenols in wastewater by organo-bentonites. Appl Clay Sci 44:151–155

Milonjic SK (2007) A consideration of the correct calculation of thermodynamic parameters of adsorption. J Serb Chem Soc 72:1363–1367

Nabais JV, Mouquinho A, Galacho C, Carrott PJM, Ribeiro Carrott MML (2008) In vitro adsorption study of fluoxetine in activated carbons and activated carbon fibres. Fuel Process Technol 89:549–555

Nunes AA, Franca AS, Oliveira LS (2009) Activated carbons from waste biomass: an alternative use for biodiesel production solid residues. Bioresour Technol 100:1786–1792

Piccin JS, Vieira MLG, Gonçalves JO, Dotto GL, Pinto LAA (2009) Adsorption of FD&C red no. 40 by chitosan: isotherm analysis. J Food Eng 95:16–20

Piccin JS, Dotto GL, Vieira MLG, Pinto LAA (2011) Kinetics and mechanism of the food dye FD&C red 40 adsorption onto chitosan. J Chem Eng Data 56:3759–3765

Piccin JS, Feris LA, Cooper M, Gutterres M (2013) Dye adsorption by leather waste: mechanism diffusion, nature studies, and thermodynamic data. J Chem Eng Data 58:873–882

Redlich OJDL, Peterson DL (1959) A useful adsorption isotherm. J Phys Chem 63:1024–1024

Rodrigues AE (2015) Simulated moving bed technology: principles, design and process applications. Elsevier, Amsterdam

Rouquerol F, Rouquerol J, Sing KSW (eds) (2014) Adsorption by powders and porous solids: principles, methodology and applications. Elsevier, Amsterdam

Russo ME, Di Natale F, Prigione V, Tigini V, Marzocchella A, Varese GC (2010) Adsorption of acid dyes on fungal biomass: equilibrium and kinetics characterization. Chem Eng J 162:537–545

Ruthven DM (1984) Principles of adsorption and adsorption processes. Wiley, New York

Saad MEK, Khiari R, Elaloui E, Moussaoui Y (2014) Adsorption of anthracene using activated carbon and *Posidonia oceanica*. Arab J Chem 7:109–113

Sahadevan R, Mahendradas DK, Shanmugasundaram V, Shanmugam K, Velan M (2009) Sorption kinetics and equilibrium analysis for the removal of Reactive Red 2 and Reactive Blue 81 dyes from synthetic effluents using dried soya bean meal. Int J Chem React Eng 7:A33

Shafeeyan MS, Daud WMAW, Shamiri A (2014) A review of mathematical modeling of fixed-bed columns for carbon dioxide adsorption. Chem Eng Res Des 92:961–988

Smith JM, Van Ness HC, Abbott MM (2005) Introduction to chemical engineering thermodynamics. McGraw-Hill, Columbus

Suzuki M (1993) Fundamentals of adsorption. Elsevier, Amsterdam

Van den Hul HJ, Lyklema J (1968) Determination of specific surface areas of dispersed materials. Comparison of the negative adsorption method with some other methods. J Am Chem Soc 90:3010–3015

Vázquez G, González-Álvarez J, Garcia AI, Freire MS, Antorrena G (2007) Adsorption of phenol on formaldehyde-pretreated Pinus pinaster bark: equilibrium and kinetics. Bioresour Technol 98:1535–1540

Vieira MLG, Esquerdo VM, Nobre LR, Dotto GL, Pinto LAA (2014) Glass beads coated with chitosan for the food azo dyes adsorption in a fixed bed column. J Ind Eng Chem 20:3387–3393

Weber CT, Collazzo GC, Mazutti MA, Foletto EL, Dotto GL (2014) Removal of hazardous pharmaceutical dyes by adsorption onto papaya seeds. Water Sci Technol 70:102–107

Wong YC, Szeto YS, Cheung W, McKay G (2004) Adsorption of acid dyes on chitosan-equilibrium isotherm analyses. Process Biochem 39:695–704

Worch E (2008) Fixed-bed adsorption in drinking water treatment: a critical review on models and parameter estimation. J Water Supply Res T 57:171–183

Yousef RI, El-Eswed B, Ala'a H (2011) Adsorption characteristics of natural zeolites as solid adsorbents for phenol removal from aqueous solutions: kinetics, mechanism, and thermodynamics studies. Chem Eng J 171:1143–1149

Yuvaraja G, Krishnaiah N, Subbaiah MV, Krishnaiah A (2014) Biosorption of Pb (II) from aqueous solution by *Solanum melongena* leaf powder as a low-cost biosorbent prepared from agricultural waste. Colloid Surf B 114:75–81

Zhang Y, Yan L, Xu W, Guo X, Cui L, Gao L, Wei Q, Du B (2014) Adsorption of Pb (II) and Hg (II) from aqueous solution using magnetic $CoFe_2O_4$-reduced graphene oxide. J Mol Liq 191:177–182

Zhang Y, Wang W, Zhang J, Liu P, Wang A (2015) A comparative study about adsorption of natural palygorskite for methylene blue. Chem Eng J 262:390–398

Zhou X, Liu H, Hao J (2012) How to calculate the thermodynamic equilibrium constant using the Langmuir equation. Adsorpt Sci Technol 30:647–649

Chapter 3
Adsorption Kinetics in Liquid Phase: Modeling for Discontinuous and Continuous Systems

Guilherme Luiz Dotto, Nina Paula Gonçalves Salau, Jeferson Steffanello Piccin, Tito Roberto Sant'Anna Cadaval Jr., and Luiz Antonio Almeida de Pinto

Abstract Adsorption is one of the most widely applied unit operations to separate molecules that are present in a fluid phase using a solid surface. Adsorption kinetic aspects should be evaluated in order to know more details about its mechanisms, characteristics, and possibilities of application. These data can determine the residence time to reach the required concentration of the adsorbate, making possible the design and operation of an adsorption equipment and defining the performance in batch and continuous systems. This chapter presents the particularities of adsorption kinetics in liquid phase. Batch and fixed-bed systems are considered. For discontinuous batch systems, diffusional mass transfer models and adsorption reaction models are discussed. For fixed-bed systems, the shape of breakthrough curves is studied on the basis of mass balance equations and empirical models. Furthermore, the design and scale up of fixed-bed columns are detailed according to the length of unused bed (*LUB*) and bed depth service time (*BDST*) concepts. Several numerical methods are presented in order to solve the required models for batch and fixed-bed systems. Some parameter estimation techniques are discussed in order to obtain the fundamental parameters for adsorption purposes, like mass transfer coefficients and empirical parameters.

Keywords Adsorption • Kinetics • Mass transfer • Liquid phase • Modeling

---

G.L. Dotto (✉) • N.P.G. Salau
Chemical Engineering Department, Federal University of Santa Maria, 1000 Roraima Avenue, 97105-900, Santa Maria, RS, Brazil
e-mail: guilherme_dotto@yahoo.com.br

J.S. Piccin
Food Engineering Department, Passo Fundo University, UPF, Br. 285, Km 171, 99052-900, Passo Fundo, RS, Brazil

T.R.S. Cadaval Jr. • L.A.A. de Pinto
Industrial Technology Laboratory, School of Chemistry and Food, Federal University of Rio Grande, km 08 Itália Avenue, 96203-900, Rio Grande, RS, Brazil

© Springer International Publishing AG 2017

A. Bonilla-Petriciolet et al. (eds.), *Adsorption Processes for Water Treatment and Purification*, DOI 10.1007/978-3-319-58136-1_3

Contents

3.1 Introduction

Adsorption is one of the most widely applied unit operations used to separate molecules that are present in a fluid phase (adsorbate) using a solid surface (adsorbent). In this chapter, the adsorption kinetics in liquid phase will be addressed. Adsorption kinetics is expressed as the rate of adsorbate removal from the fluid phase to the adsorbent or the time involving the mass transfer of one or more components contained in a liquid to the adsorbent (Qiu et al. 2009). When adsorption is studied, kinetic aspects should be evaluated in order to know more details about its mechanisms, characteristics, and possibilities of application. These data can determine the residence time to reach the required concentration of the adsorbate, making possible the design and operation of an adsorption equipment and defining the performance in batch and continuous systems (Ruthven 1984).

Several mathematical models have been suggested to describe adsorption operations, which are classified as *diffusional models* and *adsorption reaction models*. Both are used to describe the kinetic peculiarities of adsorption operation; however, they are completely uneven in its essence (Qiu et al. 2009). Adsorption diffusion models are constructed based on three successive steps: external mass transfer, intraparticle diffusion, and adsorption on active sites. On the other hand, adsorption reaction models, originating from chemical reaction kinetics, are based on the whole process of adsorption, without considering the adsorption diffusion steps previously mentioned. They are widely used to represent the adsorption data in the literature. The more extensively adsorption reaction models utilized are the pseudo-first-order, pseudo-second-order, and Elovich equation (Wan and Hanafiah 2008; Cadaval et al. 2013; Dotto et al. 2013).

Here, we reviewed diffusional models and adsorption reaction models to be applied in discontinuous and continuous adsorption operations. The models and its mathematical solutions are detailed.

3.2 Adsorption Kinetics in Discontinuous Batch Systems

In a discontinuous batch adsorption system, the kinetic profile is fundamental due to several reasons. The kinetic profile provides information about the adsorption rate, equilibrium time, and effectiveness of the adsorbent. Furthermore, based on the kinetic curve, it is possible to infer the mass transfer mechanisms, which are the rate-limiting steps of the adsorption process (Ruthven 1984; Suzuki 1993; Qiu et al. 2009). The kinetic profile in a discontinuous batch adsorption system is normally represented by curves of C_t (adsorbate concentration in the bulk solution) versus t (contact time) or q_t (amount of adsorbate adsorbed on the adsorbent) versus t. In order to obtain information about the adsorption process from these curves, several empirical, semiempirical, theoretical, and diffusion-based models are employed (Do 1998; Qiu et al. 2009; Plazinski and Rudzinski 2009; Piccin et al. 2011; Baz-Rodríguez et al. 2012; Ocampo-Pérez et al. 2012; Dotto et al. 2014, 2016). In this section, these models are divided in two main classes, which are normally employed in the majority of adsorption articles: diffusional mass transfer models and adsorption reaction models.

3.2.1 *Diffusional Mass Transfer Models*

Figure 3.1 shows the representation of the main mass transfer mechanisms that occur in a discontinuous batch adsorption operation. In this case, the adsorbent particle with radius R is defined as volume of control, and three consecutive steps of mass transfer are considered: external mass transfer, intraparticle diffusion, and adsorption on active sites (Ruthven 1984; Do 1998; Ocampo-Pérez et al. 2012; Dotto et al. 2016). The external mass transfer mechanism is relative to the movement of the adsorbate (molecules/ions) from the bulk solution (with concentration C_t) to the external surface of the adsorbent particle (with concentration $C_S(t)$). This mass transfer step is governed by the external mass transfer coefficient (k_f). The intraparticle diffusion mechanism in turn is relative to the movement of the adsorbate (molecules/ions), inside the adsorbent particle. The intraparticle diffusion mechanism occurs by effective pore volume diffusion, surface diffusion, or a combination of both mechanisms. The effective pore volume diffusion describes the transport of the adsorbate (molecules/ions) in the liquid phase inside of the particle and is represented by D_P, the effective pore volume diffusion coefficient. The surface diffusion is relative to the transport of the adsorbate over the surface of the adsorbent particles, from sites of higher energy to sites of lower energy. This mechanism is represented by the surface diffusion coefficient D_S. Finally, the adsorption on active sites is relative to the interaction of the adsorbate with the active sites of the adsorbent. In this context, the diffusional mass transfer models are constructed on the basis in the three abovementioned consecutive steps: external mass transfer, intraparticle diffusion (effective pore volume diffusion, surface

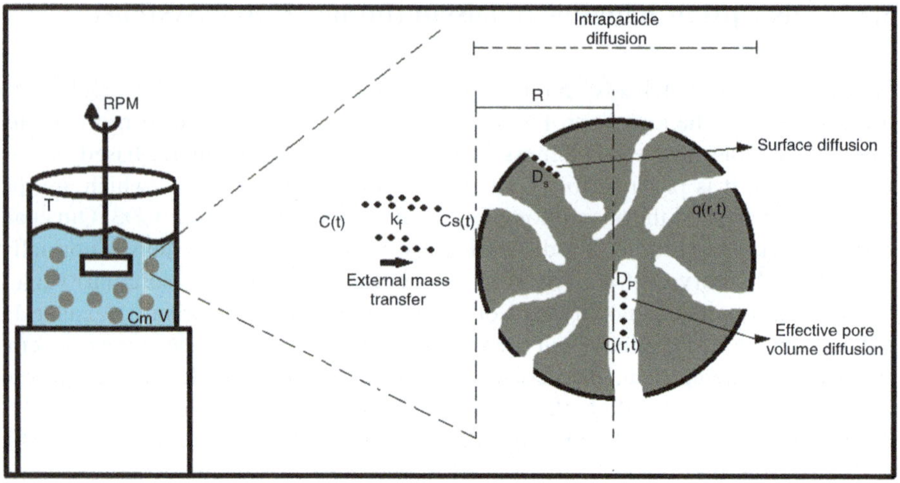

Fig. 3.1 Representation of the main mass transfer mechanisms which occur in a discontinuous batch adsorption operation

diffusion, or a combination of both mechanisms), and adsorption on an active site; and they represent realistically the adsorption kinetics (Ocampo-Pérez et al. 2012; Nieszporek 2013; Dotto et al. 2014).

One of the most complete diffusional models is the pore volume and surface diffusion model (PVSDM). This model is based on the following assumptions: batch system adsorption occurs at constant temperature; the particles are spherical; the mass transport by convection within the pores is negligible; the intraparticle diffusion can occur by pore volume diffusion and surface diffusion or both; the values of effective pore volume diffusion coefficient (D_p) and effective surface diffusion coefficient (D_s) are constant; and the adsorption rate on an active site is instantaneous (Leyva-Ramos and Geankoplis 1985; Ocampo-Pérez et al. 2010). This model is defined as

$$V\frac{dC_t}{dt} = -mSk_F\left(C_t - C_{s(t)}\big|_{r=R}\right) \tag{3.1}$$

$$t = 0, C_t = C_0 \tag{3.2}$$

$$\varepsilon_p\frac{\partial C_r}{\partial t} + \rho_p\frac{\partial q}{\partial t} = \frac{1}{r^2}\frac{\partial}{\partial r}\left[r^2\left(D_p\frac{\partial C_r}{\partial r} + \rho_p D_s\frac{\partial q}{\partial r}\right)\right] \tag{3.3}$$

$$t = 0, 0 \le r \le R, C_r = 0 \tag{3.4}$$

$$\left.\frac{\partial C_r}{\partial r}\right|_{r=0} = 0 \tag{3.5}$$

$$D_p \frac{\partial C_r}{\partial r}\bigg|_{r=R} + \rho_p D_s \frac{\partial q}{\partial r}\bigg|_{r=R} = k_F \left(C_t - C_{s(t)}\big|_{r=R} \right) \tag{3.6}$$

In Eqs. (3.1), (3.2), (3.3), (3.4), (3.5), and (3.6), V is the volume of solution, m is the amount of adsorbent, ε_p is the void fraction of the adsorbent, ρ_p is the apparent density of the adsorbent, S is the external surface area per mass of the adsorbent, C_0 is the initial adsorbate concentration in the bulk solution, C_r is the adsorbate concentration varying with the position and time, and q is the mass of adsorbate per mass of adsorbent varying with the position and time, respectively. The other symbols were already defined.

The PVSDM model can be simplified by considering that the sole intraparticle diffusion mechanism may be either pore volume diffusion (PVDM) ($D_p \neq 0$, $D_s = 0$) or surface diffusion (SDM) ($D_p = 0$, $D_s \neq 0$). Furthermore, to solve this model, it is considered that there exists a local equilibrium between the adsorbate concentration of the pore solution, C_r, and the mass of adsorbate adsorbed on the pore surface, q. This equilibrium relationship between C_r and q is represented by the adsorption isotherm:

$$q = f(C_r) \tag{3.7}$$

The isotherm studies are detailed in Chap. 2 of this book. Also, if V is constant, it is evident that C_t and q_t are always related by Eq. (3.8):

$$q_t = \frac{(C_0 - C_t)V}{m} \tag{3.8}$$

For illustration, the overall adsorption rate of Reactive Black 5 dye (RB5) on chitosan-based materials (powder and film) was investigated by Dotto et al. (2016) using the PVSDM model. The geometry of the adsorbents and swelling effects were evaluated. The authors found that the surface diffusion was the intraparticle diffusion mechanism that governed the adsorption, since its contribution was higher than 92% regardless the position and time. The D_s values ranged from 2.85×10^{-11} to 12.1×10^{-11} cm^2/s. The swelling effect was most pronounced for the chitosan films, providing an increase of about 65 times in the D_s value. On the other hand, Flores-Cano et al. (2016) studied the adsorption rate of metronidazole, dimetridazole, and diatrizoate on activated carbons using the PVSDM model. The results revealed that the surface diffusion contributed >90% of the total intraparticle diffusion, confirming that surface diffusion is the mechanism that controls the intraparticle diffusion of these pollutants on activated carbons. Finally, Largitte and Laminie (2015) evaluated the concentration decay curves for the adsorption of lead on a granular activated carbon using the PVSDM model. The results showed that the PVSDM model fitted the data reasonably well and the values of D_P were higher than those of D_S. k_L was around 10^{-4} cm s^{-1}, whereas D_P and D_S were around 10^{-6} and 0 cm^2/s. Therefore, the overall rate of adsorption was controlled by intraparticle diffusion, which was exclusively due to the pore volume diffusion.

Three models can be derived from the PVSDM model: external mass transfer model (EMTM), pore volume diffusion model (PVDM), and surface diffusion model (SDM) (Costa and Rodrigues 1985; Leyva-Ramos and Geankoplis 1985; Garcia-Reyes and Rangel-Mendez 2010).

The external mass transfer model (EMTM) assumes that the movement of solute from the liquid phase to the adsorbent is only due to external mass transfer. In this way, the intraparticle diffusion is instantaneous, so there is not a concentration gradient inside the particle. Therefore, the intraparticle diffusion resistance is considered to be insignificant. EMTM model is given by Eqs. (3.9), (3.10), (3.11), and (3.12) (Dotto et al. 2016):

$$V\frac{dC_t}{dt} = -mSk_F\left(C_t - C_{s(t)}\big|_{r=R}\right) \tag{3.9}$$

$$t = 0, C_t = C_0 \tag{3.10}$$

$$\frac{m\varepsilon_p}{\rho_p}\frac{dC_r}{\partial t}\bigg|_{r=R} + m\frac{dq_t}{dt} = mSk_F\left(C_t - C_{s(t)}\big|_{r=R}\right) \tag{3.11}$$

$$t = 0, C_t = 0, q_t = 0 \tag{3.12}$$

The pore volume diffusion model (PVDM) with external resistance is a simplification of the PVDSM model, used when the intraparticle diffusion is controlled only by effective pore diffusion ($D_p \neq 0$, $D_s = 0$). This simplification leads to (Garcia-Reyes and Rangel-Mendez 2010)

$$V\frac{dC_t}{dt} = -mSk_F\left(C_t - C_{s(t)}\big|_{r=R}\right) \tag{3.13}$$

$$t = 0, C_t = C_0 \tag{3.14}$$

$$\varepsilon_p\frac{\partial C_r}{\partial t} + \rho_p\frac{\partial q}{\partial t} = \frac{1}{r^2}\frac{\partial}{\partial r}\left[r^2\left(D_p\frac{\partial C_r}{\partial r}\right)\right] \tag{3.15}$$

$$t = 0, 0 \leq r \leq R, C_r = 0 \tag{3.16}$$

$$\frac{\partial C_r}{\partial r}\bigg|_{r=0} = 0 \tag{3.17}$$

$$D_p\frac{\partial C_r}{\partial r}\bigg|_{r=R} = k_F\left(C_t - C_{s(t)}\big|_{r=R}\right) \tag{3.18}$$

Based on the same analogy, the SDM model is used when the intraparticle diffusion mechanism is only controlled by surface diffusion; then the set of governing equations are given by

$$V\frac{dC_t}{dt} = -mSk_F\left(C_t - C_{s(t)}\big|_{r=R}\right) \tag{3.19}$$

$$t = 0, C_t = C_0 \tag{3.20}$$

$$\varepsilon_p\frac{\partial C_r}{\partial t} + \rho_p\frac{\partial q}{\partial t} = \frac{1}{r^2}\frac{\partial}{\partial r}\left[r^2\left(\rho_p D_s\frac{\partial q}{\partial r}\right)\right] \tag{3.21}$$

$$t = 0, 0 \le r \le R, C_r = 0 \tag{3.22}$$

$$\left.\frac{\partial C_r}{\partial r}\right|_{r=0} = 0 \tag{3.23}$$

$$\rho_p D_s\left.\frac{\partial q}{\partial r}\right|_{r=R} = k_F\left(C_t - C_{s(t)}\big|_{r=R}\right) \tag{3.24}$$

Furthermore, if the external mass transfer is negligible, Eq. (3.18) (PVDM) and Eq. (3.24) (SDM) can be replaced by the adsorption isotherm, or by other boundary condition.

Another important model used in adsorption systems is named homogeneous surface diffusion model (HSDM). This model considers a dual mass transport mechanism across the hydrodynamic boundary layer surrounding the adsorbent particle and intraparticle resistance within the particle in the form of surface diffusion (Leyva-Ramos and Geankoplis 1985). Mathematical equations of HSDM are

$$m\frac{dq_t}{dt} = mAk_F\left(C_t - C_{s(t)}\big|_{r=R}\right) \tag{3.25}$$

$$t = 0, C_t = C_0 \tag{3.26}$$

$$m\frac{dq_t}{dt} = V\frac{dC_t}{dt} \tag{3.27}$$

$$\rho_p\frac{\partial q}{\partial t} = \frac{1}{r^2}\frac{\partial}{\partial r}\left[r^2\left(\rho_p D_s\frac{\partial q}{\partial r}\right)\right] \tag{3.28}$$

$$t = 0, 0 \le r \le R, q = 0 \tag{3.29}$$

$$\left.\frac{\partial q}{\partial r}\right|_{r=0} = 0 \tag{3.30}$$

$$\rho_p D_s\left.\frac{\partial q}{\partial r}\right|_{r=R} = k_F\left(C_t - C_{s(t)}\big|_{r=R}\right) \tag{3.31}$$

In the same way of the other models, HSDM requires a relation between the amount of adsorbate adsorbed on the adsorbent and the amount of the adsorbate in the solution, which is given by the adsorption isotherm.

3.2.2 Adsorption Reaction Models

Adsorption reaction models originating from chemical reaction kinetics are based on the adsorption as a single phenomenon, unlike diffusive model. Adsorption reaction models are widely utilized to describe the kinetic process of adsorption. In batch systems, the more applied models are the pseudo-first-order, pseudo-second-order, and Elovich equation (Qiu et al. 2009; Largitte and Pasquier 2016).

3.2.2.1 Pseudo-First-Order Model

The pseudo-first-order model presented by Lagergren is based on the solids' capacity to adsorb and is given by Eq. (3.32) (Lagergren 1898):

$$\frac{dq_t}{dt} = k_1(q_e - q_t) \tag{3.32}$$

where q_e and q_t (mg/g) are the adsorption capacities at equilibrium and time t (min), respectively, and k_1 (min^{-1}) is the pseudo-first-order rate constant of the kinetic model. Integrating Eq. (3.32) using the initial conditions of $q_t = 0$ at $t = 0$ and $q_t = q_t$ at $t = t$ leads to Eq. (3.33) (Ho 2004):

$$\ln\left(\frac{q_e}{q_e - q_t}\right) = k_1 t \tag{3.33}$$

which can be rewritten as

$$q_t = q_e(1 - \exp(k_1 t)) \tag{3.34}$$

Several scientific papers present the pseudo-first-order model as the most suitable to represent adsorption kinetics. For the adsorption of FD&C yellow 5 onto chitosan film, the pseudo-first-order model was the more satisfactory (Cadaval et al. 2015). In this study, a little effect of the stirring rate was confirmed by the little variation in the k_1 values at different stirring rates. This model is normally used when the adsorption operation is fast, attaining the equilibrium within 20–30 min.

3.2.2.2 Pseudo-Second-Order Model

Ho described a kinetic process of the adsorption of divalent metal ions onto peat (Ho and McKay 2000), in which the chemical bonding among divalent ions and functional groups on peat, were responsible for the ionic exchange. Therefore, the peat-metal interaction can be presented as Eq. (3.35), which occurs in the adsorption of Cu^{2+} ions onto the adsorbent (Coleman et al. 1956):

$$2P^{-} + Cu^{2+} \leftrightarrow CuP_2 \tag{3.35}$$

where P^- represents the active sites on the peat surface. In the above reaction, the main assumptions were that the adsorption followed a second-order behavior and the interaction adsorbent adsorbate was chemical.

The adsorption rate described by Eq. (3.36) is dependent of divalent ion concentration on the surface of peat at time t and at equilibrium. Thus, the rate can be expressed as

$$\frac{d(P)_t}{dt} = k_2\left[(P)_0 - (P)_t\right]^2 \tag{3.36}$$

where $(P)_0$ is the amount of equilibrium sites available on the peat, $(P)_t$ is the amount of active sites occupied on the peat at time, and k_2 (g/(mg min)) is the pseudo-second-order rate constant of adsorption.

Since the driving force $(q_e - q_t)$ is proportional to the available fraction of active sites, then it can be written as

$$\frac{dq_t}{dt} = k_2(q_e - q_t)^2 \tag{3.37}$$

and integrating Eq. (3.37) using the initial conditions of $q_t = 0$ at $t = 0$ and $q_t = q_t$ at $t = t$, yields

$$\frac{1}{(q_e - q_t)} = \frac{1}{q_e} + k_2 t \tag{3.38}$$

This equation can be rewritten as follows:

$$q_t = \frac{t}{\left(\frac{1}{h_0}\right) + \left(\frac{t}{q_e}\right)} \tag{3.39}$$

and

$$h_0 = k_2 q_e^2 \tag{3.40}$$

where h_0 (mg/(g min)) is the initial adsorption rate.

The second-order rate equation has been successfully applied to the adsorption of metal ions, dyes, and organic substances from aqueous solutions. Several studies for adsorption of divalent metals reported that the majority of the sorption kinetics follows pseudo-second-order mechanism (Ho 2006; Aydin and Aksoy 2009). For instance, the adsorption of Cu (II) from copper mine water by chitosan films and the matrix effects were studied, and the pseudo-second-order model showed better fit than the other ones (Frantz et al. 2017).

3.2.2.3 Elovich Model

The Elovich equation was developed by Zeldowitsch (1934) and was used to describe the adsorption rate of carbon monoxide on manganese dioxide, which decreased exponentially with an increase of the gas adsorbed. Thus, Elovich equation is applied to determine the kinetics of chemisorption of gases onto heterogeneous surface (Rudzinski and Panczyk 2000) and was obtained from the following differential equation:

$$\frac{dq}{dt} = ae^{-\beta q} \tag{3.41}$$

where q represents the amount of gas adsorbed at time t, a is the desorption constant, and β is the initial adsorption rate. With the assumption of $a\beta t >> 1$, Eq. (3.41) was integrated with the initial conditions of $q = 0$ at $t = 0$ and $q = q$ at $t = t$ resulting in (Chien and Clayton 1980)

$$q_t = \beta \ln (a\beta) + \beta \ln t \tag{3.42}$$

or

$$q_t = \left(\frac{1}{a}\right) \ln (1 + a\beta t) \tag{3.43}$$

This equation has been applied to describe the adsorption process of different molecules from liquid medium. For example, Elovich model was the more suitable to fit the kinetic data for the adsorption of FD&C red 2 onto chitosan film (Cadaval et al. 2015) and carotenoids and chlorophylls in rice bran oil bleaching (Pohndorf et al. 2016).

3.3 Fixed-Bed Adsorption

In operational level, a fixed-bed column presents a certain working time to adsorb contaminants, such that the effluent outlet can comply with allowable levels of concentration. This working time can be expressed by the so-called breakthrough curve, and this behavior is shown in Fig. 3.2.

Considering a vertical upward flow, initially the adsorbent is completely free of the solute, while the flow of liquid is initiated by the column. The solute is gradually transferred to the adsorbent until at a point above the bed (point I in Fig. 3.2a). As the process progresses, the initial portions of the adsorbent are completely saturated (point III in Fig. 3.2a), and the zone where the solute is removed advances toward the final part of the column. The portion of the column in which the solute reduction occurs is called the mass transfer zone (point II in Fig. 3.2a).

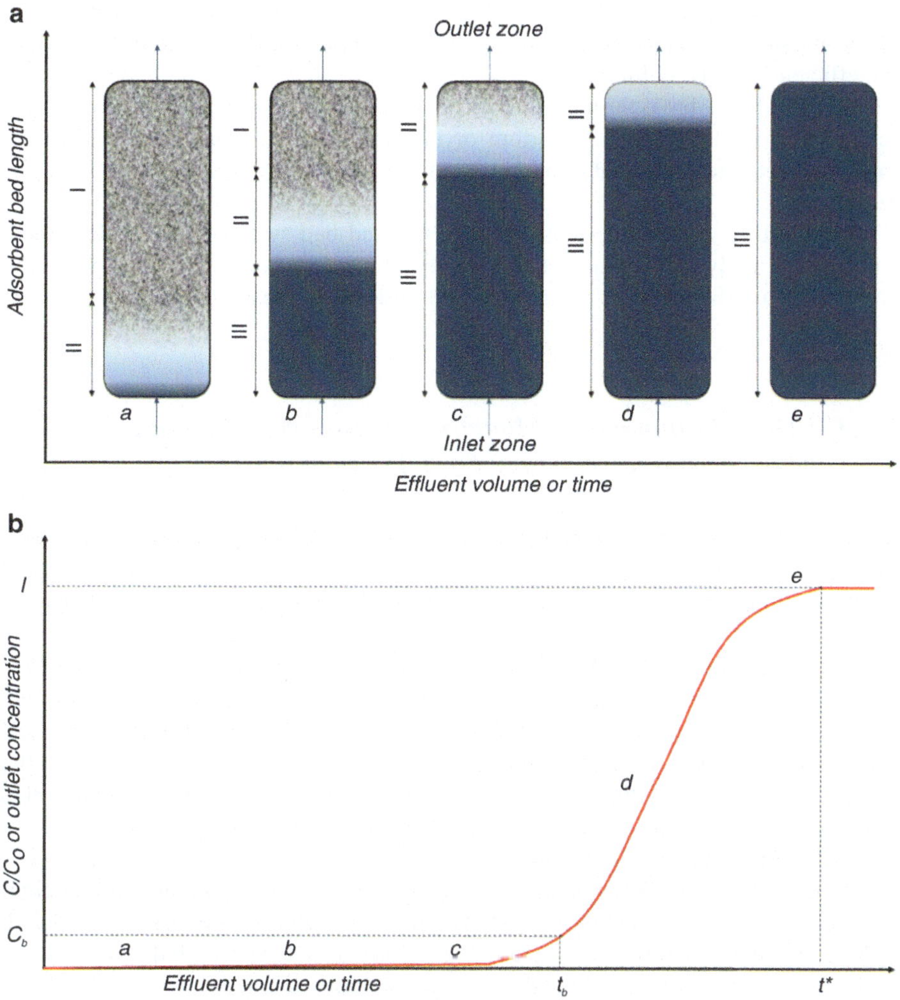

Fig. 3.2 (**a**) Progression of the mass transfer through fixed-bed column and (**b**) concentration profile of solute concentration at adsorbent bed outlet during the adsorption process

When the mass transfer zone reaches the end of column, the concentration of adsorbate in the liquid gradually increases (point c in Fig. 3.2b), since not all of the solute can be removed. Then, the start of the mass transfer zone reaches the end of the column (point e in Fig. 3.2b) and the whole column is saturated, not occurring more solute removal.

The portion of the curve between the points c and e (Fig. 3.2b) of the column is called the breakthrough curve, and the point at which the concentration of adsorbate at the output of the column reaches the maximum limit is called the breakpoint.

In this way, the shape of the rupture curve provides information about the length of the mass transfer zone (point II in Fig. 3.2a), and the smaller this zone, the greater the efficiency of the column.

In general, the breakthrough curve can be affected by thermodynamic factors related to the adsorption equilibrium or isotherms, kinetic factors related to the mass transfer rate, and fluid dynamic factors related to the flow velocity (Cooney 1999).

Thus, in this session, the mass balance in fixed-bed systems will be described, showing the analytical solutions that represent the kinetics of the column adsorption and presenting the methods of analysis and scale-up for real systems.

3.3.1 Mass Balance and Modeling of the Breakthrough Curves Based on Mass Transfer Mechanism

The differential mass balances for an elementary volume of a fixed-bed column, including the fluid phase and the adsorbent within this elementary volume, are used for the development of a mathematical model, which describes the kinetic behavior of the system.

In the mathematical modeling of fixed-bed adsorption, the following considerations must be made: (i) the system is isothermal; (ii) there is only one adsorbate soluble in the liquid; (iii) the concentration of the solute in the liquid is so small that, if the whole has been adsorbed, there will be no change in the flow rate of the fluid; (iv) there is not radial velocity and; (v) from this latter consideration, it is also considered that there is not variation in adsorbate concentration in both phases in the radial directions (Cooney 1999).

For the differential mass balance in the column, a control volume with a height Δz and a circular session identical to the column diameter (or area A) is considered. A fluid stream containing the species to be adsorbed passes through the voids of the bed (ε). Then, the volume of solid (V_s) in the volume of control is

$$V_s = (1 - \varepsilon)A\Delta z \tag{3.44}$$

Applying the mass conservation law, the mass balance of any solute i in the control volume is given by

$$\varepsilon A\Delta z\frac{\partial C_i}{\partial t} = \varepsilon AN_i \mid_z - \varepsilon AN_i \mid_{z+\Delta z} - (1 - \varepsilon)A\Delta z\frac{\partial q}{\partial t} \tag{3.45}$$

where $\varepsilon A\Delta z\frac{\partial C_i}{\partial t}$ is the solute accumulation in the control volume, $\varepsilon AN_i|_z$ is the mass rate of solute that enters the control volume in the z direction, $\varepsilon AN_i|_{z+\Delta z}$ is the mass rate of solute that leaves volume control in the z direction, and $(1 - \varepsilon)A\Delta z\frac{\partial q}{\partial t}$ is the mass rate of solute adsorbed, respectively.

Dividing Eq. (3.45) by $\varepsilon A \Delta z$ and applying the limit when Δz tends to zero yields

$$\frac{\partial C_i}{\partial t} = \frac{\partial N_i}{\partial z} - \frac{(1-\varepsilon)}{\varepsilon}\frac{\partial q}{\partial t} \tag{3.46}$$

The mass flux of solute in the fluid phase in the z direction (N_{iz}) is given by a convective portion, due to fluid movement, and another diffusive, due to a gradient of concentration caused by the adsorption along the column, according to the following equation:

$$\frac{\partial N_i}{\partial z} = v_z \frac{\partial C_i}{\partial z} - D_L \frac{\partial^2 C_i}{\partial z^2} \tag{3.47}$$

where v_z is interstitial velocity of the fluid in the z-direction and D_L is the coefficient of axial dispersion or diffusion. Then, substituting in Eq. (3.46), and rearranging the equation, the mass balance in the adsorption column can be described as

$$v_z \frac{\partial C_i}{\partial z} + \left(\frac{\partial C_i}{\partial t}\right)_{\bar{Z}} + \frac{(1-\varepsilon)}{\varepsilon}\left(\frac{\partial q}{\partial t}\right)_{\bar{Z}} = D_L \left(\frac{\partial^2 C_i}{\partial z^2}\right)_t \tag{3.48}$$

The adsorption rate $\left(\frac{\partial q}{\partial t}\right)_{\bar{Z}}$ is described by the mass transfer in the particle (previous section), and then Eq. (3.48) can be solved numerically using the following initial and boundary conditions:

$$t = 0; C = 0; q = 0 \tag{3.49}$$

$$t > 0; Z = 0; C = C_0 \tag{3.50}$$

$$t \to \infty; \frac{\partial C_i}{\partial t} = 0 = \frac{\partial q}{\partial t} = 0 \tag{3.51}$$

3.3.2 *Empirical Models for Breakthrough Curves*

The breakthrough curve behavior prediction is fundamental for the analysis and design of fixed-bed adsorption systems. From this curve, parameters, such as the breakthrough time and the saturation time of the column, are obtained, giving an idea of the length of the mass transfer zone. For this reason, several models were developed from analytical solutions of the differential mass balance in the fixed bed or by empirical solutions.

3.3.2.1 Bohart-Adams Model

Bohart-Adams model was developed considering the surface reaction theory which assumes that the equilibrium is not instantaneous, and the rate of adsorption is proportional to the adsorption capacity and the concentration of solute (Bohart and Adams 1920). This model is suitable for adsorption systems with high affinity equilibrium behavior (or irreversible isotherm) and is expressed as (Cooney 1999)

$$\frac{C_t}{C_0} = \exp\left(k_{AB}C_0 t - k_{AB}q_0\frac{z}{v_z}\right) \tag{3.52}$$

where C_0 and C_t are the input and output solute concentrations, respectively, k_{AB} is the Bohart-Adams model kinetic constant, q_0 is the stoichiometric capacity of the bed (related to the adsorption capacity predicted by the equilibrium isotherm for $C_e = C_0$, in units of mass per volume of adsorbent), and z is the length of bed.

3.3.2.2 Thomas Model

Thomas (1944) solved the differential mass balance for a system with adsorption isotherms of the Langmuir type, no axial dispersion, and kinetic described by pseudo-second-order model. Thomas model is one of the widely used models, and this model is based on the plug flow behavior in the bed, i.e., no axial dispersion, expressed as

$$\frac{C_t}{C_0} = \frac{1}{1 + \exp\left(\frac{k_{Th}q_0 m}{Q} - k_{Th}C_0 t\right)} \tag{3.53}$$

where k_{Th} is the Thomas kinetic constant, m is the mass of adsorbent, and Q is the operating flow rate. In this equation, q_0 is expressed in units of mass of solute per mass of adsorbent.

3.3.2.3 Wolborska Model

Wolborska model (1989) was based on the general equation of diffusional mass transfer for low concentration range; see Eq. (3.48). For an external diffusion with a constant coefficient, it is possible to derive

$$\frac{\partial q}{\partial t} = -v_m\frac{\partial q}{\partial z} = \beta_a(C_b - C_i) \tag{3.54}$$

where v_m is the migration rate of the solute through the fixed bed, C_i is the interface solid/liquid concentration, and β_a is the kinetic coefficient of external diffusion.

Using Eq. (3.54), and assuming that $C_i << C_b$, $v_m << v_z$, and neglecting the axial dispersion, the breakthrough curves can be described as

$$\ln\left(\frac{C_t}{C_0}\right) = \frac{\beta_a C_0}{q_0} t - \frac{\beta_a z}{v_z} \tag{3.55}$$

3.3.2.4 Yoon-Nelson Model

Yoon-Nelson model (1984) was proposed to describe the nature of breakthrough curves of adsorbate gases on activated charcoal. This model is based on the assumption that the rate of adsorption degradation for each molecule is proportional to the adsorption rate and the yield curve of the adsorbed material, represented as

$$\frac{C_t}{C_0} = 1 + \exp(\tau k_{YN} - k_{YN} t) \tag{3.56}$$

where k_{YN} is the Yoon-Nelson kinetic constant and τ is the predict time to the advance of 50% of adsorption front.

3.3.3 Design of Fixed-Bed Adsorption Systems

In fixed-bed adsorption tests (laboratory scale), the adsorption capacity of the bed is related to the area above the rupture curve, as can be seen in Fig. 3.2b. Thus, the adsorption capacity is described by

$$q = \frac{QC_0 \int_0^t \left(1 - \frac{C_t}{C_0}\right) dt}{m} \tag{3.57}$$

The stoichiometric capacity of the bed (q_{eq}) can be calculated by integrating Eq. (3.57) to the point that the concentration of the effluent outlet (C_t) is identical to the inlet concentration (C_0). In this case, this adsorption capacity is associated with complete bed saturation. Theoretically, the time stoichiometric tends to infinity, and its precise location involves trial and error. The stoichiometric capacity of the column can also be obtained from equilibrium isotherm, as described in Chap. 2.

Already the column useful capacity (q_b) occurs when the concentration of effluent outlet (C_t) reaches the concentration of break of the column (C_b). Then, q_b is obtained by integrating Eq. (3.57) to the breakthrough time (t_b).

For the design of fixed-bed adsorption systems, the useful length of the bed (L) can be calculated from a mass balance, considering a convex isotherm (i.e., Langmuir or Freundlich), using the mass of solute fed into the column with respect to the adsorption capacity of the adsorbent (in units of mass per volume of adsorbent, or q_0/ρ), described by

$$L = \frac{QC_0t_s}{S_0(1-\varepsilon)^{q_0}/_{\rho}} \tag{3.58}$$

In Eq. (3.58), the dividend represents the mass of solute fed in a service time (t_s) and the divisor the amount of accumulated mass per unit of bed length. Introducing the concept of hydraulic load (H), denoted by the ratio between the volumetric flow rate (Q) and the cross-sectional area (S_0) of the bed, or $H = \varepsilon v_z = Q/S_0$, the equation can be rewritten as

$$L = \frac{HC_0t_s}{(1-\varepsilon)^{q_0}/_{\rho}} \tag{3.59}$$

However, the useful length of the bed (L) considers a sharp breakthrough curve. As previously discussed, the shape of the breakthrough curve changes according to the rates of mass transfer and reaction with the adsorption sites. Thus, in order for the system to meet the design conditions, an extra length must be added to the bed.

For the calculation of the extra length of the bed, two methodologies will be presented: the length of unused bed (LUB) and the bed depth service time (BDST). These methods, performed on a laboratory scale, are intended to calculate the fraction of the column required for the reduction of the initial concentration to the acceptable design conditions and are related to the length of the mass transfer zone.

Moreover, in the scale-up of bed from laboratory data (isotherms or breakthrough curves), it is fundamental that the conditions like hydraulic load, pH, temperature, and concentration are similar.

3.3.3.1 LUB Concept

In this way, the length of unused bed (*LUB*) is a relation between the stoichiometric capacity (q_{eq}) and the useful capacity of the column (q_b). Associating the two equations, it has the following:

$$LUB = \left(1 - \frac{q_b}{q_{eq}}\right)z = \left(1 - \frac{t_b}{t^*}\right)z \tag{3.60}$$

3.3.3.2 Bed Depth Service Time (BDST)

BDST model was used to describe the fixed-bed adsorption column operation (Hutchins 1973). This approach becomes useful if the bed depth-breakthrough time data determined from a set of curves with different bed sizes is analyzed using the irreversible isotherm model. The BDST model is expressed as

$$\ln\left(\frac{C_0}{C_t}-1\right)=\frac{kq_0D(1-\varepsilon)}{\varepsilon v_z}-kC_0t^0 \tag{3.61}$$

Note that $q_0(1-\varepsilon)$ is the adsorption capacity of the bed per volumetric unit, denoted by N_0. Thus, rearranging the equation, the breakthrough time (t_b) is

$$t_b=\frac{N_0}{\varepsilon vC_0}D-\frac{\ln\left(\frac{C_0}{C_b}-1\right)}{kC_0} \tag{3.62}$$

where C_b is the breakthrough concentration. As can be seen, this equation suggests that the rupture time (t_b) has a linear relationship with the length of the column (D). For this case, the critical situation for operation is $t_b = 0$. Thus, the critical bed depth of the column (D_0) can be calculated by

$$D_0=\frac{\varepsilon v}{kN_0}\ln\left(\frac{C_0}{C_b}-1\right) \tag{3.63}$$

or the service time (t_s) of a bed with length D can be expressed by

$$t_s=\frac{N_0}{\varepsilon vC_0}(D-D_0) \tag{3.64}$$

3.4 Numerical Methods and Parameters Estimation

Numerical solution of diffusional mass transfer models requires the application of the numerical method of lines to solve partial differential equations (Ocampo-Pérez et al. 2010; Souza et al. 2017) which can be expressed as a system of simultaneous ordinary differential equations. For batch systems, the boundary conditions can be expressed as implicit single algebraic equations, and the Newton method can be used to solve those (Souza et al. 2017). In this case, the mass transport parameters can be either calculated, such as external mass transfer coefficient, molecular diffusivity, effective pore volume diffusion coefficient, and tortuosity factor (Leyva-Ramos et al. 2012; Ocampo-Pérez et al. 2012), or estimated by nonlinear least-square optimization, such as surface diffusion coefficient and external mass transfer coefficient (Ocampo-Pérez et al. 2010; Souza et al. 2017). For fixed bed, the mass transport parameters can also be either calculated, such as fluid interstitial velocity and voids of the bed, or estimated by nonlinear least-square optimization, such as axial dispersion or diffusion coefficient.

The prediction of adsorption reaction models and empirical models for breakthrough curves depends on parameters whose values must be estimated from available experimental (Largitte and Pasquier 2016).

3.4.1 Solving Diffusional Mass Transfer Models

The numerical method of lines utilizes ordinary differential equations for the time derivative and finite differences on the spatial derivatives (Schiesser 1991). In finite difference method, the derivatives in the partial differential equation are approximated by linear combinations of function values at the grid points. The derivatives in the partial differential equation of diffusional transfer models are discretized into $N + 1$ points on the spatial derivatives (radius), where N is the number of grid points.

Figure 3.3 shows the illustrative representation of discretization (grid points) of the transport of adsorbate molecules from the bulk solution to the spherical particle. In this way, it has C_r in different points i where $i = 0$ is the grid point at $r = 0$ and $i = N + 1$ is the grid point at $r = R$. The grid points are spaced equally with the step size given by following equation:

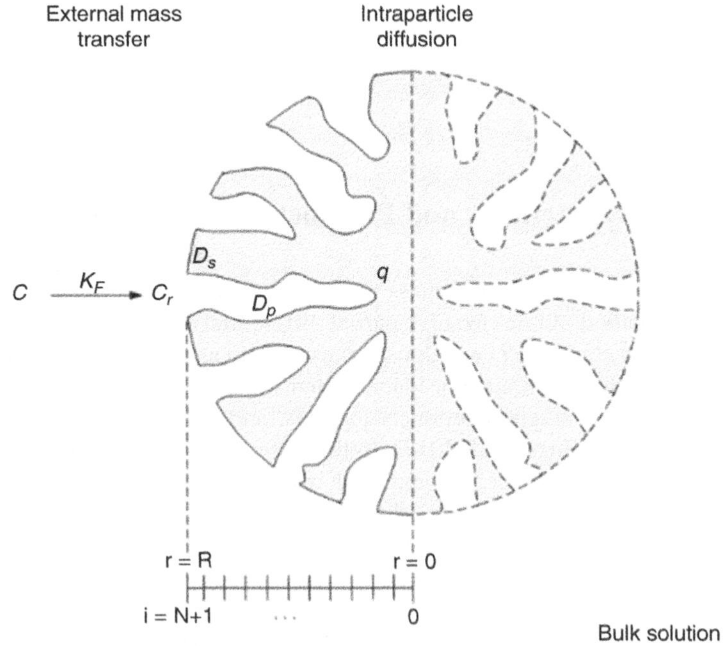

Fig. 3.3 Discretization of the transport of the adsorbate molecules from the bulk solution to the spherical particle (illustrative representation)

Table 3.1 Finite approximation functions used to solve the partial differential equations

Difference	First-order formula	Second-order formula
Forward difference for first derivative	$\frac{d}{dx}f(x_i) = \frac{f(x_{i+1})-f(x_i)}{\Delta x}$	$\frac{d}{dx}f(x_i) = \frac{-3f(x_i)+4f(x_{i+1})-f(x_{i+2})}{2\Delta x}$
Central difference for first derivative		$\frac{d}{dx}f(x_i) = \frac{f(x_{i+1})-f(x_{i-1})}{2\Delta x}$
Backward difference for first derivative	$\frac{d}{dx}f(x_i) = \frac{f(x_i)-f(x_{i-1})}{\Delta x}$	$\frac{d}{dx}f(x_i) = \frac{3f(x_i)-4f(x_{i-1})+f(x_{i-2})}{2\Delta x}$
Forward difference for second derivative	$\frac{d^2}{dx^2}f(x_i) = \frac{f(x_i)-2f(x_{i+1})+f(x_{i+2})}{\Delta x^2}$	$\frac{d^2}{dx^2}f(x_i) = \frac{2f(x_i)-5f(x_{i+1})+4f(x_{i+2})-f(x_{i+3})}{\Delta x^2}$
Central difference for second derivative		$\frac{d^2}{dx^2}f(x_i) = \frac{f(x_{i+1})-2f(x_{i1})+f(x_{i-1})}{\Delta x^2}$
Backward difference for second derivative	$\frac{d^2}{dx^2}f(x_i) = \frac{f(x_i)-2f(x_{i-1})+f(x_{i-2})}{\Delta x^2}$	$\frac{d^2}{dx^2}f(x_i) = \frac{2f(x_i)-5f(x_{i-1})+4f(x_{i-2})-f(x_{i-3})}{\Delta x^2}$

$$H = \frac{R}{N+1} \tag{3.65}$$

The partial differential equation of diffusional transfer models can be rewritten using the second-order central difference approximations of the first and second derivative. The second-order forward difference approximations of the first derivative can be used to solve the boundary condition at $r = 0$, and the second-order backward difference approximations of the first derivative can be used to solve the boundary condition at $r = R$ (Souza et al. 2017). Useful finite approximations are shown in Table 3.1.

Implicit single algebraic equations are resulting from solving the boundary conditions with finite difference approximations. The most common method for solving nonlinear algebraic equations is the Newton method (Edgar and Himmelblau 2001). Trust-region modification of Newton method (Sorensen 1982) is the basis for the MATLAB built-in routine *fsolve* (Beers 2007).

Discretized partial differential equations yield ordinary differential equation systems that are very stiff; therefore, to avoid a very small time step, an implicit method can be used, as the backward difference formula (BDF) (Gear 1971). The MATLAB built-in routine *ode15s* is a variable order solver based on the numerical differentiation formulas (NDFs). Optionally, it uses the backward differentiation formulas (BDF, also known as Gear method) that are usually less efficient (Shampine and Reichelt 1997).

Nonlinear least-square optimization can be used to estimate the mass transport parameters. The nonlinear least-square objective function to be minimized is defined as the sum of the differences between the experimental data (C) and model data ($\hat{C}$) of adsorbate concentration, given as

$$\min|f(p)| = \sum_{j=1}^{NE}\sum_{i=1}^{NY}\left[C_{(i,j)} - \hat{C}_{(i,j)}\right]^2 \tag{3.66}$$

where *NE* is the number of experiments, *NY* is the number of experimental data points, and *p* is the mass transport parameter.

The mass transport parameter (p) can represent:

(a) *Surface diffusion coefficient*: using pore volume and surface diffusion model (PVSDM), surface diffusion model (SDM), and homogeneous surface diffusion model (HSDM) for batch systems
(b) *External mass transfer coefficient*: using the external mass transfer model (EMTM) for batch systems
(c) *Axial dispersion or diffusion*: using the mass transfer model for fixed-bed column

The MATLAB built-in routine *lsqnonlin* solves nonlinear least-square (nonlinear data-fitting) problems with optional lower and upper bounds on the parameters (p). By default, *lsqnonlin* chooses the trust-region-reflective algorithm that is a subspace trust-region method and is based on the interior-reflective Newton method (Coleman and Yi 1994, 1996). Each iteration involves the approximate solution of a large linear system using the method of preconditioned conjugate gradients (PCG) (Barrett et al. 1994).

The mass transport parameter is estimated with intervals of 95% of confidence. Simulations are performed using the diffusional transfer models with the estimated parameter to compare to each set of experimental data and, hence, to check the fit accuracy. The Student t-test can be performed to find out if the means of the adsorbate concentration experimental curve and of the adsorbate concentration curve predicted by the model are significantly different. Further, the χ^2 and the Fisher exact test are also useful to verify if variances of the experimental and model concentration data differ in any interesting way (Schwaab and Pinto 2007).

3.4.2 Solving Adsorption Reaction Models and Empirical Models for Breakthrough Curves

It is common to use linear regression (also known as linear least-square analyses) to estimate the values of the parameters of an adsorption reaction model. However, the sorption data are better simulated when the adsorption reaction models are fitted by nonlinear regression (Largitte and Pasquier 2016).

Whereas an adsorption reaction model can be directly used in nonlinear regression, it must be linear with respect to the parameters to be used in linear regression. Very often, a linear relationship is hypothesized between a log transformed of the dependent variable and the model parameters.

The nonlinear least-square objective function to be minimized is defined as the sum of the differences between the experimental data (y) and model data ($\hat{y}$) of adsorbate adsorbed at time t, given as below:

$$\min|f(p)| = \sum_{j=1}^{NE} \sum_{i=1}^{NY} \left[y_{(i,j)} - \hat{y}_{(i,j)}\right]^2 \qquad (3.67)$$

In case of adsorption reaction models, y is the adsorbate adsorbed at time t. Here, p is the kinetic parameter that can represent:

(a) *Adsorption rate constant*: using pseudo-first-order model and pseudo-second-order model
(b) *Desorption rate constant and initial adsorption rate*: using Elovich equation

In the case of empirical models for breakthrough curves, y is the ratio of concentration of the effluent outlet (C_t) and the inlet concentration (C_0). Here, p represents the parameters set of the chosen empirical model for breakthrough curves.

The MATLAB built-in routine *nlinfit* solves nonlinear regression (nonlinear data-fitting). For non-robust estimation, *nlinfit* uses the Levenberg-Marquardt nonlinear least-square algorithm (Levenberg 1944; Marquardt 1963; Moré 1977).

The kinetic parameter can be estimated with intervals of 95% of confidence. Simulations are performed using the adsorption kinetic models with the estimated parameter to compare to each set of experimental data and, hence, to check the fit accuracy. The coefficient of determination (R^2) and Akaike information criterion (AIC) are also calculated. The Student t-test is used to find out if the means of the experimental data and of the predicted data by the model are significantly different. Further, the χ^2 and the Fisher exact test can be also performed to verify if variances of the experimental and model data differ (Schwaab and Pinto 2007).

3.5 Conclusion

This chapter presented a general description of the adsorption kinetics in liquid phase, considering batch and fixed-bed systems. For discontinuous batch adsorption systems, it was verified that the kinetic curves can be represented by diffusional mass transfer models and adsorption reaction models. Diffusional mass transfer models are based on mass transfer steps, while adsorption reaction models consider adsorption as a reaction. In this way, we believe that diffusional mass transfer models are more suitable. For the breakthrough curves obtained in fixed-bed systems, the same analogy can be made, being more useful the mass transfer-based models. Furthermore, it was reviewed that LUB and BDST concepts are adequate for design and scale-up of adsorption systems. Finally, several numerical methods and parameter estimation techniques were presented and discussed in order to better understand the treatment of the experimental adsorption data using batch and fixed-bed columns.

References

Aydin YA, Aksoy ND (2009) Adsorption of chromium on chitosan: optimization, kinetics and thermodynamics. Chem Eng J 151:188–194

Barrett R, Berry M, Chan TF (1994) Templates for the solution of linear systems: building blocks for iterative methods. Siam, Philadelphia

Baz-Rodríguez SA, Ocampo-Pérez R, Ruelas-Leyva JP, Aguilar-Madera CG (2012) Effective transport proper-ties for the pyridine-granular activated carbon adsorption system. Braz J Chem Eng 29:599–611

Beers KF (2007) Numerical methods for chemical engineering. Cambridge University Press, Cambridge

Bohart GS, Adams EQ (1920) Some aspects of the behavior of charcoal with respect to chlorine. J Am Chem Soc 42:523–529

Cadaval TRS Jr, Camara AS, Dotto GL, Pinto LAA (2013) Adsorption of Cr (VI) by chitosan with different deacetylation degrees. Desalin Water Treat 51:7690–7699

Cadaval TRS, Dotto GL, Pinto LAA (2015) Equilibrium isotherms, thermodynamics, and kinetic studies for the adsorption of food azo dyes onto chitosan films. Chem Eng Commun 202:1316–1323

Chien SH, Clayton WR (1980) Application of Elovich equation to the kinetics of phosphate release and sorption in soils. Soil Sci Soc Am J 44:265–268

Coleman TF, Li Y (1994) On the convergence of reflective Newton methods for large-scale nonlinear minimization subject to bounds. Math Program 67:189–224

Coleman TF, Li Y (1996) An interior, trust region approach for nonlinear minimization subject to bounds. SIAM J Optim 6:18–445

Coleman NT, McClung AC, Moore DP (1956) Formation constants for Cu (II)-peat complexes. Science 123:330–331

Cooney DO (1999) Adsorption design for wastewater treatment. Lewis Publishers, Boca Raton

Costa C, Rodrigues AE (1985) Intraparticle diffusion of phenol in macroreticular adsorbents: modelling and experimental study of batch and CSTR adsorbers. Chem Eng Sci 40:983–993

Do DD (1998) Adsorption analysis: equilibria and kinetics. Imperial College Press, London

Dotto GL, Moura JM, Cadaval TRS Jr, Pinto LAA (2013) Application of chitosan films for the removal of food dyes from aqueous solutions by adsorption. Chem Eng J 214:8–16

Dotto GL, Buriol C, Pinto LAA (2014) Diffusional mass transfer model for the adsorption of food dyes on chitosan films. Chem Eng Res Des 92:2324–2332

Dotto GL, Ocampo-Pérez R, Moura JM, Cadaval RTS Jr, Pinto LAA (2016) Adsorption rate of Reactive Black 5 on chitosan based materials: geometry and swelling effects. Adsorption 22:973–983

Edgar TF, Himmelblau DM (2001) Optimization of chemical processes. McGraw-Hill, New York

Flores-Cano JV, Sánchez-Polo M, Messoud J, Velo-Gala I, Ocampo-Pérez R, Rivera-Utrilla J (2016) Overall adsorption rate of metronidazole, dimetridazole and diatrizoate on activated carbons prepared from coffee residues and almond shells. J Environ Manag 169:116–125

Frantz TS, Silveira N Jr, Quadro MS, Andreazza R, Barcelos AA, Cadaval TRS Jr, Pinto LAA (2017) Cu (II) adsorption from copper mine water by chitosan films and the matrix effects. Environ Sci Pollut Res. doi:10.1007/s11356-016-8344-z

Garcia-Reyes RB, Rangel-Mendez JR (2010) Adsorption kinetics of chromium(III) ions on agro-waste materials. Bioresour Technol 101:8099–8108

Gear CW (1971) The automatic integration of ordinary differential equations. ASM Commun 14:176–179

Ho YS (2004) Citation review of Lagergren kinetic rate equation on adsorption reactions. Scientometrics 59:171–177

Ho YS (2006) Review of second-order models for adsorption systems. J Hazard Mater 136:681–689

Ho YS, McKay G (2000) The kinetics of sorption of divalent metal ions onto sphagnum moss peat. Water Res 34:735–742
Hutchins RA (1973) New method simplifies design of activated carbon systems. Am J Chem Eng 80:133–138
Lagergren S (1898) About the theory of so-called adsorption of soluble substances. Kungliga Svenska Vetenskapsakademiens, Handlingar 24:1–39
Largitte L, Laminie J (2015) Modelling the lead concentration decay in the adsorption of lead onto a granular activated carbon. J Environ Chem Eng 3:474–481
Largitte L, Pasquier R (2016) A review of the kinetics adsorption models and their application to the adsorption of lead by an activated carbon. Chem Eng Res Des 109:495–504
Levenberg K (1944) A method for the solution of certain nonlinear problems in least squares. Q Appl Math 2:164–168
Leyva-Ramos R, Geankoplis CJ (1985) Model simulation and analysis of surface diffusion of liquids in porous solids. Chem Eng Sci 40:799–807
Leyva-Ramos R, Ocampo-Perez R, Mendoza-Barron J (2012) External mass transfer and hindered diffusion of organic compounds in the adsorption on activated carbon cloth. Chem Eng J 183:141–151
Marquardt D (1963) An algorithm for least-squares estimation of nonlinear parameters. SIAM J Appl Math 11:431–441
Moré JJ (1977) The Levenberg-Marquardt algorithm: implementation and theory, numerical analysis. In: Watson GA (ed) Lecture notes in mathematics, vol 630. Springer, Berlin, pp 105–116
Nieszporek K (2013) The balance between diffusional and surface reaction kinetic models: a theoretical study. Sep Sci Technol 48:2081–2089
Ocampo-Perez R, Leyva-Ramos R, Alonso-Davila P, Rivera-Utrilla J, Sánchez-Polo M (2010) Modeling adsorption rate of pyridine onto granular activated carbon. Chem Eng J 165:133–141
Ocampo-Pérez R, Rivera-Utrilla J, Gómez-Pacheco C, Sánchez-Polo M, Lopez-Peñalver JJ (2012) Kinetic study of tetracycline adsorption on sludge-derived adsorbents in aqueous phase. Chem Eng J 213:88–96
Piccin JS, Dotto GL, Vieira MLG, Pinto LAA (2011) Kinetics and mechanism of the food dye FD&C Red 40 adsorption onto chitosan. J Chem Eng Data 56:3759–3765
Plazinski W, Rudzinski W (2009) Kinetics of adsorption at solid/solution interfaces con-trolled by intraparticle diffusion: a theoretical analysis. J Phys Chem C 113:12495–12501
Pohndorf RS, Cadaval TRS Jr, Pinto LAA (2016) Kinetics and thermodynamics adsorption of carotenoids and chlorophylls in rice bran oil bleaching. J Food Eng 185:9–16
Qiu H, Pan LL, Zhang Q, Zhang W, Zhang Q (2009) Critical review in adsorption kinetic models. J Zhejiang Univ Sci A 10:716–724
Rudzinski W, Panczyk T (2000) Kinetics of isothermal adsorption on energetically heterogeneous solid surfaces: a new theoretical description based on the statistical rate theory of interfacial transport. J Phys Chem 104:9149–9162
Ruthven DM (1984) Principles of adsorption and adsorption processes. Wiley, New York
Schiesser WE (1991) The numerical method of lines: integration of partial differential equations. Academic, San Diego
Schwaab M, Pinto JC (2007) Experimental data analysis I. Fundamentals of statistics and parameter estimation. E-papers, Rio de Janeiro
Shampine LF, Reichelt MW (1997) The MATLAB ODE suite. SIAM J Sci Comp 18:1–22
Sorensen DC (1982) Newton's method with a model trust region modification. SIAM J Num Anal 19:409–426
Souza PR, Dotto GL, Salau NPG (2017) Detailed numerical solution of pore volume and surface diffusion model in adsorption systems. Chem Eng Res Des 122:298–307
Suzuki M (1993) Fundamentals of adsorption. Elsevier, Amsterdam
Thomas HC (1944) Heterogeneous ion exchange in a flowing system. J Am Chem Soc 66:1466–1664

Wan WS, Hanafiah MAKM (2008) Adsorption of copper on rubber (*Hevea brasiliensis*) leaf powder: kinetic, equilibrium and thermodynamic studies. Biochem Eng J 39:521–530

Wolborska A (1989) Adsorption on activated carbon of *p*-nitrophenol from aqueous solution. Water Res 23:85–91

Yoon YH, Nelson JH (1984) Application of gas adsorption kinetics: part 1: a theoretical model for respirator cartridge service time. Am Ind Hyg Assoc J 45:509–516

Zeldowitsch J (1934) Über den mechanismus der katalytischen oxydation von CO an MnO2. Acta Phys Chim URSS 1:364–449

Chapter 4
Hydrothermal Carbonisation: An Eco-Friendly Method for the Production of Carbon Adsorbents

Carlos Javier Durán-Valle, Almudena B. Botet-Jiménez, and Delia Omenat-Morán

Abstract The hydrothermal carbonisation process has attracted great interest in recent years, as it is an eco-friendly method for obtaining carbonaceous materials. This method consists of heating an organic material in the presence of water in a closed vessel at temperatures above 100 °C. The pressure generated by the water vapour next to a heat supply gives rise to the reactions necessary to form hydrothermal carbon (HTC). This chapter describes the manufacture of these materials and their application as adsorbents of contaminants in the aqueous phase. The manufacturing method differs from the classic pyrolysis method because of its lower power consumption, higher yield, and drastic reduction of pollutant emissions. HTCs are chars with high oxygen content, a large number of functional groups, and low porous development. They have been used to remove numerous contaminants of all types, but they are especially suitable for the removal of heavy metals.

Keywords Hydrothermal carbonization • Adsorbents • Water pollution • Heavy metals

Contents

C.J. Durán-Valle (✉)
Departamento de Química Orgánica e Inorgánica, Universidad de Extremadura, Avda. de Elvas, s/n, 06006, Badajoz, Spain

Instituto Universitario de Investigación del Agua, Cambio Climático y Sostenibilidad, Universidad de Extremadura, Avda. de Elvas, s/n, 06006, Badajoz, Spain
e-mail: carlosdv@unex.es

A.B. Botet-Jiménez • D. Omenat-Morán
Departamento de Química Orgánica e Inorgánica, Universidad de Extremadura, Avda. de Elvas, s/n, 06006, Badajoz, Spain

© Springer International Publishing AG 2017

A. Bonilla-Petriciolet et al. (eds.), *Adsorption Processes for Water Treatment and Purification*, DOI 10.1007/978-3-319-58136-1_4

4.1 Introduction

Carbonaceous materials have numerous applications that have been known for some time. The two most important purposes (depending on the amount used) are as fuel or as adsorbent. As an adsorbent, the employment of activated carbons is important, which are usually obtained from charcoal (or biochar) that is obtained from the biomass.

The traditional method of carbonisation is pyrolysis of the biomass. In general terms, it consists of heating a lignocellulosic biomass at elevated temperatures (400–1000 °C) for several hours, or even days. This method poses some problems:

- The use of high temperatures requires a large amount of energy, in addition to the hazardous process and the wear on the materials employed (by both high temperatures and corrosive by-products).
- Biomass drying also requires energy, as the presence of liquid water considerably increases energy consumption.
- An environmental problem is the production of high amounts of pyroligneous acid, tar, and contaminating gases, which do not allow the process to be classified as environmental friendly.
- Lastly, pyrolysis yield, though variable, is usually very low.

This is where hydrothermal carbonisation comes into play. Hydrothermal carbons (HTCs) can be obtained from a biomass and water at mild temperatures (typically 200 °C) under water-generated pressure. Their advantages are:

- The raw material does not need to be dried because of the fact that it will be mixed with liquid water. Together with mild temperatures, both factors represent significant savings on energy expenditure.

- Consequently, a "wet" biomass, such as grass, leaves, or a glucose solution can be used.
- No gases are released, no tar is produced, and the liquid obtained is not particularly polluting and very easy to handle later.
- In addition, the yield is higher, about twice as high as in pyrolysis (Kumar et al. 2011).

Therefore, it is a method of carbonisation that will be used frequently in the future. This process has been known, as the early 1900s (Bergius and Specht 1913), for making synthetic coal at low temperatures and was recently rediscovered for the preparation of chars.

The product obtained has only one property in which it is expected to be inferior to the pyrolytic coal: its heat capacity, which will be somewhat less per unit mass. But there are many applications (Hu et al. 2008; Titirici and Antonietti 2010; Titirici 2012; Titirici et al. 2012) for which this carbonaceous material has been successfully tested. In this chapter, we will focus on obtaining HTCs and their use in the purification of contaminated water.

4.2 Hydrothermal Carbon Preparation

The properties of an HTC depend on factors such as the precursor selected, hydrothermal process conditions, and the use of templates or solids to coat. Subsequently, HTCs can be activated or functionalized in several ways. Figure 4.1 shows a diagram of the possible processes taken into account in the manufacture of HTCs.

4.2.1 Precursors

Several precursors have been employed as a source of carbon in the preparation of these materials, as shown in Fig. 4.2. A typical option is to use pure compounds from the biomass, such as glucose (Sevilla and Fuertes 2009; Chen et al. 2011; Yu et al. 2014; Martín-Jimeno et al. 2015; Roldán et al. 2016), starch, sucrose (Sevilla and Fuertes 2009), or fructose (Alatalo et al. 2016). The selection of these compounds is because of the simplicity of the process, as the reactions involved can be studied more easily. There are numerous bibliographical references that try to understand the hydrothermal process (Titirici 2012), and they usually use a simple compound, such as glucose. One advantage of these substances is that they are very abundant and inexpensive. For example, glucose is the most abundant carbohydrate unit in biomass. When using glucose, HTC is obtained in spherical forms, which grow in size with increasing carbonisation time. Chen et al. (2011) tested several compounds (glucose, sucrose, starch, and cellulose) as precursors, but glucose was finally selected because the result has more carboxylic groups.

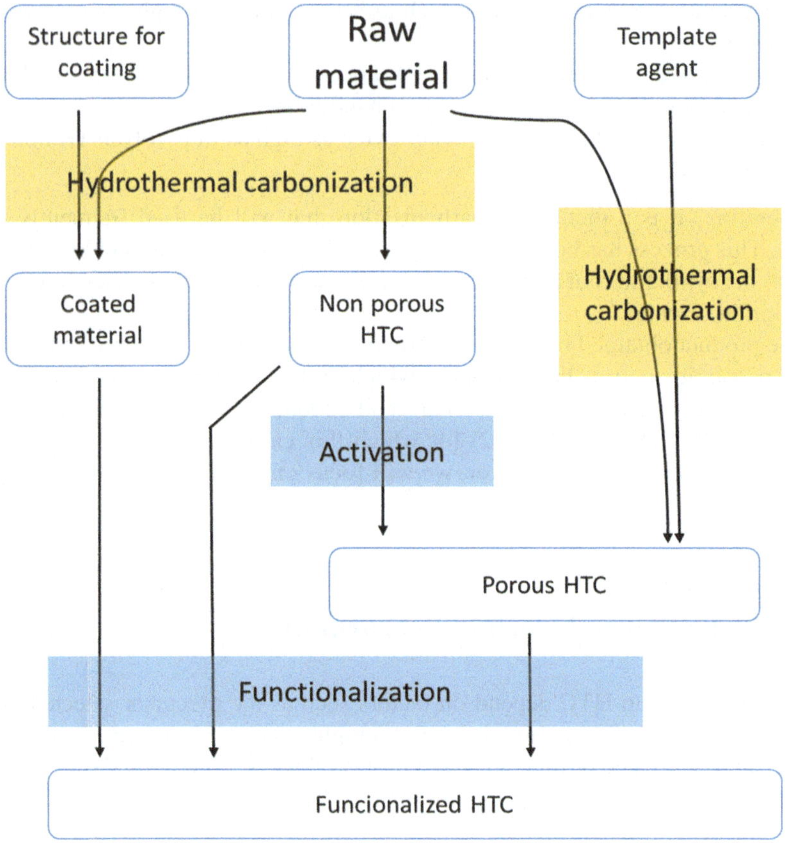

Fig. 4.1 Diagram of processes in hydrothermal carbons synthesis

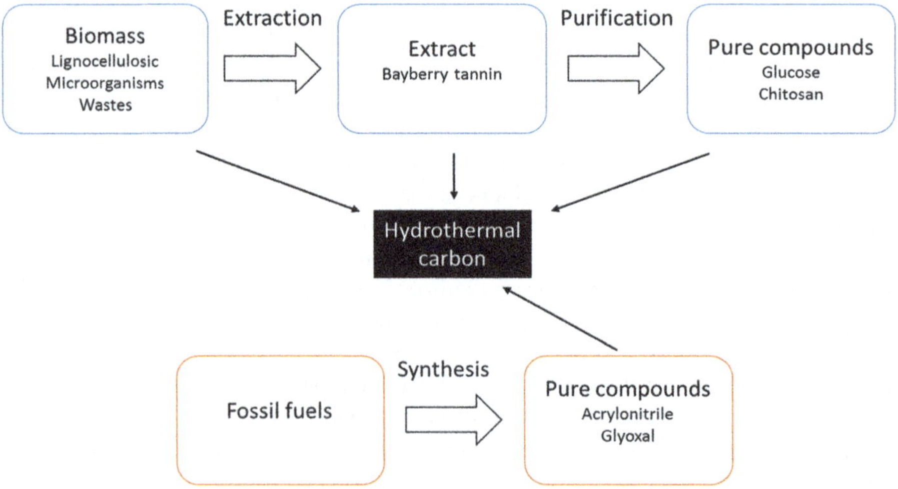

Fig. 4.2 Main origins of hydrothermal carbons, with examples

An intermediate choice between the use of pure compounds and an unmodified biomass is the use of biomass extracts. For example, bayberry tannin has also been used (Li et al. 2014). Lignocellulosic materials are very common as raw materials for charcoals or activated carbon manufacturing. They are composed of three main components: cellulose, hemicellulose, and lignin. As cited in Sect. 4.1, one advantage of HTC is its ability to utilise biomass feedstock without drying before or during the process. It has an edge in terms of energy savings over pyrolytic carbonisation. Obviously, lignocellulosic materials have been widely used in HTC preparation (Hu et al. 2008). Other materials that have been used include switchgrass *Panicum virgatum* (Kumar et al. 2011; Regmi et al. 2012), pine needles *Salix psammophila K* (Liu et al. 2013; Zhu et al. 2014), wood (Laube and Reza 2016), pinewood (Liu et al. 2010), pinewood sawdust, rice husks (Liu and Zhang 2009, 2011), walnut shells, sunflower stems, olive stones (Román et al. 2013), rice straw (Dai et al. 2014), wheat straw, sawdust, corn stalks (Sun et al. 2015), and peanut hulls (Xue et al. 2012). The effect of removing lignin (Liu and Guo 2015) has also been studied, and it has been concluded that lignin, more stable than cellulose and hemicellulose, protects the other two from heat treatment. A sample of lignin-free wood carbonises faster than normal wood. It has been observed (Hu et al. 2008) that soft plant tissues, without an extended crystalline cellulose scaffold, yield globular carbonaceous nanoparticles; however, hard plant tissues with structural crystalline cellulose scaffolds can preserve their outer shape.

A less common option is the use of microorganisms. An industrial microorganism (*Saccharomyces cerevisiae*) has been used to obtain HTC (Li et al. 2016). Another source of carbonisable biomass is waste products. These can be industrial, such as beer waste (Hao et al. 2014), or livestock waste, such as swine manure (Cao et al. 2011). Another biomass residue of animal origin is prawn shells, which are a source of chitin (poly-β-(1-4)-*N*-acetyl-D-glucosamine). Chitin can be deacetylated to give rise to chitosan (Zhang et al. 2014a; Shen et al. 2016), which can also be carbonised. These cases have the advantages of eliminating and valuing wastes, which are usually considered zero-cost raw materials, and of creating an extra environmental profit. We can also work with compounds that do not come from a biomass, such as glyoxal and acrylonitrile (Yang et al. 2014).

4.2.2 Hydrothermal Process

The hydrothermal process is characterised by moderate temperatures (<300 °C) and high pressure, as the water included in the reactor increases the pressure when it exceeds 100 °C (autogenous pressure). This reaction can also be performed by raising the pressure externally, for instance, by using a pressurised gas. But if it is done in equipment without a pressure measurement appliance, the temperature must be monitored because of the fact that pressure increases very quickly. Table 4.1 shows pressure data obtained autogenously as a function of temperature, calculated according to the Clausius-Clapeyron Eq. (4.1).

Table 4.1 Autogenous pressure (vapour pressure) of water

Temperature (°C)	Pressure (MPa)
100	0.10
120	0.20
140	0.36
160	0.62
180	1.03
200	1.62
220	2.46
240	3.62
260	5.18
280	7.22
300	9.83
320	13.11
340	17.16
360	22.08
380	27.97
400	34.94

$$\ln \frac{P1}{P2} = \left(\frac{\Delta H_{\text{vap}}}{R}\right)\left(\frac{1}{T2} - \frac{1}{T1}\right) \tag{4.1}$$

where $P1$ and $P2$ are the water vapour pressures at temperatures $T1$ and $T2$, respectively, ΔH_{vap} is the water vaporisation enthalpy (40,650 J mol^{-1}), and R is the gas constant (8.314 J K^{-1} mol^{-1}).

If 374 °C and 22.1 MPa are exceeded, water will behave as a supercritical fluid. In this situation, mainly liquid and gaseous fuels are obtained, and it is denominated by hydrothermal liquefaction or hydrothermal gasification. At very high temperatures (Kruse et al. 2013), the formation of gases like H_2 and CH_4 is favoured. At intermediate temperatures, hydrolysis of the biomass occurs, and medium molecules, such as ketones, carboxylic acids, or phenols, are obtained, from which bio-oil can be prepared. At lower temperatures, the hydrolysis reaction decreases and dehydration is favoured, so the main product is a solid, HTC. As the solid product (char) will be analysed in this chapter, we will limit ourselves to studying the subcritical conditions of hydrothermal carbonisation below 300 °C.

It should be kept in mind that if substances other than water are used that can reach the vapour state (Yang et al. 2014), the partial pressures of each of these substances will be added together to give rise to a greater pressure than expected. This fact must be considered in order to avoid accidents and to compare different reagents under the same conditions. One factor that can explain this water reactivity, besides the energy supplied by the heating, is the increase of the water ionisation constant when the temperature increases (Bandura and Lvov 2006), causing a higher concentration of H^+ and OH^- that can catalyse the dehydration reaction. This effect is slightly amplified by the increase in pressure. In Table 4.2, some of the parameters of hydrothermal carbonisation used in published works are shown.

Table 4.2 Main experimental parameters of hydrothermal carbonisation

Temp (°C)	Heating rate (°C/min)	Time (h)	Biomass (g)	Rate (biomass/ water)	References
140–200	–	10	10	1/5	Shen et al. (2016)
160	–	18	1	1/40	Zhao et al. (2016)
180	–	24	2	1/7.5	Yu et al. (2014)
180	5	24	20	1/3.5	Chen et al. (2011)
180	1.5	12	5	1/6	Li et al. (2016)
180	2.5	48	16	1/4	Song et al. (2012)
180	–	16	2	1/7.5	Zhang et al. (2013), Zhou et al. (2015)
180	1.5	24	20	1/3.75	Wang et al. (2012)
180	1.5	24	12.7	1/4.72	Yang et al. (2014)
180	1.5	24	20.8	1/2.40	Yang et al. (2014)
180	–	12	2	1/9	Zhang et al. (2014b)
190	–	16	–	1/10	Demir-Cakan et al. (2009)
200	–	16	3	1/10	Liu et al. (2013)
200	–	20	5	1/12	Sun et al. (2015)
200	–	16	2	1/9	Martín-Jimeno et al. (2015)
200	–	4–14	3	1/10[a]	Dai et al. (2014)
220	–	20	5	1/20	Román et al. (2013)
230	2	6	1000	1/8	Laube and Reza (2016)
250	0.7	20	–	1/4	Cao et al. (2011)
250	–	20	11.6	1/4	Sun et al. (2011)
300	7	0.5	–	1/7	Regmi et al. (2012), Kumar et al. (2011)
300	10	0.33	5	1/6	Liu and Zhang (2009, 2011), Liu et al. (2010)
300	–	5	60	1/6.7	Xue et al. (2012)
300	–	–	–	1/12.5	Zhu et al. (2014)

[a]Water solutions, but not pure water

As shown in Table 4.2, in most cases, temperatures of 180–200 °C are reached, with varying times between 10 and 24 h. When heating to higher temperatures (Liu and Zhang 2009, 2011; Liu et al. 2010; Kumar et al. 2011; Regmi et al. 2012), the preparation time can be considerably reduced. Although this can be considered an advantage, it must be taken into account that the autogenous pressure produced by raising the temperature from 200 °C to 300 °C is multiplied several times (Table 4.1) and that certain commonly used materials, such as Teflon, cannot be used at that temperature. Higher temperatures mean more energy expenditure, which can be compensated for by the brevity of the process. These issues should be considered if HTC production is intended to reach an industrial scale, but are not the most relevant for laboratory research. More important could be the influence of temperature on the porous and chemical structures of the HTC.

The experiments shown in this study have been done on a scale of a few grammes, except Laube and Reza (2016). Apparently, this kind of study is still in the laboratory phase, and there does not seem to be any interest in obtaining medium- to large-scale adsorbents.

A parameter with greater variation is the biomass/water ratio, which varies from 1:2.4 to 1:40. No exhaustive studies have been done on the influence of this proportion, although there is one reference (Sevilla and Fuertes 2009) in which a higher yield and larger size of the spheres obtained were observed by increasing the concentration of glucose. But this result does not allow conclusions to be drawn when using other materials, such as a ligand-cellulosic biomass. It should be remembered that the presence of water is crucial to this synthesis method, as it generates autogenous pressure and reacts with the parting material to produce carbonisation.

One problem posed by HTCs at the end of the hydrothermal carbonisation process is that they have very low porosity. Several strategies have been put into place to introduce micro- and mesopores and thus increase the specific surface area. The most common method and the one used with pyrolysed carbons is activation (see Sect. 4.2.5), but templates have also been used to improve this property (see Sect. 4.2.3).

Some authors add a small amount of acid, as it favours the dehydration processes of some organic molecules, such as carbohydrates (Cao et al. 2011; Liu et al. 2013; Zhang et al. 2014b). On the other hand, Dai et al. (2014) observed that increasing the hydrothermal treatment time from 4 to 14 h hardly changed the composition of the final product, although it must be considered that they used a solution of $LaCl_3$ and KOH instead of distilled water. The presence of these compounds can catalyse the hydrothermal process (Hu et al. 2008) in such a way that extended time does not make a big difference. Slight carbon enrichment and a decrease in the content of hydrogen and oxygen can be observed. There is a large difference (Dai et al. 2014) when varying the composition of the raw material, but there are only small differences between HTCs prepared at different temperatures from the same precursor.

Sevilla and Fuertes (2009) carried out an extensive study on the effect of preparation conditions (temperature, time, and concentration) on the properties of HTC, using glucose, sucrose, and starch. The yield obtained in the hydrothermal process increased with increased temperature (in the range of 170–240 °C), with longer treatment time (from 0.5 to 15 h), and with higher concentration of the carbohydrate in water. Similarly, the average size of the HTC spheres obtained varied. However, Shen et al. (2016) have reported a decrease in yield from 78.8% to 35.2% when increasing the temperature from 140 °C to 220 °C. The influence of hydrothermal treatment conditions on the properties of the HTC is a complex subject and seems to depend heavily on the raw material used.

Sevilla and Fuertes (2009) detected the existence of a large number of functional groups, but they were unevenly distributed. The nucleus of the spheres is highly aromatic with stable oxygen groups (ether, quinone, pyrone, etc.), while the shell contains a high density of functional groups (hydroxyl, phenol, carboxylic, etc.).

Thus, the exterior would be hydrophilic, while the interior would be hydrophobic. This hydrophilic outer surface favours the dispersion of the HTC particles in water.

At the end of the hydrothermal treatment, the product must be ovendried before being used. Numerous authors have previously performed washing with acetone to remove organic substances soluble in this solvent. The reason comes from the fact that the hydrothermal treatment was initially used for gasification and liquefaction of the organic matter, and the liquid produced when washing it with acetone is a biofuel. Except for a decrease in the contribution of functional groups (Cao et al. 2011), no major structural changes in the HTCs thus treated have been described.

4.2.3 Templates

One of the main disadvantages of the use of HTCs in adsorption is that they present both limited porosity and surface area. This is also true of pyrolytic carbons, and activation is used to improve their properties. Activation can also be used in HTCs and will be discussed in Sect. 4.2.5.

Another method to increase porosity is the use of templates in the hydrothermal process. One of the templates employed (Martín-Jimeno et al. 2015) is graphene oxide. It presents several interesting characteristics: good dispersibility in water and organic solvents, greater reactivity than non-oxidized graphene, and a tendency to form monolithic gels. Thus, a suspension of graphene oxide was prepared in which glucose was dissolved. This mixture was heated in an autoclave at 200 °C for 16 h, yielding a carbonaceous hydrogel that was dried and subsequently activated. This method allows to obtain a monolith instead of the usual powder formed by the spheres characteristic of this process. The density of the monolith increases as the glucose concentration increases, indicating that the carbonisation must be incorporated into the template structure. It could also be considered to be an example of coating (see Sect. 4.2.4).

A reagent frequently dealt with in chemical activation is $ZnCl_2$. Roldán et al. (2016) introduced it into the mixture before the hydrothermal process, employing the "salt-templating" methodology. The basic concept is that the hypersaline conditions stabilise the surface of the first small nanoparticles, avoiding Ostwald ripening or excessive particle growth. These particles turn collectively unstable at sufficiently high concentrations, undergoing spinodal phase separation and cross-linking towards the final porous carbon gels. The more salt is added, the smaller the primary particles are and, hence, the higher the surface area is. The HTC prepared with $ZnCl_2$ presents a greater specific surface once the hydrothermal process is finished. But when washing the carbon to remove the Zn and when performing the pyrolysis of the HTC to activate it thermally, the specific surface increases considerably, whether $ZnCl_2$ has been used or not, but with narrower pores in this case. There is a big difference when the HTC is not washed before the pyrolysis because, in that case, a much higher specific surface is obtained because of the activation process. Note that it has already been mentioned that $ZnCl_2$ is usually used in the chemical activation of pyrolytic carbons. This, in fact, does not imply an improvement in the adsorption capacity of dyes, which are the pollutants with which they have been tested, being more effective the introduction of N and S heteroatoms.

The same method of "salt templating" was exercised with a LiCl-$ZnCl_2$ mixture (Alatalo et al. 2016). Micro-/mesoporous HTCs were thus achieved with a specific surface area greater than 400 m^2/g obtained with BET method. Another common template (Hu et al. 2008) for pyrolytic carbons that could be brought into play here is silica. By modifying the polarity of the silica surface, porous structures of various sizes and shapes could be synthesised.

4.2.4 Coating

One possibility offered by the HTC method, when employing water-soluble substances such as glucose, is the possibility of coating carbonaceous shells on a structured surface. The coating process is dependent (Titirici 2012) on the surface properties, hydrothermal carbonisation time, and concentration. It is mainly used to coat nanostructures and in applications related to energy storage. This technique is not usually applied to make adsorbent materials, although there are some examples.

Zhao et al. (2016) used polyacrylonitrile fibres as a template. First, they reinforced the bond between the fibres with ethylenediamine and ethylene glycol. Then, they added these fibres to a glucose solution and performed the hydrothermal carbonisation of the carbohydrate, which was deposited on the fibre structure.

The opposite process can also be performed: depositing certain compounds on the HTC, for which the existence of oxygenated functional groups is very helpful. This process should not be confused with functionalisation, and sometimes it is difficult to establish a limit between the two. We can consider the process to be coating when it forms a species with its own crystallinity and functionalisation if a molecule binds to the HTC by a covalent bond and acts as an organic functional group of the corresponding graphene molecule. An example of coating is the deposition of noble metal nanoparticles on carbonaceous nanofibers (Quian et al. 2007). These materials are often used in catalysis.

4.2.5 Activation

The activation process is added to improve the porous structure of the carbonaceous materials, i.e., increase the volume of the pores and the specific surface. There are two main types: chemical activation and physical activation. The main difference between the two is the reactivity of the compounds used as activating agents. The activation process is important when HTC adsorbents are prepared because their capacity is often related to their specific surface and their porous structure. This aspect is more important in adsorption than in other applications of HTCs.

4.2.5.1 Chemical Activation

One of the methods used is chemical activation. The compounds employed are more reactive than those employed in physical activation, and the temperature range is wide, ranging from room temperature to about 800 °C. Activation time is usually from 1 to 2 h. It must be taken into account that the use of high temperatures (>400 °C) can destroy existing functional groups, so in many applications, it is necessary to evaluate if it is convenient to increase porosity and specific surface area by modifying the surface chemistry.

One of the most common reagents is KOH. Thus, Martín-Jimeno et al. (2015) activated an HTC obtained with templates of graphene oxide, treating it with KOH at temperatures of 700 °C and 800 °C. Regmi et al. (2012) also performed a treatment with a KOH solution at room temperature. Although the treatment entails an improvement in the adsorption of Cu(II) and Cd(II), it hardly implies a change when the concentration of the KOH solution oscillates between 0.5 and 2.0 N. Activation under these conditions gives rise to very high yields (Sun et al. 2015) with little loss of mass. But it does change the PZC of the HTCs obtained from acidic to slightly basic. This implies a change in the electrical charge of the surface as a function of pH, which is different from the usual behaviour of the HTCs. Also, a NaOH 1 M solution can be used as an activating agent (Zhao et al. 2016).

On the other hand, Hao et al. (2014) used H_3PO_4 to activate an HTC obtained from beer waste. In this case, they studied the influence of different variables of the activation process on the product obtained. These variables were the activation temperature (600–700 °C), the activation time, the concentration of H_3PO_4, and the flow rate of N_2.

4.2.5.2 Physical and Thermal Activation

The other habitual method of activation is physical activation, which consists of heating the sample in the presence of a gas, the most common gases being air, carbon dioxide, and water vapour. The observed changes are those usually described in the activation of pyrolytic carbons; however, lower performance (greater burn-off) has been reported as compared to the activation of pyrolytic carbons. This is not a surprise, as hydrothermal carbonisation results in a carbonaceous material richer in oxygen than pyrolytic carbon. This implies that it contains more volatile matter and that a greater proportion is lost than the fixed carbon in the activation treatment.

Thus, by activating an HTC with carbon dioxide at 800 °C during a variable period of time (Liu and Zhang 2011), there is an increase in the ash content and a decrease in the yield as the activation time is increased. At the same time, there is an increase of the specific surface. An increase in the proportion of wide pores (meso- and macropores) also occurs (and this does not always happen with pyrolytic carbons).

When comparing the activation of HTCs from different materials with air and carbon dioxide (Román et al. 2013), it is observed that the specific surface and volume of micropores, which are two related parameters in activated carbon, are always higher when using CO_2. In contrast, different results are obtained in the volume of meso- and macropores. In both activation processes, micropores are created, and the specific surface area increases.

A variation of this method is thermal activation, which consists of heating the HTC in the presence of an inert gas. In this case, the activation is produced by the temperature and not by the reaction with the gas, as it only serves as a carrier of the by-products formed. Although some authors (Zhu et al. 2014) use it as an activation method (N_2, from 300 °C to 700 °C), one should consider whether a second pyrolysis-type carbonisation or an actual activation is actually being performed. As mentioned above, HTCs differ from pyrolytic carbons because of their low porous development. In the work of Zhu et al. (2014), the evolution of the composition of the chars according to temperature is as expected: an increase of the carbon content and a reduction of the hydrogen and (especially) oxygen content. This variation is important when starting the heat treatment (up to 300 °C) and also between the temperatures of 400 °C and 600 °C. In the first step, it can be assumed that changes in composition are mainly because of water loss (desorption or dehydration). And between 400 °C and 600 °C, changes are because of the pyrolysis process. A consequence of this variation in the oxygen ratio is that the pH changes from slightly acidic in the HTC (5.2) to very alkaline in the sample treated at 700 °C (pH 11.8).

Table 4.3 summarises some of the activation methods found in the literature. Chemical activation methods (with liquid or solid reagents) predominate over physical activation methods (using gaseous reagents). The activation times are, in general, shorter than those of hydrothermal carbonisation. There may be large variations in the conditions under which the activation is carried out.

4.2.6 Functionalisation

There are many methods of changing the surface chemistry of a carbonaceous material, and some of them have been used with HTC. Activation can modify the functional groups of the carbonaceous material, but its main function is to improve the porosity. For this reason, activation methods will not be dealt with in this section.

Before modifying the surface chemistry of an HTC, its properties must be taken into account. If it is obtained from a biomass, it must be assumed that the raw material contains a large amount of oxygen in its composition. Most of the starting materials (carbohydrates and lignocellulosic materials) have an oxygen content approaching 50% by weight, excluding moisture. As high temperatures have not been used, many of the original functional groups have not been destroyed. We have, therefore, a carbonaceous material with a high number of functional groups.

Table 4.3 Main experimental parameters of activation of HTCs

Temp (°C)	Heating rate (°C/min)	Isotherm time (h)	Activating agent	Rate (agent/HTC)	References
Room	–	1	KOH 2M	250 mL/g	Regmi et al. (2012), Sun et al. (2015)
700–800	–	0.5	KOH[a]	1/1 and ¼ weight	Martín-Jimeno et al. (2015)
600–700	10	3	H_3PO_4 45–80%	4 mL/g	Hao et al. (2014)
Room	–	2	NaOH 1M	750 mL/g	Zhao et al. (2016)
800	10	0.5–2	CO_2	30 mL/min	Liu and Zhang (2011)
850	–	0.5	CO_2	40 mL/min	Román et al. (2013)
250	–	0.5	air	100 mL/min	Román et al. (2013)
300–700	4	4[b]	N_2	1 L/min	Zhu et al. (2014)

[a]A concentrated solution of KOH. Incipient wetness procedure
[b]Total time

4.2.6.1 Functionalisation During the Hydrothermal Process (One Step)

Functionalisation can begin during the hydrothermal process. Thus, Roldán et al. (2016) used glucose as a source of carbon and used either pyrrole-2-carboxaldehyde to add nitrogen atoms or 2-thiophenecarboxaldehyde atoms to add sulphur atoms. Nitrogen atoms may be included in the HTC if they already exist in the raw material, for example, in chitin (White et al. 2009). Functionalised HTC with fructose and 2-thiophenecarboxaldehyde (Alatalo et al. 2016) has also been prepared. Note that glucose has been mixed with phosphoric acid (Zhang et al. 2014a). We can observe another example in the use of an alkaline solution of $LaCl_3$ instead of water to obtain a lanthanide-doped HTC (Dai et al. 2014). Demir-Cakan et al. (2009) used acrylic acid mixed with glucose to obtain a high number of carboxyl groups in the HTC.

Functionalisation can be carried out by choosing the appropriate starting material. A work has been previously cited (Chen et al. 2011) comparing HTCs prepared from glucose, sucrose, starch, and microcrystalline cellulose. HTCs with different numbers of carboxyl groups were obtained, despite using the same conditions in the hydrothermal process.

4.2.6.2 Post-functionalisation (Two Steps)

Chemical modification of the surface can be done after obtaining the HTC. The methods followed are similar to those employed for activated carbons of pyrolytic origin.

A simple method is oxidation or calcination under air. atmosphere It is usually carried out at temperatures below 400 °C (Chen et al. 2011; Zhang et al. 2013; Yu et al. 2014) because, at this temperature, most of the functional groups begin to be destroyed. But the temperature must be higher than 200 °C, as it is necessary for sufficient oxygen chemisorption of the air to occur so that the functional groups can be formed. The objective in this temperature range is to achieve a greater number of oxygen functional groups, which in many cases improve the adsorption capacity of the HTC. Zhou et al. (2015) have studied the effects of temperature (from 100 °C to 300 °C) and time (from 1 to 5 h) in detail. The oxygen content and the number of carboxyl groups increase linearly with temperature and time, except in the change from 250 °C to 300 °C where the number of carboxyl groups increases significantly. Chen et al. (2011) studied the effect of temperature between 100 °C and 400 °C and analysed the functional groups using the Boehm method. As the temperature increased, the weaker acidic functional groups (type phenol) decreased from 100 °C, while the strongly acidic ones (type carboxyl) increased with temperature. The variation in both cases is especially important when the temperature exceeds 200 °C. In contrast, the increase in oxidation time, although increasing the number of carboxyl groups, does it very weakly. These results have been confirmed using FTIR and XPS data.

Another method of oxidation (Li et al. 2016) is oxidation with potassium permanganate in a strongly acidic medium, according to the well-known Hummers method (Hummers and Offeman 1958), to obtain graphite oxide. Two other oxidation methods frequently used with carbonaceous materials are the use of nitric acid (Liu et al. 2013) and the use of hydrogen peroxide (Xue et al. 2012). In both cases, the oxygen content increases, and at the same time, a decrease of PZC is verified.

As a general rule, mild oxidation methods give rise to the adsorption of oxygen to create functional groups, increasing the mass of the carbon. Aggressive methods produce carbon oxidation and the formation of carbon oxides, reducing the total mass. It is not strange, therefore, that in the described conditions the mass yield exceeds 100% in some cases.

Some uses of functionalisation are highly specific. One example is the use of *o*-phosphoethanolamine and *N,N′*-dicyclohexylcarbodiimide (Yu et al. 2014) with the aim of improving the adsorption of $(UO_2)^{2+}$ by the phosphate groups, because the phosphate group has strong tendencies to form complexes with uranyl ions.

Other functionalisations require several stages. For example, Song et al. (2012) used 5-azacytosine to adsorb U(VI), as this multidentate ligand had demonstrated a high affinity for dissolution towards the actinide metals. To bind the 5-azacytosine to the carbon, they first performed calcination in air at 300 °C. The product obtained was reacted with ethane-1,2-diamine in the presence of *N,N′*-dicyclohexylcarbodiimide. And in the third step, amino groups grafted on the HTC surface reacted with 5-azacytosine.

Another similar case is functionalisation with salicylideneimine (Wang et al. 2012). The first two steps are similar, but in the third step, salicylaldehyde reacts with the amino groups previously formed to anchor to the surface of the HTC as salicylideneimine, which is a good ligand of cations.

Functionalisation may sometimes depend on the raw material. Thus, Liu and Zhang (2011) prepared two HTCs activated from pinewood dust and rice husks. The last is known to have a lot of silicium that converts to silica. When phenol adsorption was probed, there was a large difference in the amount adsorbed between the two coals, which can be explained by the greater specific surface area of the pinewood dust HTC. However, the difference in Cu(II) adsorption was lower, which can be explained by the presence of silicon forming silanol groups, which easily bind Cu, increasing the adsorption capacity.

A two-stage process whose functionalisation begins in the hydrothermal process is HTC preparation (Yang et al. 2014) from glyoxal and acrylonitrile to obtain cyano (-CN) functional groups. These react with hydroxylamine hydrochloride to obtain amidoxime grafted to the HTC surface.

4.2.7 Hydrothermal Versus Pyrolytic Carbonisation

It is interesting to compare the two methods of carbonisation, although there are not many papers dedicated to this question. The first differences we can find are in the preparation conditions. The hydrothermal process is carried out at moderate temperatures (180–300 °C) and high pressure (1–10 MPa), while pyrolytic carbonisation is carried out at a higher temperature (400–1000 °C) and at atmospheric pressure (0.10 MPa). In fact, out of these two factors, temperature seems to be the one that most influences the final properties of the material obtained.

Some papers have been published comparing materials obtained by the two methods of carbonisation. Liu et al. (2010) have studied the properties of HTC obtained at 300 °C and pyrolytic carbon obtained at 700 °C in an atmosphere of N_2 from pinewood. Cao et al. (2011) used dried swine manure as raw material. The pyrolytic char was obtained at 620 °C after 2 h, and the HTC was obtained at 250 °C after 20 h. Falco et al. (2011) prepared HTC at 180 °C for 12 h and pyrolytic char in an atmosphere of nitrogen at various temperatures. In this experiment, the HTCs were later pyrolysed.

The composition obtained by elemental analysis is very different between the two carbons used in the work of Liu et al. (2010). Pyrolytic carbon contains a large amount of carbon while reducing the content of hydrogen, nitrogen, and oxygen. This is because the higher the temperature, the degraded the heteroatom-containing functional groups (i.e., the more labile groups). The composition obtained suggested (Durán-Valle 2006) that the aromatisation process progressed more in the pyrolytic carbon than in the HTC.

The proximate analysis is related to these data. A larger amount of heteroatoms is related to functional groups less stable when heated, and therefore, the HTC contains a greater amount of volatile matter and less fixed carbon. The amount of ash is also lower in the HTC, but this can be mainly attributed to the higher yield of this carbonisation, as it can be considered that the amount of ash in the raw material is usually kept approximately stable during both processes of carbonisation. That is, the variation observed is because of the decrease in the fixed carbon content and especially in a volatile matter. Some authors also proposed (Liu et al. 2010) that the presence of water in the hydrothermal method can dissolve part of the mineral matter.

Another difference found (Liu et al. 2010) is in the acidic/basic properties. As it contains more oxygen, the HTC has more acidic functional groups, and its zero point charge (PZC) is lower than in pyrolytic carbons.

As for the surface structure, both carbons have poor porosity development, which is even lower in HTC. The specific surface obtained is low, mainly because of the small volume of micropores formed. For many applications in adsorption of these materials, it is necessary to develop porosity, especially when apolar pollutants are removed. This can be done by using templates in the hydrothermal process or by activating both types of carbons, as discussed above.

The degree of graphitisation is usually very low in HTCs. Demir-Cakan et al. (2009) have measured this parameter using Raman spectroscopy. The data of the proximate and ultimate analyses and of the physical structure support this conclusion.

Sun et al. (2011) compared two pyrolytic carbons (obtained at 400 °C) and two HTCs (obtained at 250 °C) from poultry litter and wheat straw. The differences between the coals obtained are not as big as has been described in other works (Liu et al. 2010). This can be explained by the low temperature and limited pyrolysis time (2–4 h) in the work of Sun et al. (2011), suggesting that pyrolytic carbons are on the limit of what can be considered a carbonaceous material and not a roasting material. Thus, the elemental composition is similar, although the small differences observed support what has been indicated in other works: higher carbon content and lower nitrogen, hydrogen, and oxygen content in the pyrolytic carbon compared to HTC. It is also observed that the specific surface is lower in the case of thermal coals than in HTCs, which is not usually the case. However, the measure of aromaticity by ^{13}C-NMR indicates that it is more developed in pyrolytic carbons.

A similar result is observed when comparing an HTC obtained at 180 °C with a pyrolytic carbon prepared at 350 °C; both are rich in oxygen and hydrogen when compared to a pyrolytic carbon obtained at 750 °C, which is most common (Falco et al. 2011). The change in composition depends on temperature, and it is observed that pyrolysis of the HTC increases the carbon content and decreases hydrogen and oxygen content, obtaining an elemental analysis similar to that of a carbon obtained by pyrolysis at the same temperature.

The yields depend on both the raw material and the carbonisation conditions and can reach very different values. In general, it can be stated that the hydrothermal method achieves a yield double that of the pyrolytic method.

4.3 Adsorption

Among the other applications cited in the introduction, HTCs have been employed in adsorption processes, especially in the removal of contaminants from aqueous solutions.

4.3.1 Dye Adsorption

As stated in Chap. 1, dyes are pollutants that are difficult to remove from wastewater because they are non-biodegradable as well as toxic to plants, animals, and humans. The greatest environmental problem with dyes is that their absorption of sunlight entering the water interferes with the growth of bacteria, which dips to an insufficient level for biologically degrading impurities in the water. Also, the growth of algae decreases, as sunlight does not reach them; some algae even die, affecting the food chain. The dye molecules are distinguished by their large size, which is why adsorbents with a larger pore size, between the broad microporous and the narrow mesoporous, are preferable.

Alatalo et al. (2016) used a $ZnCl_2$-LiCl template, using fructose as the carbon source organic material. When 2-thiophene carboxaldehyde was added, a solid with a larger surface area and pore volume was obtained. Both materials were tested as methylene blue adsorbents. The material without sulphur is more effective, probably because of higher oxygen content. In addition, it maintains its adsorption capacity better when performing several consecutive cycles.

Martín-Jimeno et al. (2015) adsorbed three dyes (rhodamine B, fuchsin basic, and methyl orange; see Fig. 4.3) on HTCs made on templates, some of them activated with KOH. The best-absorbed dye was methyl orange, because it is the narrowest of the three, and these basically prepared templates show mainly narrow microporosity. Those HTCs activated with high KOH ratios developed broader

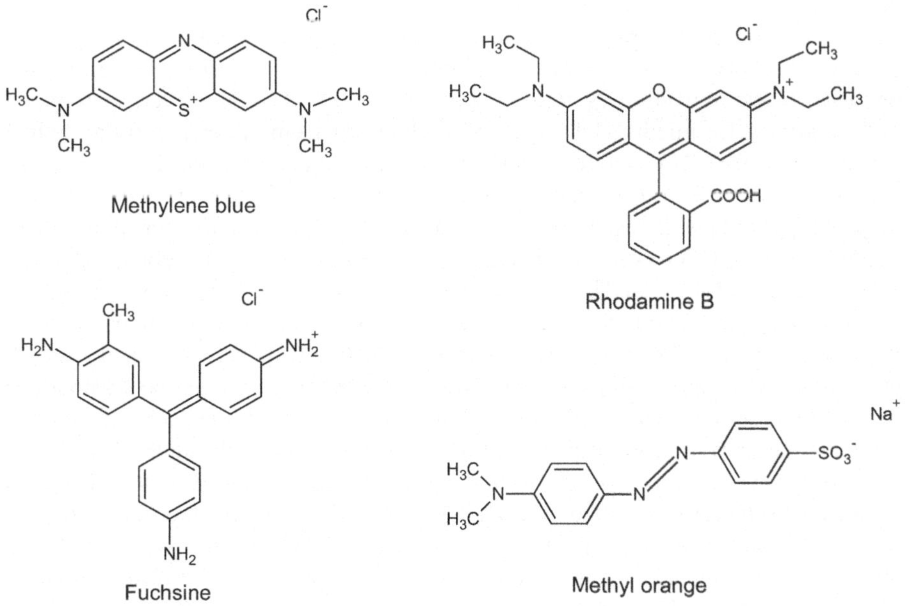

Fig. 4.3 Molecules of dyes cited in this section

porosity (broad micropores and narrow mesopores), so they adsorbed a greater amount of rhodamine B and fuchsin basic than the other HTCs.

Roldán et al. (2016) prepared HTC using glucose, doping the HTC with N or S by adding pyrrolecarboxaldehyde, thiopenecarboxaldehyde, or both. These change the porosity before pyrolysis, with a higher percentage of mesopores. Also, the sample contains $ZnCl_2$. In this experiment, these authors did not wash the salts, because $ZnCl_2$ is an activation agent activated to develop microporosity. These HTCs were used as adsorbents of methylene blue and rhodamine B. This dye is adsorbed to a lesser extent, and the authors explain it as having a somewhat larger size than methylene blue. It is interesting to note that non-pyrolysed HTCs, despite their smaller surface, adsorb dyes better. This can be explained by the lower oxygen content of the pyrolysed materials, which interact less well with the cationic dyes.

Hao et al. (2014) studied the adsorption of methylene blue dye from aqueous solutions using activated carbons produced by chemical activation of hydrothermally carbonised (HTC) beer waste. By studying the adsorption isotherms, it was concluded that activated carbon has a good adsorption capacity for the dye studied (341 mg/g). The pH value affects the adsorption process, so at a low pH, the adsorption is lower than at a high pH. This is attributed to the fact that at pH values above the PZC, the surface of the adsorbent is negatively charged, which increases the adsorption capacity of the positively charged methylene blue molecule.

4.3.2 Pesticides

One of the activities that produces a lot of water pollution is agriculture, for example, with the discharge of pesticides by agricultural irrigation systems. There are few published works about the adsorption of pesticides with HTCs. However, the removal of paraquat (1,1′-dimethyl-4,4′-bipyridyl dichloride) from water has been studied (Zhao et al. 2016). This herbicide is banned in several countries because of its high toxicity but is still used in many regions of the world. To test the removal, they prepared PAN fibres that were mixed with a solution of glucose to later submit them to the hydrothermal process. An HTC in the form of fibres was thus obtained, which was activated with 1M NaOH. The specific surface area obtained was low ($<12\ m^2/g$). The hydrothermal treatment increases the tensile strength, and although it decreases the elongation at break, it has good flexibility. Such enhanced mechanical strength is beneficial for applications in wastewater treatment. The adsorbed amount is practically zero, only with PAN fibres. It increases about 50 times after the hydrothermal treatment and is multiplied by three after activation with NaOH. This shows that adsorbent capacity comes from the presence of a porous material and is enhanced by the creation of oxygenated functional groups. As the pH decreases, the adsorption capacity of paraquat decreases. This is because of electrostatic interactions. Paraquat is a cation (Fig. 4.4) and is repelled by the positive charge of the protonated HTC surface at a low pH. As the pH increases, this repulsion decreases, and the effect disappears at a pH above 7. Also, the adjustment to kinetic

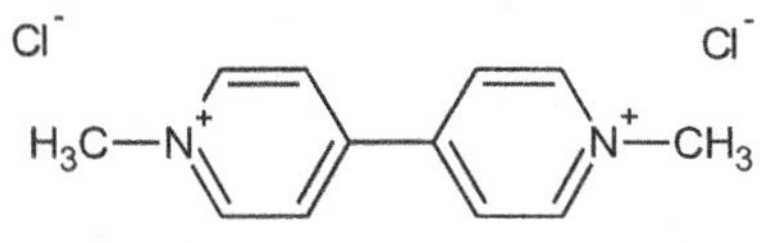

Fig. 4.4 Structure of paraquat (1,1′-dimethyl-4,4′-bipyridyl dichloride)

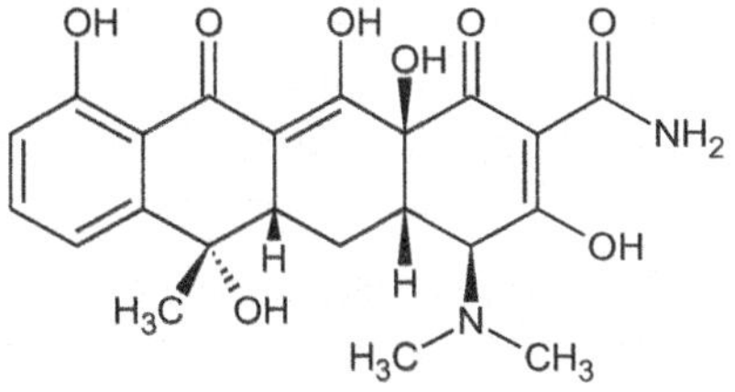

Fig. 4.5 Molecular structure of tetracycline

models (pseudo second order) and isotherms (Langmuir) is in agreement with the idea that the limiting step of adsorption occurs at the surface, which could be explained by the electrostatic interactions.

4.3.3 Drugs

There are not any publications on the adsorption of drugs with HTCs. In particular, tetracycline (Fig. 4.5) is an antibiotic that is widely used in aquaculture and veterinary medicine to improve growth rates and feed efficiencies. It is excreted as an unmetabolized compound, and the greatest known risk is the development of multiresistant bacterial strains that will no longer be able to be treated with current antibiotics. Zhu et al. (2014) have studied their adsorption with heat-activated HTC, obtaining the best results for treated char at the highest temperatures (700 °C). In this case, it must be assumed that because functional groups are not abundant in HTC, it is the specific surface that drives adsorption.

4.3.4 Endocrine Disrupting Chemicals

These compounds, as drugs, have biological activity. They can mimic the activity of natural hormones. They have been found in the effluents of municipal sewage treatment plants. For example, Sun et al. (2011) have tested HTCs as adsorbents of bisphenol A, 17α-ethinylestradiol, and phenanthrene. They deduced that compared to pyrolytic carbons, more polar HTC favours adsorption of the more polar compounds.

4.3.5 Metal Ions

Water pollution caused by heavy metals is a serious worldwide environmental problem with significant impacts on human health and on nature. One common method of removing these contaminants is adsorption. Although HTCs have been employed to adsorb many different contaminants, they have certain properties that make them useful for the adsorption of metals. In fact, they can compete with activated carbons with a specific surface area 10 times greater. Among these properties are the following:

- They have a large number of oxygenated functional groups originating from the raw material. These functional groups are polar and, therefore, more favourable for ion adsorption.
- These functional groups in the HTC can be used to give rise to other distinct functionalities, increasing or making the interaction with the metallic ions more specific.
- They are acidic. This means that in most situations where there is a metal in a solution, it is found as a cation, and the HTC has given protons to the solution and is negatively charged. The negative charge on its surface attracts positively charged metal ions (see Fig. 4.6).
- Some metals, especially in a high oxidation state, are dissolved as anions. The high number of functional groups on the HTC makes it possible to easily change the charge on its surface by modifying the pH of the medium. Thus, at a pH < PZC, the surface will be positively charged and will prefer to adsorb anions.

As mentioned above, the pH of the solution may affect the adsorption capacity, as it modifies the surface of the HTC. An example with a metal cation (M^{n+}) and a surface with a determined PZC are shown in Fig. 4.6. When pH < PZC, the HTC accepts protons, which positively charge its surface, repelling the cations. When pH > PZC, the HTC yields protons from its functional groups, remaining negatively charged and attracting the cations. The situation may be more complex than shown in Fig. 4.6, as the metal may form other species, such as anions or precipitated solids.

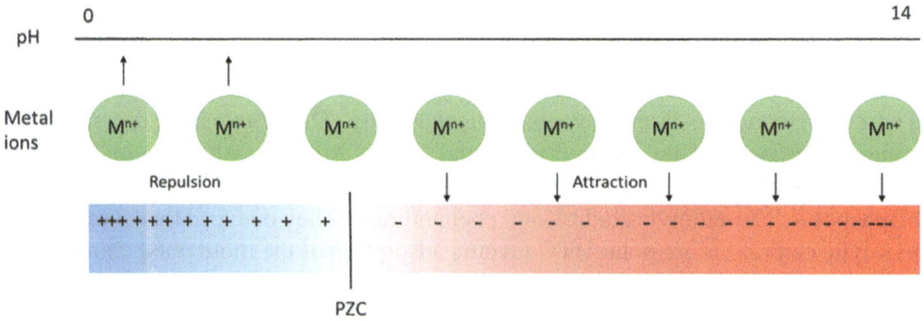

Fig. 4.6 Effect of pH on cation adsorption

4.3.5.1 *p*-Block and *d*-Block Metals

Cr(VI) is considered to be one of the top priority toxic pollutants because of its mutagenicity and carcinogenicity properties. Chromium mainly comes from processes such as textile dyeing, leather tanning, and the chromic salt industry. Shen et al. (2016) prepared HTC from chitosan obtaining better results than those published with other adsorbent materials. Chitosan is a copolymer of 2-glucosamine and *N*-acetyl-2-glucosamine. Hydroxyl and amine groups are interesting substances to adsorb metals, and chitosan is a good scavenger from Pb(II) or Cd(II) at a high pH. Cr(VI) adsorption is favoured at low pH, but chitosan is not stable in an acidic solution. In principle, chitosan cannot be used to remove Cr(VI). But a chitosan carbonaceous adsorbent can exhibit good chemical stability. To maintain amino and hydroxyl groups, carbonisation must be carried out at low temperatures, which is possible with the HTC method. HTCs were prepared at different temperatures (140–220 °C). The HTC prepared at 140 °C does not appear to be completely charred, and those prepared at 200 °C and 220 °C had lower adsorption capacity than those prepared at 160 °C and 180 °C. This may be because of the destruction of functional groups, as lower N and O contents were observed when raising the preparation temperature. As with other metallic cations, the adsorption is fast, as equilibrium is reached in 30 min. The kinetic model that best fits the experimental results is that of pseudo-first order.

Cr(VI) adsorption is strongly pH dependent. This is because this metal is in the form of anions (CrO_4^{2-}, $HCrO_4^{-}$, $Cr_2O_7^{2-}$), which are attracted electrostatically to the positive charges of the surface. These are more abundant as pH decreases in the solution. In addition, the presence of amino groups favours this increase in positive charge, as they protonate more easily than the oxygenated functional groups.

The maximum adsorption capacity obtained by Shen et al. (2016) at a pH of 4.0 was 343 mg/g and showed a saturation effect, which can be explained by the existence of a limited number of active sites on the adsorbent. The adsorption capacity of the non-carbonized chitosan is slightly higher, but, as stated above, it cannot be used in highly acidic solutions. This also supports the idea that there is a limited number of active sites, so the isotherm data are better fitted to the Langmuir model than to Freundlich equation. The enthalpy and entropy values are negative, as well as free energy, which indicates an exothermic and spontaneous process. The adsorbent is easy to recycle: a centrifugation and a wash with NaOH is enough. After five cycles, more than 92% of the initial adsorption capacity is still maintained.

Copper and cadmium are heavy metals that are involved in several industrial processes. Regmi et al. (2012) have studied their adsorption obtaining excellent results in some experimental conditions. They observed, as with other cations in solutions, that at an acidic pH level (less than 5), the adsorption is very low, as the surface of the HTC is protonated and repels the positive charges of the cations (see Fig. 4.6). The adsorption increases above a pH of 5 because the surface groups are not protonated; therefore, the negative charge that attracts the cations predominates.

It also influences the precipitation of metals. Thus, copper precipitates at a pH of 7 and cadmium at a pH of 10. But both are removed at a pH of 5, which can be due only to adsorption. There are big differences in the adsorption capacities whether we use HTC activated with KOH, inactivated HTC, or the starting biomass (switchgrass).

Liu and Zhang (2011) have also studied the adsorption of Cu(II) on activated HTC. Likewise, they have observed that the adsorbed amount increases with pH. This occurs for both pyrolytic carbons and HTCs, but with the difference that in the latter, having a PZC in the study range (pH of 1–6), the influence of pH is greater, rapidly increasing the adsorbed amount when pH $>$ PZC, as in that situation the predominant charge at the surface is negative (Liu et al. 2010). It should be added that the HTCs present a greater number of functional groups, so the number of charges of any type on the surface will be greater than in pyrolytic carbons. A negative charge attracts the cation Cu(II).

Sun et al. (2015) studied the adsorption of Cd(II) with non-modified HTCs and HTCs modified with KOH. The modified carbons showed a higher adsorption capacity, as they reduced about 90% of the Cd(II) in the solution, while the unmodified HTCs barely removed 10%. This is because of the greater number of oxygen groups on the surface of the modified carbon. These isotherms are better fitted to the Langmuir model than the Freundlich equation, which indicates homogeneity at the surface. As described in previous paragraphs (Regmi et al. 2012), when the pH is lower than the PZC, the surface of the HTC is positively charged, as is the cadmium cation, so adsorption is low. By using a pH higher than the PZC, the surface charge is negative and the adsorption increases sharply.

Demir-Cakan et al. (2009) studied the adsorption of Cd(II) and Pb(II) in HTCs prepared from glucose and acrylic acid to obtain a large number of carboxyl functional groups. The greater the amount of acrylic acid used in the hydrothermal carbonisation, the greater the adsorption of both metals. The capacity of the best HTC is superior to that of other adsorbents commonly used for the adsorption of Pb (II) and similar to them in the adsorption of Cd(II). The adsorption isotherm of both metals is better fitted to the Freundlich model than Langmuir equation, which the authors attribute to the heterogeneity of the surface. A similar work has been published on the adsorption of Cd(II) and Pb(II), but the carboxyl groups were obtained by oxidation of the HTC at 300 °C (Chen et al. 2011). The adsorbed amount is increased by releasing Pb(II) instead of Cd(II), and if oxidised HTC is used, the adsorbed amount is multiplied several times. In addition, in this experiment, the oxidised HTC was found to be superior to other common adsorbents when Pb(II) is used and competitive with them when using Cd(II).

The adsorption of Pb(II) with HTCs has been studied by Liu and Zhang (2009). The adsorbed amount is low at very acidic pH levels, but improves, reaching a maximum near pH $= 5$ and decreasing with a higher pH. Above pH $= 5$, hydroxylated compounds of Pb(II) are formed in which the interaction with the negatively charged surface weakens as the charge decreases. Pb(II) adsorption can be greatly improved by oxidising the HTC with hydrogen peroxide (Xue et al. 2012). In addition, the adsorption equilibrium is reached faster. Note that a change in adsorption has been observed: the HTC isotherm fits better to the Freundlich model and the modified HTC to the Langmuir model. The authors proposed that this is because the oxidation decreases the heterogeneity of the surface. They also pointed out that in previous work, they detected that Pb(II) is removed when a pyrolytic coal is used, by the precipitation mechanism. Crystalline compounds are formed which can be detected by XRD. But with the HTCs, these crystals are not observed. From published studies on Pb(II) adsorption, it can be concluded that the predominant interaction is that of the metal with oxygen atoms on the surface of the HTC.

4.3.5.2 *f*-Block Metals

A metallic element that is very interesting to adsorb is uranium. And it is probably this element that is the most studied using HTCs as adsorbents. Interest is because of two reasons. Uranium is used for the generation of electricity and is a nonrenewable resource, so it is valuable and worth the time and effort to recover it. But it is also an element that can affect the ecosystem because of the radiochemical and toxic effects that it presents. Carbonaceous materials are suitable adsorbents, as they have a great stability against radiation, heat, acids, and bases. Specifically, against radiation, the effect of gamma irradiation with doses from 1 to 100 kGy has been studied (Yang et al. 2014). The adsorption capacity decreases slightly, and the selectivity towards the U(VI) is hardly affected, even at the highest doses of radiation. The only problem it presents is that they usually have few functional groups so that unmodified carbons exhibit relatively low selectivity and poor adsorption capacity towards uranium. An advantage of HTCs is that, although they have a lower specific surface area than other carbon materials, they have a greater number of oxygenated surface groups, which is an advantage in the case of adsorbates such as U(VI), as we will see below.

Several HTCs have been used to adsorb U(VI). It is soluble in moderate concentrations only at acidic pH levels, up to pH = 4.5. Under these conditions, it is found as uranyl cation (UO_2^{2+}).

With respect to synthesis of HTCs for uranium removal, Li et al. (2016) prepared carbon microspheres from *Saccharomyces cerevisiae* cells by hydrothermal methods, followed by chemical modification using the Hummers method to increase the number of fungal groups. Li et al. (2014) used another approach. Some of the compounds used as an antidote in cases of acute poisoning by uranyl cations are polyphenols. As a result of this, they used a type of polyphenols (bayberry tannins) with glyoxal introduced as a cross-linking agent, with the target of obtaining a large amount of phenol groups on a stable material. They obtained more than 10 mmol/g between phenol and carboxyl groups. In addition, they achieved greater porous development when using glyoxal than when using only tannins. Other authors have also used glyoxal (Yang et al. 2014), not as a cross-linking agent but as a carbon source. They mixed glyoxal with acrylonitrile to have -CN groups that could react with $NH_2OH{\cdot}HCl$ to obtain amidoxime groups at the surface and avoid the formation of carboxyl groups. Amidoxime is known to be an effective bidentate ligand of U(VI). Yu et al. (2014) tested another strategy (taking advantage of the tendency of uranium to form complexes with phosphates) to prepare HTCs with these functional groups. For this, they obtained HTC from glucose and oxidised them in air at moderate (i.e., 300 °C). They then reacted the HTC with *o*-phosphoethanolamine and *N,N′*-dicyclohexylcarbodiimide to functionalize the surface. Even simpler was the method of Zhang et al. (2014a), who prepared HTC with phosphate groups in a single step mixing glucose and phosphoric acid. Song et al. (2012) followed a similar strategy opting for a multidentate ligand that usually has good selectivity towards actinides in the 5-azacytosine column-actinide/lanthanide separation processes. After the hydrothermal process and calcination in air at a moderate temperature, the HTC was first reacted with ethylenediamine and subsequently with 5-azacytosine. The oxidation process increased the number of carboxyl groups, unlike for Li et al. (2014). Wang et al. (2012) chose another ligand, the salicylideneimine, known for its ability to coordinate with hexavalent cations. For this, they performed a synthesis similar to that of Song et al. (2012). The difference is in using salicylaldehyde after the reaction with ethylenediamine. Liu et al. (2013) opted for a two-stage synthesis, in which after obtaining HTC from pine needles, this was oxidised with 2 M HNO_3 to obtain a greater number of carboxyl groups. In this case, HTC porosity dramatically decreases, but in spite of this, a considerably better adsorption of U(VI) is obtained. Zhang et al. (2013) prepared HTC from glucose and performed oxidation in air at temperatures between 150 °C and 300 °C, without further surface modification. The number of carboxyl groups obtained depended mainly on the oxidation temperature and not on time. This amount of functional groups influences the adsorption capacity. It is also possible to use simpler processes such as those proposed by Kumar et al. (2011) that used lignocellulosic material (switchgrass, *Panicum virgatum*) to obtain HTCs in a single stage without further chemical modification. A synthesis was also performed with chitosan (Zhang et al. 2014b), a material chosen for its abundance, low cost, and numerous amino groups in its composition.

When adsorption of U(VI) was carried out, a great influence of pH (from 1 to 4.5) was observed, with a low adsorption at very acidic pH levels and higher adsorption at a pH of 4. It was explained (Fig. 4.6) by the fact that at low pH levels, functional groups are protonated, and so there is electronic repulsion with U (VI) in its UO_2^{2+} form. But as the pH increases, the positive charge on the surface decreases, as well as the repulsion. Some authors increased the range of study to pH = 8. It is possible that the uranyl cation is soluble if it forms a CO_3^{2-} coordination compound (Kumar et al. 2011). In this case, the maximum amount of adsorption is obtained at pH = 6.0. It should be noted that above this pH, the U predominant species in a solution have a negative charge, whereas at pH levels <6.0, they have a positive charge. As for ionic strength, its effect is not clear, and with some of the adsorbents cited, it does not produce an effect, while with some others it does. Overall, the adsorption process of U(VI) is very fast. Some of the published results are shown in Table 4.4. Two factors might be responsible: the absence of developed porosity, which prevents the existence of a prolonged

Table 4.4 Some published data about the adsorption rate of U (VI)

Maximum capacity (%)	Time (min)	References
80	10	Li et al. (2016)
88	1	Li et al. (2014)
90	5	Song et al. (2012)
80	1	Wang et al. (2012)

diffusion process, and the strong interaction between the cation and the HTC's active sites. This quickness in adsorption is habitual when metallic cations are used.

The kinetics conform to a pseudo-second-order model, which is based on the fact that the limiting step of the reaction is because of a process of chemisorption on the surface, which includes the formation of bonds. In addition, the intraparticle diffusion model presents a poor fit, confirming that the slow step proceeds on the surface. This adsorption mechanism also confirms the adsorption isotherm data, which are better fitted to the Langmuir model than the Freundlich model. The Langmuir model assumes monolayer adsorption, which would be closer to a chemisorption process. It also supports the model that shows an increase of the quantity adsorbed with an increase in temperature. Li et al. (2014) and Wang et al. (2012) adjusted the isotherm data to the Dubinin-Radushkevich model and obtained a binding energy whose positive value (+12.54 kJ/mol) also suggesting chemisorption. This data is supported by positive values of enthalpy and entropy and the negative value of the free energy. The adsorption of U(VI), therefore, is spontaneous and endothermic. Liu et al. (2013) prepared two materials with very different specific surfaces and found that the one with a larger surface area saturated before the other material, demonstrating that the number of active sites for U(VI) adsorption is limited and does not depend directly on the value of the specific surface.

The results obtained for the adsorption of U(VI) in several adsorption works about this metal cation are shown in Table 4.5. It can be seen that the amount adsorbed strongly depends on the relative concentrations of HTC and U(VI), presenting very different values. Better results are obtained when the pH level is less than 5.0. It should be noted that when both data have been measured and published, there is a close relationship between the capacity measured experimentally and that indicated for the monolayer by the Langmuir model. This fact indicates that adsorption, as indicated above, must be in a single layer and probably as a chemisorption.

Another similar ion (actinide and presenting environmental problems resembling U(VI)) is Th(IV). It appears in many minerals for industrial use (Syed 1999), so it is frequently found in sewage. In addition, it is an element used as a nuclear fuel and in alloys, so it is also interesting to recover. Zhou et al. (2015) used a similar char that they had already used to adsorb U (VI) (Zhang et al. 2013), which was oxidised in air at a moderate temperature to increase the number of oxygenated functional groups. The maximum amount removed from Th (IV) is achieved at pH = 3.5, but the authors speculated that it may be because of two factors: higher adsorption (because of the decrease in positive charge on the HTC) and the beginning of precipitation as $Th(OH)_4$. The effect of the treatment temperature in

Table 4.5 The adsorption capacity of U (VI) by HTCs

Sample name	HTC conc. (mg/L)	U(VI) conc. (mg/L)	Adsorption (mg/g)		Temperature (°C)	pH	Time (h)	References
			Exper.[a]	Langmuir[b]				
Oxy-HTCM	400	100	160.0	184.8	25	4.5	12	Li et al. (2016)
HTC-btg	400	400	307.3	332.2	25	4.5	2	Li et al. (2014)
HTC	200	–	–	80.0	25	6.0	2	Yu et al. (2014)
HTC-PO_4	200	–	–	434.8	25	6.0	2	Yu et al. (2014)
HTC	5000	30	4.0	2.1	25	3.9	34	Kumar et al. (2011)
HTC	200	50	55.9	62.7	35	6.0	1	Liu et al. (2013)
HTC-COOH	200	50	176.0	205.8	35	6.0	1	Liu et al. (2013)
HCSs250	200	50	169.9	–	–	7.0	25 min	Zhang et al. (2013)
HCSs300	200	50	180.0	–	–	7.0	25 min	Zhang et al. (2013)
HCSs	200	≥110	64.9	80.0	25	6.0	1	Zhang et al. (2014b)
HCSs-$PO_4$3	200	≥110	261.8	285.7	25	5.0	1	Zhang et al. (2014a)
HTC-Sal	500	119	261.8	292.8	–	4.3	–	Wang et al. (2012)
HTC-AO	50	300	1021.6	–	25	4.5	2	Yang et al. (2014)
HCC	–	–	–	264.6	–	–	–	Zhang et al. (2014b)
HTC-Acy	500	–	408.4	–	60	4.5	2	Song et al. (2012)

[a]Experimental data of maximum adsorption
[b]Capacity of the monolayer according to the Langmuir isotherm model

air is clearer. As temperature rises, the HTCs have higher oxygen content, and this increases the amount of adsorbed thorium. The data on the effects of contact time (fast adsorption), application of mathematical models to kinetics (adjusted to the pseudo-second-order model) and isotherms (where Langmuir model represents the phenomenon better than the Freundlich model), and the effect of temperature (increasing adsorption by raising the temperature) indicate that the adsorption process, as with U(VI), is chemical adsorption.

4.3.5.3 Mixture of Metals

Few studies have been done on simultaneous metal adsorption with HTCs (Sun et al. 2015). It has been observed that the smaller the radii of the hydrated cations are, the greater the adsorption; this is also described with other adsorbents. Adsorption is greater when using HTCs with high oxygen content. In addition, it has been measured that the adsorption of a metal is lower if it is adsorbed from a mixture than if it is adsorbed from a pure solution. This is because of the competition between several metals for active adsorption sites; in other words, the number of active adsorption sites on the surface of the HTC is limited.

Xue et al. (2012) have studied the adsorption of a mixture of heavy metals in an aqueous solution using HTC in columns. Conclusions are similar to work on batch adsorption (Sun et al. 2015), which reinforces the idea that there are a limited number of active sites in HTCs.

As general remark for metal adsorption, it can be deduced that this process is governed by the existence of active sites, where a specific bond occurs. This fact is supported by kinetic, adsorption isotherms and thermodynamics, in which the Langmuir model, which supposes a limited number of active sites, is more adequate than the Freundlich model. In several cases, the existence of chemisorption may be assumed. It can also be generally deduced that the adsorption of metals depends heavily on pH levels. This can influence the charge of the species containing the metal and also the charge of the HTC surface. The priority mechanism seems to be similar to ion exchange.

4.3.6 Phosphorus

Phosphorus-containing minerals are a limited resource, and they are essential for agriculture as a nutrient. In addition, when phosphorus compounds enter an aquatic environment, they may cause eutrophication problems. Therefore, there is a high level of interest in removal (to reduce environmental problems) and recovery (to use in agriculture) of phosphorus. HTCs doped with lanthanum have been proposed as phosphorus adsorbents (Dai et al. 2014). Some authors have found that efficiency of this HTC is higher than $La(OH)_3$ and that HTC without lanthanum showed no phosphorus removal activity. The lanthanum content and phosphorus removal efficiency were increased with the hydrothermal carbonisation time. Lanthanum must be bonded to functional surface groups, because no lanthanum compounds were detected with XRD. The pH levels of the solution and coexisting anions have a limited effect on the activity of La-HTC, which can be used in a wide range of situations. As is the case for other adsorbates that bind to specific surface sites, the Langmuir model better describes the adsorption isotherm than the Freundlich model.

4.3.7 Phenols

Liu and Zhang (2011) have studied the adsorption of phenol on activated HTC. They have worked with different pH levels, and the adsorption has not deflected. This result was attributed to the fact that the phenol is in molecular form in that range of pH and is not affected by the positive charges on the surface of the HTC.

4.3.8 Wastewater

In general, adsorption studies have been performed in simple solutions. But it is also possible to study the adsorption on HTCs in more complex mixtures. The purification of water resulting from fermentation of lactic acid has been studied (Laube and Reza 2016). In this work, several different adsorbents were used, and it was concluded that HTCs prepared from willow and poplar are the most effective, combining good adsorption capacity and high physical and chemical endurance.

4.3.9 Reusability

To reduce the overall cost of practical applications, adsorbent reusability is significant. Depending on the adsorbent and adsorbate, regeneration methods are diverse. To date, some studies have analysed the reusability of HTCs.

For adsorption of U(VI: Li et al. 2016), the adsorbed amount was found to be very low at pH = 1.0 and higher when the pH level was raised to pH = 4.5. A strong acid treatment method was used to regenerate the HTC, and after five cycles no significant loss of adsorbent capacity had been observed. The same method was employed by Zhou et al. (2015) for the regeneration of HTC used to adsorb Th (IV). The same technique was used to regenerate an HTC fibre on which paraquat had been adsorbed. In this case, the manipulation produced the breakage of the external HTC separating it from the internal fibre, which resulted in a loss of efficiency (Zhou et al. 2015). Even so, it remained at 83% after five cycles. Yu et al. (2014) compared this method with the use of EDTA to recover the adsorbed U (VI) and concluded that the use of EDTA is preferable, even at concentrations lower than those of the acid. Shen et al. (2016) studied the adsorption of an anion (CrO_4^{2-}) in place of the cations studied by the previously cited authors so that their regeneration method was the opposite: they used a 2 M NaOH solution. The loss of the adsorption capacity was 7.7% after five cycles, indicating that it is a good method for regeneration. This ease of regeneration is because of the operation of HTCs as ion exchangers, thanks to a large number of functional groups on their surfaces.

4.4 Conclusions

The hydrothermal carbonisation method is an alternative to the classical pyrolysis method. It has several advantages, among which are a lower production of pollutants, better yield, and lower energy consumption, as a lower temperature is used and wet raw material can be used. These characteristics allow for the inclusion of this procedure within Green Chemistry.

The product obtained is characterised by low porous development, high oxygen content, and a large number of functional groups on its surface. It can be activated or prepared by using templates to increase its porosity. It can also be functionalized in very different ways, which is favoured by the abundance of preexisting functional groups, as well as being used to coat other materials.

It has been used in several applications, for instance, the adsorption of contaminants in water. In this field, it can be said that research is just beginning, as the number of bibliographical references found in the preparation of this document (year: 2016) is low.

HTCs have been used as adsorbents for many different substances, with good results. But to date, they appear to be especially useful for the adsorption of ions in a solution, whether organic or inorganic. This is because the existence of a large number of functional groups also allows for a large number of positive or negative charges as a function of the pH of the solution. These charges attract the ions with the opposite charge. Functional groups also facilitate the regeneration of the adsorbent by using acids or alkalis.

In summary, HTCs are materials with a wide range of possibilities for employment, and they are expected to be investigated and used frequently in the future.

Acknowledgements Authors acknowledge the support provided by Universidad de Extremadura and Junta de Extremadura/FEDER ref. GRU15123.

References

Alatalo SM, Mäkilä E, Repo E, Heinonen M, Salonen J, Kukk E, Sillanpääa M, Titirici MM (2016) Meso- and microporous soft templated hydrothermal carbons for dye removal from water. Green Chem 18:1137–1146

Bandura AV, Lvov SN (2006) The ionization constant of water over wide range of temperature and density. J Phys Chem Ref Data 35:15–30

Bergius F, Specht H (1913) Die Anwendung hoher Drucke bei chemischen Vorgngen und eine Nachbildung des Entstehungsprozesses der Steinkohle, vol 58. Verlag Wilhelm Knapp, Halle an der Saale

Cao X, Ro KS, Chappell M, Li Y, Mao J (2011) Chemical structures of swine-manure chars produced under different carbonization conditions investigated by advanced solid-state 13C nuclear magnetic resonance (NMR) spectroscopy. Energ Fuel 25:388–397

Chen Z, Ma L, Li S, Geng J, Song Q, Liu J, Wang C, Wang H, Li J, Qin Z, Li S (2011) Simple approach to carboxyl-rich materials through low-temperature heat treatment of hydrothermal carbon in air. Appl Surf Sci 257:8686–8691

Dai L, Wu B, Tan F, He M, Wang W, Qin H, Tang X, Zhu O, Pan K, Hu Q (2014) Engineered hydrochar composites for phosphorus removal/recovery: lanthanum doped hydrochar prepared by hydrothermal carbonization of lanthanum pretreated rice straw. Bioresour Technol 161:327–332

Demir-Cakan R, Baccile N, Antonietti M, Titirici MM (2009) Carboxylate-Rich carbonaceous materials via one-step hydrothermal carbonization of glucose in the presence of acrylic acid. Chem Mater 21:484–490

Durán-Valle CJ (2006) Geometrical relationship between elemental composition and molecular size in carbonaceous materials. Appl Surf Sci 252:6097–6101

Falco C, Perez Caballero F, Babonneau F, Gervais C, Laurent G, Titirici MM, Baccile N (2011) Hydrothermal carbon from biomass: structural differences between hydrothermal and pyrolyzed carbons via 13C solid state NMR. Langmuir 27:14460–14471

Hao W, Björkman E, Lilliestråle M, Hedin N (2014) Activated carbons for water treatment prepared by phosphoric acid activation of hydrothermally treated beer waste. Ind Eng Chem Res 53:15389–15397

Hu B, Yu SH, Wang K, Liu L, Xu XW (2008) Functional carbonaceous materials from hydrothermal carbonization of biomass: an effective chemical process. Dalton Trans 40:5414–5423

Hummers WS, Offeman RE (1958) Preparation of graphitic oxide. J Am Chem Soc 80:1339–1339

Kruse A, Funke A, Titirici MM (2013) Hydrothermal conversion of biomass to fuels and energetic materials. Curr Opin Chem Biol 17:515–521

Kumar S, Loganathan VA, Gupta RB, Barnett MO (2011) An assessment of U (VI) removal from groundwater using biochar produced from hydrothermal carbonization. J Environ Manag 92:2504–2512

Laube H, Reza MT (2016) Application of biosorbents for ion removal from sodium lactate fermentation broth. J Environ Chem Eng 4:10–19

Li B, Ma L, Tian Y, Yang X, Li J, Bai C, Yang X, Zhang S, Li S, Jin Y (2014) A catechol-like phenolic ligand-functionalized hydrothermal carbon: one-pot synthesis, characterization and sorption behaviour toward uranium. J Hazard Mater 271:41–49

Li F, Li D, Li X, Liao J, Li S, Yang J, Yang Y, Tang J, Liu N (2016) Microorganism-derived carbon microspheres for uranium removal from aqueous solution. Chem Eng J 284:630–639

Liu F, Guo M (2015) Comparison of the characteristics of hydrothermal carbons derived from holocellulose and crude biomass. J Mater Sci 50:1624–1631

Liu Z, Zhang FS (2009) Removal of lead from water using biochars prepared from hydrothermal liquefaction of biomass. J Hazard Mater 167:933–939

Liu Z, Zhang FS (2011) Removal of copper (II) and phenol from aqueous solution using porous carbons derived from hydrothermal chars. Desalination 267:101–106

Liu Z, Zhang FS, Wu J (2010) Characterization and application of chars produced from pinewood pyrolysis and hydrothermal treatment. Fuel 89:510–514

Liu YH, Wang YQ, Zhang ZB, Cao XH, Nie WB, Li Q, Hua R (2013) Removal of uranium from aqueous solution by a low cost and high-efficient adsorbent. Appl Surf Sci 273:68–74

Martín-Jimeno FJ, Suárez-García F, Paredes JI, Martínez-Alonso A, Tascón JMD (2015) Activated carbon xerogels with a cellular morphology derived from hydrothermally carbonized glucose-graphene oxide hybrids and their performance towards CO_2 and dye adsorption. Carbon 81:137–147

Quian HS, Antonietti M, Yu SH (2007) Hybrid "golden fleece": synthesis and catalytic performance of uniform carbon nanofibers and silica nanotubes embedded with a high population of noble-metal nanoparticles. Adv Funct Mater 17:637–643

Regmi P, Garcia Moscoso JL, Kumar S, Cao X, Maob J, Schafran G (2012) Removal of copper and cadmium from aqueous solution using switchgrass biochar produced via hydrothermal carbonization process. J Environ Manag 109:61–69

Roldán L, Marco Y, García-Bordejé E (2016) Bio-sourced mesoporous carbon doped with heteroatoms (N,S) synthesised using one-step hydrothermal process for water remediation. Micropor Mesopor Mat 222:55–62

Román S, Valente Nabais JM, Ledesma B, González JF, Laginhas C, Titirici MM (2013) Production of low-cost adsorbents with tunable surface chemistry by conjunction of hydrothermal carbonization and activation processes. Micropor Mesopor Mat 165:127–133

Sevilla M, Fuertes AB (2009) Chemical and structural properties of carbonaceous products obtained by hydrothermal carbonization of saccharides. Chem Eur J 15:4195–4203

Shen F, Su J, Zhang X, Zhang K, Qi X (2016) Chitosan-derived carbonaceous material for highly efficient adsorption of chromium (VI) from aqueous solution. Int J Biol Macromol 91:443–449

Song Q, Ma L, Liu J, Bai C, Geng J, Wang H, Li B, Wang L, Li S (2012) Preparation and adsorption performance of 5-azacytosine-functionalized hydrothermal carbon for selective solid-phase extraction of uranium. J Colloid Interf Sci 386:291–299

Sun K, Ro K, Guo M, Novak J, Mashayekhi H, Xing B (2011) Sorption of bisphenol A, 17-α-ethinyl estradiol and phenanthrene on thermally and hydrothermally produced biochars. Bioresour Technol 102:5757–5763

Sun K, Tnag J, Gong Y, Zhang H (2015) Characterization of potassium hydroxide (KOH) modified hydrochars from different feedstocks for enhanced removal of heavy metals from water. Environ Sci Pollut R 22:16640–16651

Syed HS (1999) Comparison studies adsorption of thorium and uranium on pure clay minerals and local Malaysian soil sediments. J Radioanal Nucl Chem 241:11–14

Titirici MM (2012) Hydrothermal carbons: synthesis, characterization, and applications. In: Tascon JMD (ed) Novel carbon adsorbents. Elsevier Ltd, Amsterdam, pp 351–399

Titirici MM, Antonietti M (2010) Chemistry and materials options of sustainable carbon materials made by hydrothermal carbonization. Chem Soc Rev 39:103–116

Titirici MM, White RJ, Falco C, Sevilla M (2012) Black perspectives for a green future: hydrothermal carbons for environment protection and energy storage. Energy Environ Sci 5:6796–6822

Wang H, Ma L, Cao K, Geng J, Liu J, Song Q, Yang X, Li S (2012) Selective solid-phase extraction of uranium by salicylideneimine-functionalized hydrothermal carbon. J Hazard Mater 229–230:321–330

White RJ, Antonietti M, Titirici MM (2009) Naturally inspired nitrogen doped porous carbon. J Mater Chem 19:8645–8650

Xue Y, Gao B, Yao Y, Inyang M, Zhang M, Zimmerman AR, Ro KS (2012) Hydrogen peroxide modification enhances the ability of biochar (hydrochar) produced from hydrothermal carbonization of peanut hull to remove aqueous heavy metals: batch and column tests. Chem Eng J 200:673–680

Yang X, Li J, Liu J, Tian Y, Li B, Cao K, Liu S, Hou M, Li S, Ma L (2014) Simple small molecule carbon source strategy for synthesis of functional hydrothermal carbon: preparation of highly efficient uranium selective solid phase extractant. J Mater Chem A 2:1550–1559

Yu XF, Liu YH, Zhou ZW, Xiong GX, Cao XH, Li M, Zhang ZB (2014) Adsorptive removal of U (VI) from aqueous solution by hydrothermal carbon spheres with phosphate group. J Radioanal Nucl Chem 300:1235–1244

Zhang ZB, Nie WB, Li Q, Xiong GX, Cao XH, Liu YH (2013) Removal of uranium(VI) from aqueous solutions by carboxyl-rich hydrothermal carbon spheres through low-temperature heat treatment in air. J Radioanal Nucl Chem 298:361–368

Zhang ZB, Zhou ZW, Cao XH, Liu YH, Xiong GX, Liang P (2014a) Removal of uranium (VI) from aqueous solutions by new phosphorus-containing carbon spheres synthesized via one-step hydrothermal carbonization of glucose in the presence of phosphoric acid. J Radioanal Nucl Chem 299:1479–1487

Zhang WL, Zhang ZB, Cao XH, Ma RC, Liu YH (2014b) Uranium adsorption studies on hydrothermal carbon produced by chitosan using statistical design method. J Radioanal Nucl Chem 301:197–205

Zhao R, Wang Y, Li X, Sun B, Li Y, Ji H, Qiu J, Wang C (2016) Surface activated hydrothermal carbon-coated electrospun PAN Fiber membrane with enhanced adsorption properties for herbicide. ACS Sustain Chem Eng 4:2584–2592

Zhou ZW, Xiong GX, Liu YH, Cao XH, Zhang ZB (2015) Removal of thorium(IV) from aqueous solutions by carboxyl-rich hydrothermal carbon spheres through low-temperature heat treatment in air. Desalin Water Treat 54:2516–2529

Zhu X, Liu Y, Zhou C, Luo G, Zhang S, Chen J (2014) A novel porous carbon derived from hydrothermal carbon for efficient adsorption of tetracycline. Carbon 77:627–636

Chapter 5
Removal of Heavy Metals, Lead, Cadmium, and Zinc, Using Adsorption Processes by Cost-Effective Adsorbents

Meng Xu and Gordon McKay

Abstract Heavy metals have been utilized by human beings for thousands of years, and they pervade nearly all aspects in modern economic activities. However, heavy metal ions such as cadmium, lead, zinc, nickel, and copper are detected in waste streams from mining operations, battery manufacturing, electroplating, and smelting industries due to their extensive use. The adverse effects of heavy metals on human health have been studied by international organizations including the WHO. Although the emissions have declined in most developed countries, in some parts of the developing world, human exposure to heavy metals is still a severe problem. Various technologies of heavy metal removal from effluent have been investigated and reported in literature including precipitation, membrane filtration, coagulation, adsorption/ion exchange, and electrochemical treatment. In this chapter, adsorption studies of cadmium, lead, and zinc onto various low-cost adsorbent materials will be reviewed. The equilibrium isotherm, batch kinetic studies, the effect of variables including contact time, initial pH, adsorbate concentrations, dosage, and the application of adsorption models will be discussed.

Keywords Heavy metals • Water treatment • Low-cost adsorbents

Contents

M. Xu
Department of Civil and Environmental Engineering, Hong Kong University of Science and Technology, Clear Water Bay, Kowloon, Hong Kong

G. McKay (✉)
Division of Sustainability, College of Science and Engineering, Hamad Bin Khalifa University, Education City, Qatar Foundation, Doha, Qatar
e-mail: gmckay@hbku.edu.qa

© Springer International Publishing AG 2017

A. Bonilla-Petriciolet et al. (eds.), *Adsorption Processes for Water Treatment and Purification*, DOI 10.1007/978-3-319-58136-1_5

5.1 Introduction

Heavy metals presented in source and treated water are a severe problem to public health (Hua et al. 2012). Due to the extensive use of various heavy metals in mining operations, battery manufacturing, electroplating, and other industries, the surrounding underground and surface water are subjected to the risk of metal contamination (Srivastava and Majumder 2008; Wang et al. 2012). Metals associated with these industrial activities such as cadmium, lead, zinc, and nickel are not biodegradable and have the tendency to accumulate in living organisms, leading to diseases and disorders to human health (Khan et al. 2008; Joseph 2009; Li et al. 2015).

During the past decades of development, to treat heavy metal-contaminated wastewater, several technologies have been developed and applied, among which adsorption/ion exchange has been recognized as a promising way to treat industrial waste effluents with the advantage of easy operation (Fu and Wang 2011). Traditional treatment processes such as chemical precipitation (Feng et al. 2000) and electrochemical treatment (Sadrzadeh et al. 2009) are effective when dealing with high metal concentrated wastewater. However, they fail in certain cases when the metal concentration is relatively low (Gautam et al. 2014). Membrane filtration (Blöcher et al. 2003) and adsorption can be applied to low concentration metal-contaminated wastewater, but membrane filtration is extremely expensive when treating large amounts of industrial effluent and thus can hardly be applied on an industrial scale (Fu and Wang 2011).

However, an adsorption process can make full use of certain types of adsorbent materials even when the metal concentration is quite low (Fu and Wang 2011). By choosing from various types of adsorbent materials, heavy metals can be removed together or selectively (Dąbrowski et al. 2004). Three types of adsorbent materials have been commercially available in water treatment: activated carbons (Paul Chen and Lin 2001), zeolites (Hui et al. 2005), and synthetic polymeric adsorbents (Kennedy 1973). Most activated carbons have a wide range of pore sizes that can accommodate large organic molecules; zeolite which refers to aluminosilicate materials with varying ratios of Al to Si tends to have very small pores; synthetic polymeric adsorbents usually have only micropores. Activated carbons, zeolite, and synthetic polymeric resins have already been applied successfully in several water treatment applications, among which activated carbon is the most commonly used adsorbent material.

Activated carbon can be manufactured from natural or carbonaceous materials such as coal (Ahmadpour and Do 1996; Chingombe et al. 2005), peat (Ho and McKay 1999a), coconuts (Sekar et al. 2004), and others by high-temperature steam activation or other processes. But the temperature needed in carbonization/

activation process is extremely high at around 600–1200 °C which means high energy consumption, whereas the yield of activated carbon from biomass is quite low, usually below 20%, resulting in the relatively high expense to treat large quantities. Thus cost-effective adsorbent materials are needed to replace traditional activated carbons.

The adsorption capacities vary a lot depending on the nature of the raw material as well as the modification process. Apart from the adsorption capability, cost is also an important parameter in comparing and choosing various adsorbent materials (Bailey et al. 1999; Júnior et al. 2009). The cost of an individual adsorbent material varies depending on several aspects such as local availability and the degree of required activation or modification process. In general, "low-cost" adsorbents should meet the requirements as abundant in nature; need as simple processing as possible are waste materials or by-products from another industrial process (Bailey et al. 1999).

Agricultural wastes are usually available in large quantities with cheap price which meets the requirement as "low-cost" materials (Demirbas 2008). The major constituents of these materials are lignin and cellulose which may also include other functional groups of lignin such as alcohols, ketones, aldehydes, phenolic, carboxylic, and ether groups (Júnior et al. 2009). To some extent, these groups have the ability to donate a lone pair of electrons to bind with heavy metals and form coordination complexes which makes these waste materials potential adsorbents for heavy metal uptake. Commonly used agricultural wastes include rice husks (Ajmal et al. 2003; Krishnani et al. 2008), peanut hull (Periasamy and Namasivayam 1996; Brown et al. 2000; Witek-Krowiak et al. 2011), sawdust (Yu et al. 2000; Argun et al. 2007), pinus bark and different bark samples (Vázquez et al. 1994; Al-Asheh et al. 2000), tea leaves (Ahluwalia and Goyal 2005), banana and orange peels (Annadurai et al. 2003), palm kernel husk (Onundi et al. 2010), coconut husk (Amuda et al. 2007), modified cotton, modified cellulose (O'Connell et al. 2008), modified corncob (Vaughan et al. 2001), wool fibers (Balköse and Baltacioğlu 1992), and other agricultural by-products.

In addition to agricultural wastes, industrial waste by-products have also been considered and researched for the application of heavy metal adsorption. Nowadays, industrial activities also generate enormous amounts of by-products and solid waste materials, some of which can be reused while the others will be sent for disposal in landfills. Therefore, recycle and reuse of these industrial waste materials to reduce the quantity of waste have become a hot topic. The most attractive aspect of these types of materials is that they are usually provided free or at very low price. During the past decades, many industrial wastes have been investigated with or without treatment processes, and they have been applied to metal adsorption systems such as fly ash from the energy combustion industry (Cho et al. 2005; Hui et al. 2005), bark and sawdust from the timber industry (Yu et al. 2001; Bulut and Tez 2007), black liquor from the paper industry (Srivastava et al. 1994), red mud from the aluminum industry (Gupta et al. 2001), and others.

Chitin is another kind of material that has been reported to have the ability to adsorb heavy metals (Wu et al. 2000). It is plentiful in nature and can also be produced from shrimp, prawn, and crab meat canning industries. Moreover, chitin's deacetylated derivative chitosan has a higher adsorption capacity than that of chitin. Chitosan can be produced from chitin using chemical methods or can also be found naturally in some fungal cell walls (Wu et al. 2001).

Finally, naturally abundant clay (Celis et al. 2000) and minerals with the potential to be used as adsorbents have gained attention. Clay has a large surface area which infers it has potential to be an adsorbent material; zeolites are naturally occurring silicate minerals and possess ion-exchange ability (Ok et al. 2007). In particular, zeolite presents a strong affinity toward lead and other heavy metals which makes it an effective adsorbent material.

5.2 Adsorption Process

Adsorption is a surface-based process in which the dissolved species are transported to the porous solid adsorbent by diffusion and are then adhered onto the extensive inner surface of the adsorbent. An adsorbate refers to the constituent that is attracted onto a surface, and adsorbent refers to the solid onto which the constituent is adsorbed. Adsorbent solids bind dissolved molecules by physical forces (physisorption), ion exchange, or chemical binding (chemisorption).

As stated in Chaps. 2 and 3, solid-liquid adsorption systems usually involve two types of experimental investigations: equilibrium isotherm adsorption testes and batch kinetic rate studies. Equilibrium isotherm model equations such as the Langmuir model and the Freundlich model are commonly used to describe experimental adsorption data, while the most widely used kinetic models are the pseudo-first-order model and the pseudo-second-order model. It is important to determine the best-fit isotherm and kinetic model in order to evaluate the material and to further develop suitable industrial adsorption system designs (Gerente et al. 2007).

5.2.1 Equilibrium Adsorption Isotherm

Establishing the most appropriate correlation for equilibrium curves is very important in the purpose of optimizing the condition of an adsorption system for metal ion uptake. The maximum capacity of the adsorbent can be predicted under a given set of conditions, whereby the adsorption capacity is dependent only on the equilibrium curve established between the adsorbed sites and metal solution concentrations. These equilibrium adsorption capacity curves can be obtained by measuring the adsorption isotherm of the metal ions onto the sorbents with fixed pH, dosage, and temperature. Thus it is necessary to develop isotherm models to fit the experimental data points in order to obtain the correlations of solid-phase concentration to the fluid-phase solute concentration. In this section, we recall some of the most used isotherm models for heavy metal adsorption.

5.2.1.1 The Langmuir Model

The Langmuir model calculates the amount of molecules adsorbed onto a solid surface and is often used to describe the solute adsorption (Langmuir 1918). This model assumes a monolayer adsorption onto a complete homogeneous surface with a negligible interaction between adjacent adsorbed molecules. The equilibrium equation is stated as:

$$q_e = \frac{K_L C_e}{1 + a_L C_e} \tag{5.1}$$

where q_e is the solid-phase adsorbate equilibrium concentration (mmol/g), C_e is the aqueous-phase adsorbate equilibrium concentration (mmol/L), K_L (L/g) and a_L (L/mmol) are Langmuir isotherm constants, and the monolayer capacity $q_{max} = K_L/a_L$.

The Langmuir equation is apposite to homogeneous adsorption systems where the adsorption of each adsorbate molecule has equal activation adsorption energy. When the concentration is very low, $a_L C_e$ is far smaller than unity; it implies $q_e \approx K_L C_e$ which obeys the Henry's law. Therefore, a linear expression of the Langmuir equation is:

$$\frac{C_e}{q_e} = \frac{1}{K_L} + \frac{a_L}{K_L} C_e \tag{5.2}$$

where a plot of C_e/q_e versus C_e gives a straight line where the slope is a_L/K_L and the intercept is $1/K_L$. The Langmuir equation gives the theoretical monolayer saturation capacity, q_{max} as K_L/a_L.

5.2.1.2 The Freundlich Model

The Freundlich equation is an empirical equation proposed by Freundlich. It is commonly used to describe organic components in solution; it assumes that the adsorption sites are not identical. It usually applies to heterogenous surfaces and multilayer adsorption on the adsorbate. The equation is:

$$q_e = a_F C_e^{b_F} \tag{5.3}$$

where a_F is the Freundlich constant, while b_F is the heterogeneity factor. When b_F equals unity, this model reduces to Henry's law. The smaller this term is, the more nonlinear the adsorption isotherm becomes. It is notable that when the value of b_F gets smaller than 0.1, the shape of the isotherm approaches a rectangular isotherm. But it is often criticized for not reducing to Henry's law at very dilute concentrations. Also, there is no finite limit for the adsorption uptake when the concentration is sufficiently increased which is contrary to the plateau observed in most real systems. Therefore this model is generally only valid in a narrow range of concentrations. A linear expression of the Freundlich equation can be obtained by taking logarithms of Eq. (5.3):

$$\ln q_e = \ln a_F + b_F \ln C_e \tag{5.4}$$

where a plot of ln q_e versus ln C_e can determine the constant a_F and exponent b_F. The Freundlich isotherm is derived by assuming and inserting an exponential decay energy distribution function in to the Langmuir equation. It is commonly used in the description of reversible adsorption having heterogeneous surfaces.

5.2.1.3 The Redlich-Peterson Model

The Redlich-Peterson isotherm (Redlich and Peterson 1959) is also an empirical equation which was proposed by Redlich and Peterson to represent equilibrium data; it combines elements from both Langmuir and Freundlich equations together to incorporate three parameters into the equation. It is applied to both homogeneous and heterogeneous systems.

This equation is widely used as a compromise between the Langmuir and Freundlich systems. The equation is expressed as follows:

$$q_e = \frac{k_{RP} c_e}{1 + a_{RP} {c_e}^{b_{RP}}} \tag{5.5}$$

where k_{PR}, a_{PR}, and b_{RP} are Redlich-Peterson isotherm constants. b_{RP} is between 0 and 1, and when b_{RP} equals to 1, it reduces to Langmuir. It approximates to Henry's law at low concentration and to the Freundlich at high concentration.

5.2.1.4 The Sips (Langmuir-Freundlich) Model

The Sips model (Sips 1948) is also known as the Langmuir-Freundlich isotherm. It considers the case of a molecule occupying two sites. The Sips equation takes the following form:

$$q_e = \frac{K_{LF} a_{LF} {c_e}^{n_{LF}}}{1 + a_{LF} {c_e}^{n_{LF}}} \tag{5.6}$$

where K_{LF} is the Sips isotherm equilibrium constant (L/g), a_{LF} is Sips isotherm constant $(\mathrm{L/mmol})^{1/n_{LF}}$, and n_{LF} is the Sips isotherm exponent, dimensionless. This equation is a combination of the Langmuir and Freundlich models, it will effectively reduce to a Freundlich isotherm at low adsorbate concentrations, and thus it does not follow Henry's law, while it predicts a monolayer adsorption capacity of the Langmuir isotherm at high adsorbate concentrations.

5.2.2 Kinetic Studies and Models

Kinetic experiments are carried out to evaluate the potential of the material for commercial applications. An ideal adsorbent shows properties such as high adsorption capacity as well as fast removal rate. Kinetic models are used to evaluate the mechanisms of heavy metal adsorption and its potential rate-controlling steps that include mass transport and chemical reaction processes (Aksu 2001). Another purpose for conducting batch kinetic studies is to optimize the operation conditions for large-scale processes. Various factors such as initial metal concentration, initial pH of the solution, adsorbent dosage, temperature, particle size, and other process parameters will affect the adsorption capacity. Kinetic models only concern the effects of observable parameters on the overall rate of the system. Predicting the overall rate in the given adsorption system is the most important since the system's kinetics determines adsorbate time and the reactor dimensions.

In order to understand the kinetics and the rate-controlling steps, several adsorption kinetic models have been established: the pseudo-first-order model, the pseudo-second-order model, the Elovich model, the first-order equation of Bhattacharya and Venkobachar, the Weber and Morris adsorption kinetic model, the Weber and Morris adsorption kinetic model, and the Ritchie equation. The two most widely used models are the pseudo-first-order and the pseudo-second-order model which can study the heavy metal adsorption kinetics and quantify the extent of uptake in adsorption kinetics. Some details of the common kinetic models applied in heavy metal adsorption are given below. Note that this discussion complements the information provided in Chaps. 2 and 3.

5.2.2.1 The Pseudo-First-Order Model

The pseudo-first-order rate equation is generally expressed as follows:

$$q_t = q_e[1 - \exp(-k_1 t)] \tag{5.7}$$

where q_t and q_e are amounts adsorbed onto the adsorbent material (mmol/g) at time t (min) and equilibrium, respectively. k_1 is the pseudo-first-order rate constant (min^{-1}).

The linear expression of this equation can be rearranged as follows:

$$\ln(q_e - q_t) = \ln q_e - k_1 t \tag{5.8}$$

From this form of this equation, it is clear that q_e should be known in order to fit the equation to the experimental data in the linear form. However, in many cases q_e is unknown after the batch kinetic experimental contact time unless the equilibrium isotherm has been determined. The amount of adsorbed metal ions is significantly smaller than the equilibrium amount when the adsorption process tends to be very slow.

Therefore in most reported literatures, the pseudo-first-order equation does not fit well with the experimental data for the whole range of contact time but is generally applicable for the initial 20–30 min of the adsorption process (Gerente et al. 2007).

Apart from the linear method of fitting, some groups use the sum of squared errors (SSE) method by minimizing the difference of calculated data with the experimental data:

$$\mathrm{SSE} = \sum \left(q_{\mathrm{exp}} - q_{\mathrm{cal}}\right)^2 \tag{5.9}$$

Excel built-in function solver can be used to obtain the unknown q_{e} and k_1 at the same time. However, some literature results show that the calculated q_{e} is still smaller than that from the isotherm experiments; thus this model equation may still underestimate the equilibrium capacity. To sum up, the pseudo-first-order model has been extensively used to describe metal uptake process. However, this model has the problem of predicting the equilibrium capacity; the calculated q_{e} from this model equation frequently does not agree with the experimental data. It only gives a good fit with the experimental data for the first 30 min. After that, the prediction of the model usually goes below the experimental data. It has been suggested that the time lag results in the difference in the q_{e} values. The time to overcome the boundary layer and external resistance controls the process at the beginning, and this time lag is difficult to quantify. This model equation has been applied to metal adsorption systems like cadmium, lead, and zinc, but most of the fits are moderate to poor.

5.2.2.2 The Pseudo-Second-Order Model

The pseudo-second-order model was developed by Ho and McKay (1999b) and is expressed by the following equation:

$$q_{\mathrm{t}} = \frac{k_2 q_{\mathrm{e}}^2 t}{1 + k_2 q_{\mathrm{e}} t} \tag{5.10}$$

where q_{e} and q_{t} are the amount of metal ions adsorbed (mmol/g) at equilibrium and at time t, respectively; k_2 is the pseudo-second-order rate constant of adsorption (g mmol^{-1} min^{-1}).

The pseudo-second-order rate constant can be determined by plotting the experimental data t/q versus t, and then q_{e} and k_2 can be calculated from the slope and intercept. In studies on the adsorption of cadmium by chitosan, the pseudo-second-order kinetic model was much more successful than the pseudo-first order. It has been suggested that in chemisorption processes, the pseudo-second-order model is superior to the pseudo-first-order model due to the consideration of adsorbent-adsorbate interaction in the pseudo-second-order model. The pseudo-second-order model has been recognized as the most widely fitted model for the metal adsorption from effluent.

Theoretically, the rate of an adsorption system should remain constant at a certain specified temperature; it only changes with the change of the temperature. However, the literatures show that the rate contact is quite different when the variables such as initial concentration, pH, or dosage are changed. This might be a problem for this useful model, and therefore it is important to develop a correlation between the pseudo-first and the pseudo-second-order rate constant with each variable.

5.3 Low-Cost Adsorbent Materials and Metal Adsorption

In order to reduce the cost of adsorption process, waste materials were taken into consideration to be recycled and reused as well as cheap raw materials. This section is mainly focused on four parts: agricultural wastes, industrial wastes, marine materials, and mineral materials.

5.3.1 Agricultural Waste

Agricultural waste materials are usually abundant and have been proven to be eco-friendly and economic sources of precursors or adsorbent materials. It has been reported that various agricultural wastes can be easily converted to value-added products. Agricultural wastes have been recycled and used in the removal of different types of pollutants from effluent. The basic components of agricultural waste materials include hemicelluloses, lignin, lipids, simple sugars, proteins, hydrocarbons, and starch which contains functional groups that are potential to become adsorbent materials for heavy metal adsorption. Agricultural waste products can be used directly or after modification. The agricultural wastes can be directly used for adsorption tests after recycling, washing, and making into the desired particle size. Another way to reuse the agricultural wastes is to use certain methods to pretreat the material before use. The pretreatment methods have been extensively studied in order to enhance the functional groups and therefore increase the number of active sites for adsorption purposes. The application to remove lead, cadmium, and zinc by agricultural waste with their experimental conditions and adsorption capacities has been listed in Table 5.1.

Sugarcane bagasse is a kind of lignocellulosic material which has a complex configuration with a variety of activated sites; modified sugarcane bagasse has been reported to be effective in terms of heavy metal removal (Orlando et al. 2002; Karnitz et al. 2007). After pretreatment with succinic anhydride, carboxylic functions are introduced to succinic anhydride, and the amide functions are formed by chemical introduction of polyamine. Due to the chelating properties of polyamines

Table 5.1 Agriculture waste applied for the removal of lead, cadmium, and zinc and their adsorption capacities

Adsorbent material	Metal ion	Initial metal concentration	pH	Adsorbent dosage (g/L)	Contact time	Temperature (°C)	Adsorption capacity (q_e)	References
Agave bagasse (raw)	Pb(II)	60 mg/L	5	2	–	25	35.60 mg/g	Velazquez-Jimenez et al. (2013)
Agave bagasse (HCl)	Pb(II)	60 mg/L	5	1	–	25	42.31 mg/g	Velazquez-Jimenez et al. (2013)
Agave bagasse (HNO_3)	Pb(II)	60 mg/L	5	1	–	25	54.29 mg/g	Velazquez-Jimenez et al. (2013)
Agave bagasse (NaOH)	Pb(II)	60 mg/L	5	1	–	25	50.12 mg/g	Velazquez-Jimenez et al. (2013)
Banana peel	Pb(II)	30–80 mg/L	5	40	20 min	25	2.18 mg/g	Anwar et al. (2010)
KCl-modified orange peel	Pb(II)	10–300 mg/L	5–5.5	5	120 min	25	141.84 mg/g	Guo et al. (2011)
Sulfured orange peel	Pb(II)	25–800 mg/L	5	5	–	30	164.00 mg/g	Liang et al. (2011)
Orange peel xanthate (XOP)	Pb(II)	100 mg/L	–	5	1.5 h	25	218.34 mg/g	Liang et al. (2010)
Mercerized garlic peel	Pb(II)	1–200 mg/L	5	0.5	30 min	25	109.05 mg/g	Liu et al. (2014)
Native garlic peel	Pb(II)	1–200 mg/L	5	0.5	30 min	25	51.73 mg/g	Liu et al. (2014)
Lentil husk	Pb(II)	50 mg/L	5	2	24 h	30	81.43 mg/g	Basu et al. (2015)
$Ca(OH)_2$-modified musk-melon peel	Pb(II)	1–5 mmol/L	4.5	5	120 min	25	0.81 mmol/g	Huang and Zhu (2013)
Ponkan peel	Pb(II)	15 mmol/L	5	2	60 min	25	112.10 mg/g	Pavan et al. (2008)
$ZnCl_2$-activated grapefruit peel	Pb(II)	100 mg/L	5.3–6.5	10	90 min	30	12.73 mg/g	Pei and Liu (2012)
Rice husk	Pb(II)	1.17 mg/L	–	0.06	–	25	87.18%	Hegazi (2013)
Olive pomace	Pb(II)	–	3–5	10	60 min	–	0.074 mmol/g	Pagnanelli et al. (2003)
Agave bagasse (raw)	Cd(II)	60 mg/L	5	2	–	25	13.27 mg/g	Velazquez-Jimenez et al. (2013)

Agave bagasse (HCl)	Cd(II)	60 mg/L	5	1	–	25	12.50 mg/g	Velazquez-Jimenez et al. (2013)
Agave bagasse (HNO_3)	Cd(II)	60 mg/L	5	1	–	25	13.50 mg/g	Velazquez-Jimenez et al. (2013)
Agave bagasse (NaOH)	Cd(II)	60 mg/L	5	1	–	25	18.32 mg/g	Velazquez-Jimenez et al. (2013)
Banana peel	Cd(II)	30–80 mg/L	3	30	20 min	25	5.71 mg/L	Anwar et al. (2010)
Sugarcane bagasse	Cd(II)	50 mg/L	6	20	60 min	25	99.5%	Garg et al. (2008)
Maize corncob	Cd(II)	50 mg/L	6	20	60 min	25	99%	Garg et al. (2008)
Jatropha oil cake	Cd(II)	50 mg/L	6	20	60 min	25	85%	Garg et al. (2008)
Citric acid-modified orange peel	Cd(II)	0.001–0.005 mol/L	6	4.3	120 min	–	0.90 mmol/g	Li et al. (2007)
KCl-modified orange peel	Cd(II)	10–300 mg/L	5–5.5	5	120 min	25	125.63 mg/g	Guo et al. (2011)
Orange peel xanthate (XOP)	Cd(II)	50 mg/L	–	5	1.5 h	25	76.57 mg/g	Liang et al. (2010)
HCl-modified *Cucumis sativus* peel	Cd(II)	5–150 mg/L	5	1	60 min	25	58.14 mg/g	Pandey et al. (2014)
Grapefruit peel	Cd(II)	10–200 mg/L	5	4	60 min	–	42.09 mg/g	Torab-Mostaedi et al. (2013)
Unmodified rice straw	Cd(II)	25–350 mg/mL	2–6	–	180 min	25	13.90 mg/g	Ding et al. (2012)
Wheat stem	Cd(II)	0.1–1.2 mmol/L	5	–	60 min	–	0.10 mmol/g	Tan and Xiao (2009)
Rice husk	Cd(II)	0.48 mg/L	–	0.06	–	25	67.92%	Hegazi (2013)
Olive pomace	Cd(II)	–	3–5	10	60 min	–	0.062 mmol/g	Pagnanelli et al. (2003)
Agave bagasse (raw)	Zn(II)	60 mg/L	5	2	–	25	7.84 mg/g	Velazquez-Jimenez et al. (2013)
Agave bagasse (HCl)	Zn(II)	60 mg/L	5	1	–	25	12.40 mg/g	Velazquez-Jimenez et al. (2013)

(continued)

Table 5.1 (continued)

Adsorbent material	Metal ion	Initial metal concentration	pH	Adsorbent dosage (g/L)	Contact time	Temperature (°C)	Adsorption capacity (q_e)	References
Agave bagasse (HNO_3)	Zn(II)	60 mg/L	5	1	–	25	14.43 mg/g	Velazquez-Jimenez et al. (2013)
Agave bagasse (NaOH)	Zn(II)	60 mg/L	5	1	–	25	20.54 mg/g	Velazquez-Jimenez et al. (2013)
KCl-modified orange peel	Zn(II)	10–300 mg/L	5–5.5	5	120 min	25	45.29 mg/g	Guo et al. (2011)
Sulfured orange peel	Zn(II)	25–800 mg/L	5	5	–	30	80.0 mg/g	Liang et al. (2011)
Orange peel xanthate (XOP)	Zn(II)	50 mg/L	–	5	1.5 h	25	49.85 mg/g	Liang et al. (2010)
Mango peel	Zn(II)	10–500 mg/L	5–6	5	120 min	25	28.21 mg/g	Iqbal et al. (2009)
Nano-porous-activated neem bark	Zn(II)	200 mg/L	–	1.2	2 days	35	11.90 mg/g	Maheshwari et al. (2015)
Rice husk ash	Zn(II)	10–100 mg/L	6	10	100 h	25	7.15 mg/g	El-Said et al. (2011)
Raw *Eucalyptus sheathiana bark*	Zn(II)	20–70 mg/L	5	0.25	2 h	30	128.21 mg/g	Afroze et al. (2016)
NaOH-treated *Eucalyptus sheathiana bark*	Zn(II)	20–70 mg/L	5	0.25	2 h	30	250.00 mg/g	Afroze et al. (2016)

toward metals such as zinc, cadmium, lead, and copper, the modified succinic anhydride becomes suitable for the treatment of metal-contaminated wastewater (Bianchi et al. 1991).

Velazquez-Jimenez et al. (2013) tested the raw and modified *Agave salmiana* bagasse for metal adsorption. HCl, HNO_3, NaOH, tartaric, citric, and oxalic acids were used to modify bagasse, respectively, to determine if its concentration of active groups could be improved. These materials were then tested for the removal of Cd(II), Pb(II), and Zn(II) ions from effluent at pH 5, and desorption studies were performed at pH 2 and 4 at 25 °C. The characterization techniques are used mainly to identify carboxyl-, hydroxyl-, sulfur-, and nitrogen-containing groups in bagasse. The carboxylic groups were mainly responsible for metal uptake. The results show that the adsorption capacity of raw bagasse is about 8, 14, and 36 mg/g for zinc, cadmium, and lead, respectively. The adsorption capacity was improved around 27–62% after modification with HNO_3 and NaOH. However, citric, oxalic, and tartaric acid did not have significant effects on improving adsorption capacity.

Recently, Demiral and Güngör (2016) used grape bagasse as raw material and phosphoric acid to activate the material to activated carbon. The obtained activated carbon has a highest surface area of 1455 m^2/g and total pore volume of 0.88 cm^3/g at 500 °C carbonization temperature with an impregnation ratio of 5/1. Then the activated carbon was used to remove copper with maximum capacity of 43.47 mg/g at 45 °C. The Langmuir and Dubinin-Radushkevich isotherm equations showed better fits, while the rate of adsorption was found to conform to the pseudo-second-order kinetic model. Gurgel and Gil (2009) have introduced two different methods to prepare two new chelating materials, MMSCB 3 and MMSCB 5, using succinylated twice-mercerized sugarcane bagasse (MMSCB 1). MMSCB 1 was activated with 1,3-diisopropyl-carbodiimide and acetic anhydride, respectively, and later both intermediate were reacted with triethylenetetramine to obtain MMSCB 3 and MMSCB 5. The capacities of MMSCB 3 and MMSCB 5 for copper, cadmium, and lead uptake from single-component solutions were evaluated at different contact times, pH, and initial metal ion concentrations. The adsorption isotherm results show good fit with the Langmuir model, and the maximum adsorption capacities of MMSCB 3 and MMSCB 5 for copper, cadmium, and lead were found to be 59.5 and 69.4, 86.2 and 106.4, and 158.7 and 222.2 mg/g, respectively. Feng et al. (2011) studied the adsorption of cadmium, lead, and nickel using the grafted copolymerization-modified orange peel (OPAA). The Langmuir model correlated with the experimental data better than the Freundlich model. From the Langmuir equation, the maximum uptake capacities for lead and cadmium were 476.1 and 293.3 mg/g, respectively. The adsorption capacity of the modified material has increased 4.2- and 4.6-fold for lead and cadmium, respectively, in comparison with the original orange peel. Moreover, the kinetics for lead and cadmium adsorption followed the pseudo-second-order model.

Afroze et al. (2016) recently studied the adsorption of zinc ions from effluent using raw and base-modified *Eucalyptus sheathiana* bark. Bark is a common by-product from the timber industry; *Eucalyptus* trees are evergreen and fast-growing trees which are abundantly available worldwide. The equilibrium adsorption studies showed that both the Freundlich and Langmuir models can be capable of describing the systems for zinc uptake using raw or base-modified eucalyptus bark. Also, the pseudo-second-order rate equation is found to well describe the adsorption of zinc for both raw and modified bark. The maximum adsorption capacity of modified eucalyptus bark for zinc removal at 30 °C was 250.0 mg/g which was quite competitive to other adsorbent materials.

Lentil husk was also recognized as a promising low-cost adsorbent with a maximum removal capacity of 81.43 mg/g for lead under the optimized condition of pH (5.0) and temperature (30 °C) with an initial metal ion concentration of 250 mg/L by Basu et al. (2015). The functional groups of lentil husk were modified by treating with different chemicals to investigate their role in adsorption; chemical modification of functional groups revealed both hydroxyl and carboxyl groups played crucial role in the binding process.

Rice husk is another category of agricultural waste; since rice is the most popular food in Asia, large quantities of rice husk were produced every year. Nowadays, rice husk is used as energy source for power plants; however, 18% of the total amount of husk remains as rice husk ash (RHA). It has been investigated that the major component of RHA is silica, while it also contains some alumina content, both of which are typically used in zeolite synthesis (Gupta et al. 2011). Santasnachok et al. (2015) synthesized zeolite from agriculture waste by-product rice husk ash (RHA). The synthesized material was applied in cadmium-contaminated wastewater in the initial concentration range of 50–500 mg/L. The maximum removal capacity of synthesized zeolite Na-A and Na-X for cadmium was found to be 736.38 and 684.46 mg/g, respectively, at dosage of 0.25 g/L.

5.3.2 *Industrial By-Products and Wastes*

Industrial by-products are usually considered as useless waste materials and cause major disposal problems. Fly ash is a waste material generated from combustion processes. Previously, fly ash is mainly used for road construction or formation of bricks and cement. However, fly ash contains high percentage of silica and alumina which makes it possible to be used as precursor of zeolite-type adsorbent material (Bhatnagar and Sillanpää 2010). The reuse of fly ash to develop low-cost and efficient adsorbents for wastewater treatment follows the concept of using waste to treat waste (Shawabkeh et al. 2004; Belviso et al. 2010). Many studies have been reported on fly ash, raw fly ash, modified fly ash, or on dispersed TiO_2-fly ash and have been tested as adsorbents for heavy metals removal from wastewater (Gupta and Ali 2004; Harja et al. 2012; Shyam et al. 2013).

Aluminum industry waste has also been studied under the concept of recycle waste for water treatment. Red mud is a waste material from the aluminum industry and has been widely investigated. It is formed during the production of alumina when bauxite ore is subjected to caustic leaching (Bhatnagar et al. 2011); sodium hydroxide solution is used in the refining process. And thus this material is highly alkaline with a pH as high as 10–13. So it needs to be neutralized before being used as an adsorbent material. Red mud is mainly composed of fine particles containing aluminum, iron, silicon, hydroxides, and titanium oxides. Zhu et al. (2007) evaluated the potential of granular red mud (GRM) to remove cadmium from aqueous solution. The pseudo-second-order model can describe the experimental kinetic data at initial pH 6.0 and 3.0. The maximum adsorption capacities for GRM were 38.2 mg/g at 20 °C, 43.4 mg/g at 30 °C, and 52.1 mg/g at 40 °C. Gupta et al. (2001) converted red mud into an inexpensive and efficient adsorbent and used it for the removal of lead and chromium from aqueous solutions which exhibits good adsorption capacities. The equilibrium data fits well with both the Freundlich and the Langmuir model. Red mud was treated with hydrogen peroxide at room temperature for 24 h in order to oxidize the adhering organic impurities and was then washed repeatedly with distilled water, followed by an activation process in a muffle furnace at 500 °C for 3 h. The adsorption results showed that the maximum uptake for lead took place at pH 4.0.

The steel industry also produces a large quantity of by-products which create serious disposal problems. Basic oxygen furnace (BOF) slag is a kind of by-product from steel-making process which is mainly applied for construction applications including road engineering (Miraoui et al. 2012) and cement production (Han et al. 2015) due to the hydration activity of BOFs. Some authors have investigated the adsorption of cadmium, lead, zinc, and copper from effluent by BOFs (Xue et al. 2009, 2013). The sieve size of BOFs was controlled under 0.6 mm first and then treated by acid (0.1 M HCl) and washed by deionized water. The equilibrium isotherms showed that copper, cadmium, and zinc had similar but relatively higher adsorption capacity compared to lead in single-element system. The adsorption affinity is found to follow the sequence Zn > Cu > Pb > Cd in single-element systems, while the uptake order in the multielement system was Pb > Cu > Zn > Cd.

The paper industry produces a large amount of black liquor, from which lignin can be extracted. The extracted lignin was characterized and used for lead and zinc removal (Srivastava et al. 1994). The uptake amount of lead was more than that of zinc, while the adsorption capacity increases with the increase of pH value.

The waste from the leather-processing industry has also been investigated by researchers in the application of metal adsorption. Fathima et al. (2005) utilized animal skins which are high in protein content to complex with iron and then used it for the removal of chromium(VI). The effect of pH and the initial concentration of chromium(VI) by iron treated fleshing is studied. The adsorption capacity by iron-treated material is shown to be greatly improved to 51 mg of chromium(VI) per gram of treated fleshing, while it is only flesh 9 mg/g for the untreated fleshing. However, this study has not been applied to other metals to date.

Recently, a newly developed industrial waste material is introduced. The material is recycled from waste printed circuit board recycling industry. With the rapid development of the electronic devices, the lifetime of such devices has become shorter and shorter resulting in the massive waste electronic devices. Printed circuit boards (PCBs) are the major component of the abandoned electronic devices. Recent commercial recycling of PCBs is based on using a mechanical-physical technique to separate the PCBs into metallic and nonmetallic fractions which is quite environmentally friendly compared to the traditional treatment methods. The metallic fraction is mainly copper which is of high value and can be marketed easily, while the nonmetallic fraction (NMF) which accounts for 70% of the total PCB weight has always been considered as a low-value by-product. Hadi et al. (2013) recycled the nonmetallic part and investigated the activation of this material for the application of heavy metal adsorption. The optimum condition to activate the material is to impregnate the raw material with activating agent KOH at weight ratio of 2 (KOH/NMF = 2) under continuous stirring for 3 h at room temperature. The resulting slurry was heated to 250 °C for 3 h in a muffle furnace under a nitrogen atmosphere. The product was then washed with DI water and subsequently dried at 110 °C. Xu et al. (2014) performed a detailed equilibrium study of this activated nonmetallic fraction for the adsorption of cadmium. The adsorption results show that the maximum uptake capacity of this material for cadmium removal reached to 2.1 mmol/g (initial pH = 4) which shows that this material can effectively remove cadmium ions from effluent and is highly competitive with the commercial resins. The Redlich-Peterson model is the best-fit model. Lead adsorption has also been studied with a maximum capacity of 3.2 mmol/g at initial pH of 4. Lead also shows higher affinity than cadmium in the binary-component system (Xu et al. 2015).

Finally, other adsorbent materials derived from industrial waste and their adsorption capacities for lead, cadmium, and zinc have been listed in Table 5.2.

5.3.3 *Marine Materials*

Adsorbents have been derived from waste seafood because of the abundant amount of chitin found the in the exoskeletons of crabs and other arthropods (Rinaudo 2006). Seafood processing produces large quantities of by-products. Chitin is second to cellulose in terms of polysaccharide quantities in nature. It is also easily found as waste product from the crab meat canning industry. It has been reported that chitin or modified chitin can remove metal ions from effluent (Benaissa and Benguella 2004; Karthik and Meenakshi 2015). Chitosan is a partially acetylated glucosamine biopolymer that can be found in the cell wall of some fungi such as *Mucorales strain*. However, it is mainly derived from deacetylation of chitin and has been known as the most important derivative of chitin due to the useful features such as hydrophilicity, biodegradability, and antibacterial property (Rinaudo 2006). Chitosan can be made by treating the chitin in the shells of shrimp and other crustaceans with an alkaline substance such as sodium hydroxide, and it is reported that chitosan chelates six times amounts of metals compared to chitin due to the exposed free amino groups during deacetylation process. Chitosan has potential for commercial biomedical applications and has been applied in the treatment of heavy metal-contaminated wastewaters (Pandey and Tiwari 2015; Chen et al. 2017).

Table 5.2 Industrial waste and their adsorption capacities for lead, cadmium and zinc removal

Adsorbent material	Metal ion	Initial metal concentration	pH	Adsorbent dosage (g/L)	Contact time	Temperature (°C)	Adsorption capacity (q_e)	References
Fly ash	Pb(II)	1.17 mg/L	–	0.06	–	25	76.06%	Hegazi (2013)
Cancrinite-type zeolite from fly ash	Pb(II)	0.5–4 mmol/L	6	0.5	72 h	25	2.53 mmol/g	Qiu and Zheng (2009)
Red mud	Pb(II)	2.41–4.83 mM	4	10	24 h	30	64.79 mg/g	López et al. (1998)
Black liquor	Pb(II)	0.097–0.97 mM	4–6	2–6	24 h	30	18.65 mg/g	Srivastava et al. (1994)
Blast furnace sludge	Pb(II)	4000 mg/L	–	50	5 h	20	64.17–79.87 mg/g	López-Delgado et al. (1998)
Activated NMF	Pb(II)	0.5–7 mmol/L	4	1	4 days	20	3.2 mmol/g	Xu et al. (2015)
Fly ash	Cd(II)	0.48 mg/L	–	0.06	–	25	73.54%	Hegazi (2013)
Bagasse fly ash	Cd(II)	14 mg/L	6	10	–	30	1.24 mg/g	Gupta et al. (2003)
Red mud	Cd(II)	0.1–0.75 g/L	4.7–6.2	1	48 h	–	10.57 mg/g	López et al. (1998)
Blast furnace slag	Cd(II)	0.889 mM	5	10	–	30	18.72 mg/g	Gupta et al. (1997)
Blast furnace sludge	Cd(II)	1500 mg/L	–	50	5 h	20	6.74–10.15 mg/g	López-Delgado et al. (1998)
Activated NMF	Cd(II)	0.2–7.5 mmol/L	4	1	4 days	20	2.1 mmol/g	Xu et al. (2014)
Sawdust	Cd(II)	10 mg/L	5	10	4 h	30	94.5%	Naiya et al. (2009)
Cancrinite-type zeolite from fly ash	Zn(II)	0.5–4 mmol/L	6	0.5	72 h	25	1.532 mmol/g	Qiu and Zheng (2009)
Red mud	Zn(II)	0.2–0.7 g/L	6.9–7.8	1	48 h	–	12.59 mg/g	López et al. (1998)
Black liquor	Zn(II)	0.01–0.1 mM	5–7	2–6	24 h	30	95.0 mg/g	Srivastava et al. (1994)
Blast furnace slag	Zn(II)	1.52 mM	6	20	–	30	17.66 mg/g	Gupta et al. (1997)
Blast furnace sludge	Zn(II)	1500 mg/L	–	50	5 h	20	4.25–9.25 mg/g	López-Delgado et al. (1998)
Activated NMF	Zn(II)	5 mmol/L	4	1	–	20	2.0 mmol/g	Hadi et al. (2014)

The high adsorption capacity by chitosan and its derivatives for the removal of heavy metal ions from aquatic systems is due to the presence of multifunctional groups; the high hydrophilicity is because a large number of hydroxyl groups on the glucose units with high chemical reactivity as well as flexible structure of the polymer chain (Bhatnagar and Sillanpää 2009).

Compared to chitin, chitosan has superior adsorption ability for heavy metals due to its higher content of amino groups. Chemical modifications of chitosan such as carboxyalkyl substitution, aldehyde cross-linking, ligand cross-linking, and polyamination are accessible to prevent it from dissolution in acidic media ($pH < 2$) or to enhance adsorption ability. Sargın and Arslan (2015) prepared chitosan microcapsules via cross-linking and evaluated their capability for the removal of copper, cadmium, zinc ions, and other metallic ions. Different pH values, metal concentrations, temperature, amount of adsorbent, and adsorption time have been studied. The adsorption equilibrium followed the Langmuir isotherm model, and the adsorption capacity of the chitosan/sporopollenin microcapsules was found to be 0.15 mmol/g for cadmium and 0.25 mmol/g for zinc.

Seaweed (marine algae) is another kind of abundantly available adsorbent material. Seaweed does not only grow in saltwater but also in freshwater. Usually, seaweed materials, especially brown algae, possess a relatively high surface area and present high binding affinity toward metals due to the functional groups such as amino, hydroxyl, carboxyl, and sulfate as binding sites in the cell wall structure (Özer et al. 2006). Metals can be attracted by an electrostatic force or can form complexes with the material (Hamdy 2000; Jalali et al. 2002).

The adsorption capacity for metal ions was significantly improved after modification. Typical modification methods include acid and acid-base treatment (Suzuki et al. 2005). The surface chemistry was changed dramatically after the treatment resulting in different adsorption preferences. Suzuki et al. (2005) used 0.1 M HCl and 0.1 M NaOH to treat *Ulva* seaweed, respectively. The adsorption isotherm of the acid-pretreated material was similar to that of the non-treated one, while the adsorption capacity of the alkali-pretreated material increased a lot. The isotherm model was well fitted by the Langmuir model with an adsorption capacity of 90 mg/g for the alkali-pretreated material. The uptake capacities of typical marine material-derived adsorbents for heavy metal removal have been listed in Table 5.3.

5.3.4 Zeolite and Clay

Zeolites, clays, sediment, and soil are readily available, inexpensive materials, and thus are good candidates as low-cost adsorbent materials or precursors. Sediment and soil are mostly used for the removal of organic pollutants such as glyphosate (Morillo et al. 2000) and herbicides (Kibe et al. 2000) from water, while zeolite and clay present good adsorption capabilities for metal removal. The removal capacities for cadmium, lead, and zinc using zeolite and clay as adsorbent materials are listed in Table 5.4.

Table 5.3 The uptake capacities of various marine material-derived adsorbents for lead, cadmium, and zinc removal

Adsorbent material	Metal ion	Initial metal concentration	pH	adsorbent dosage (g/L)	Contact time	Temperature (°C)	Adsorption capacity (q_e)	References
Cross-linked chitosan with epichlorohydrin	Pb(II)	0–15 ppm	6	0.1	4 h	25	34.13 mg/g	Chen et al. (2008)
Chitosan	Pb(II)	5–100 mg/L	4–7	1	24 h	25	0.036 mmol/g	Rangel-Mendez et al. (2009)
Seaweed	Pb(II)	–	5.5	–	–	22–24	1.78 mmol/g	Ahmady-Asbchin et al. (2009)
Sargassum hystrix (brown algae)	Pb(II)	200 mg/L	4.5	2	3 h	30	265 mg/g	Jalali et al. (2002)
S. natans (brown algae)	Pb(II)	200 mg/L	4.5	2	3 h	30	224 mg/g	Jalali et al. (2002)
Padina pavonica (brown algae)	Pb(II)	200 mg/L	4.5	2	3 h	30	210 mg/g	Jalali et al. (2002)
Sargassum sp.	Pb(II)	–	5	1	6 h	22	1.16 mg/g	Sheng et al. (2007)
Chemically modified chitosan	Cd(II)	100 mg/L	8	5	16 h	25	357.14 mg/g	Sankararamakrishnan et al. (2007)
Chitosan	Cd(II)	100 mg/L	8	5	16 h	25	85.47 mg/g	Sankararamakrishnan et al. (2007)
Chitosan	Cd(II)	5–100 mg/L	4–7	1	24 h	25	0.016 mmol/g	Rangel-Mendez et al. (2009)
Chitin	Cd(II)	10–300 mg/L	5.7–6.02	2	24 h	25	13 mg/g	Benguella and Benaissa (2002)
Seaweed	Cd(II)	–	5.5	–	–	22–24	0.85 mmol/g	Ahmady-Asbchin et al. (2009)
Ceramium virgatum (red algae)	Cd(II)	10–400 mg/L	5	10	120 min	20–50	39.7 mmol/g	Sarı and Tuzen (2008)
Alkali-pretreated *Ulva onoi* (green algae)	Cd(II)	10–500 mg/L	7.8	1	12 h	20	90.70 mg/g	Suzuki et al. (2005)

(continued)

Table 5.3 (continued)

Adsorbent material	Metal ion	Initial metal concentration	pH	adsorbent dosage (g/L)	Contact time	Temperature (°C)	Adsorption capacity (q_e)	References
Ulva onoi (green algae)	Cd(II)	10–500 mg/L	7.8	1	12 h	20	61.90 mg/g	Suzuki et al. (2005)
Sargassum sp.	Cd(II)	–	5	1	6 h	22	0.76 mg/g	Sheng et al. (2007)
Laminaria japonica (brown algae)	Cd(II)	–	5	1	180 min	25	1.85 mmol/g	Liu et al. (2009)
Cross-linked chitosan with epichlorohydrin	Zn(II)	0–15 ppm	6	0.1	4 h	25	10.21 mg/g	Chen et al. (2008)
Seaweed	Zn(II)	–	5.5	–	–	22–24	0.71 mmol/g	Ahmady-Asbchin et al. (2009)
Ulva fasciata sp. (green algae)	Zn(II)	20–100 mg/L	5	3.33–16.67	20 min	30	13.5 mg/g	Prasanna Kumar et al. (2007)
Laminaria japonica (brown algae)	Zn(II)	–	5	1	180 min	25	1.42 mmol/g	Liu et al. (2009)
Bifurcaria bifurcate (brown algae)	Zn(II)	75–100 mg/L	5	–	–	25	30.3 mg/g	Freitas et al. (2008)
Fucus spiralis (brown algae)	Zn(II)	75–100 mg/L	5	–	–	25	34.3 mg/g	Freitas et al. (2008)

Table 5.4 The uptake capacities of zeolite and clay materials for lead, cadmium, and zinc removal

Adsorbent material	Metal ion	Initial metal concentration	pH	Adsorbent dosage (g/L)	Contact time	Temperature	Adsorption capacity (q_e) or removal %	References
Kaolinite	Pb(II)	50 mg/L	5.7	2	180 min	30	5.3 mg/g	Gupta and Bhattacharyya (2008)
Montmorillonite	Pb(II)	50 mg/L	5.7	2	180 min	30	21.7 mg/g	Gupta and Bhattacharyya (2008)
Natural zeolite	Pb(II)	4.9 mg/L	2.2	20	–	–	99.5%	Wingenfelder et al. (2005)
Natural zeolite	Pb(II)	9.7 mg/L	5.6	20	–	–	99.9%	Wingenfelder et al. (2005)
Natural zeolite	Pb(II)	8.94 mmol/L	5.2	10	3 days	23	1.71 mmol/g	Perić et al. (2004)
Clinoptilolite	Pb(II)	10–800 mg/L	–	2.5	48 h	–	27.7 mg/g	Sprynskyy et al. (2006)
Raw clinoptilolite	Pb(II)	400 mg/L	4.5	2.5	4 h	22	80.933 mg/g	Günay et al. (2007)
NaCl-treated clinoptilolite	Pb(II)	400 mg/L	4.5	2.5	4 h	22	122.400 mg/g	Günay et al. (2007)
Illite	Cd(II)	0–250 mg/L	2.5–8	5	–	25	350 mmol/kg	Echeverría et al. (2002)
Kaolinite	Cd(II)	50 mg/L	5.5	2	240 min	30	4.0 mg/g	Gupta and Bhattacharyya (2008)
Montmorillonite	Cd(II)	50 mg/L	5.5	2	240 min	30	21.6 mg/g	Gupta and Bhattacharyya (2008)
ZrO-derivative kaolinite	Cd(II)	10–50 mg/L	5.5	2	240 min	30	5.27 mg/g	Gupta and Bhattacharyya (2006)

(continued)

Table 5.4 (continued)

Adsorbent material	Metal ion	Initial metal concentration	pH	Adsorbent dosage (g/L)	Contact time	Temperature	Adsorption capacity (q_e) or removal %	References
ZrO-derivative montmorillonite	Cd(II)	10–50 mg/L	5.5	2	240 min	30	36.63 mg/g	Gupta and Bhattacharyya (2006)
TBA-derivative kaolinite	Cd(II)	10–50 mg/L	5.5	2	240 min	30	6.31 mg/g	Gupta and Bhattacharyya (2006)
TBA-derivative montmorillonite	Cd(II)	10–50 mg/L	5.5	2	240 min	30	43.47 mg/g	Gupta and Bhattacharyya (2006)
Natural zeolite	Cd(II)	5.3 mg/L	2.2	20	–	–	21.0%	Wingenfelder et al. (2005)
Natural zeolite	Cd(II)	5.3 mg/L	5.6	20	–	–	85.8%	Wingenfelder et al. (2005)
Clinoptilolite	Cd(II)	80 mg/L	–	2.5	48 h	–	4.22 mg/L	Sprynskyy et al. (2006)
Bentonite	Cd(II)	25–100 mg/L	6.5	2.5	90 min	30	0.084 mmol/g	Purna Chandra Rao et al. (2006)
Natural zeolite	Zn(II)	49.1 mg/L	2.2	20	–	–	23.4%	Wingenfelder et al. (2005)
Natural zeolite	Zn(II)	41 mg/L	5.6	20	–	–	93.6%	Wingenfelder et al. (2005)
Natural zeolite	Zn(II)	10.08 mmol/L	5.3	10	3 days	23	0.812 mmol/g	Perić et al. (2004)
Bentonite	Zn(II)	25–100 mg/L	6.5	2.5	90 min	30	0.141 mmol/g	Purna Chandra Rao et al. (2006)

Zeolites are naturally occurring, highly porous aluminosilicate minerals with different cavity structures, which can also be produced commercially. The porous structure of zeolite can accommodate various cations including sodium, potassium, calcium, magnesium ions, and others. These positive-charged ions are loosely held in the structure and can be easily exchanged with other metal ions in solution and therefore possess ion-exchange abilities. Zeolites consist of more than 40 natural species, among which clinoptilolite is the most abundant and frequently studied one. Sprynskyy et al. (2006) studied the adsorption behavior on raw and pretreated clinoptilolite for lead and cadmium removal. The maximum adsorption capacity is 4.22 mg/g and 27.7 mg/g for Cd^{2+} and Pb^{2+}, respectively, at an initial concentration of 80 mg/L. The experimental data is well fitted with both the Langmuir and the Freundlich models; the Freundlich model exhibits better fitting at high metal concentrations. Wingenfelder et al. (2005) investigated the removal of lead, cadmium, and zinc from synthetic mine waters by a natural zeolite. Lead can be efficiently removed from neutral as well as from acidic solutions, whereas the uptake of zinc and cadmium decreased when the pH is low and the iron concentrations are high.

Clay usually possesses large surface area and negative charge on the structure of fine-grain silicate minerals. The negative charge can be neutralized by the adsorption of positive-charged cations which might be the reason that it presents good adsorption capacities for metal removal (Ali et al. 2012). Various types of clay have been investigated and modified to enhance their efficiency for heavy metal removal. Kaolinite and montmorillonite are two of the most important and commonly used clay minerals that have been reported in heavy metal removal applications.

Gupta and Bhattacharyya (2008) used kaolinite and montmorillonite for cadmium and lead removal from water. The uptake capacities increase with the increase of pH values until the formation of insoluble metal hydroxides. The adsorption reaction is rapid with maximum uptake amount being observed within 180 min for lead and 240 min for cadmium. Both kaolinite and montmorillonite follow the Langmuir monolayer model with respect to lead and cadmium; the removal capacity values are in the range of 6.8–11.5 mg/g (kaolinite) and 21.1–31.1 mg/g (montmorillonite). Different modification methods have also been developed in order to improve the quality and characteristics of the clay. Two commonly used techniques are intercalation/pillaring and acid activation. Intercalation is the insertion of a guest species in the interlayer region of a clay mineral with preservation of the layered structure (Liu and Wu 2001), while acid treatments of clay minerals control mineral weathering and genesis (Panda et al. 2010). These treatments can replace exchangeable cations with H^+ ions, Al^{3+}, and other cations. The attached ions can escape out from both tetrahedral and octahedral sites leaving SiO_4 groups largely intact (Theocharis et al. 1988). Gupta and Bhattacharyya (2006) use kaolinite, montmorillonite, and their poly(oxo zirconium) and tetrabutylammonium derivatives to remove cadmium from aqueous solution. The uptake of cadmium was very fast at the beginning of contact and slowed down gradually indicating diffusion into the inner surface of the adsorbent. The pseudo-second order can describe the experimental data better than the pseudo-first order. And the equilibrium isotherm data was fitted well by both the Langmuir model and the Freundlich model. The adsorption behavior was poor in strongly acidic solution but improved in an alkaline medium, and the uptake amount continuously increased with the increase of pH values.

5.4 Conclusion

Adsorption plays an important role in removing heavy metals from effluent. Activated carbon is the most popular adsorbent material and has been used widely in wastewater treatment for decades. However, commercial activated carbons have drawbacks in terms of heavy metal removal efficiency and cost problems. So the overall idea is to find suitable cost-effective alternatives to replace traditional activated carbon. Under these circumstances, new adsorbent materials, especially cost-effective adsorbent materials, need to be developed. The low-cost adsorbent materials usually have the problem of low adsorption capacity; thus cheap and effective modification methods should be introduced to improve the adsorption capacity while controlling the cost of activation at the same time.

In this chapter, a wide range of low-cost adsorbent materials have been reviewed and discussed. Low-cost materials have been divided into four categories based on the nature of the materials. The common characteristics of these materials are abundance either naturally occurring or low-value by-products. By analyzing the composition and developing low-cost treatment methods, they become value-added materials with good adsorbent properties.

References

Afroze S, Sen TK, Ang HM (2016) Adsorption removal of zinc (II) from aqueous phase by raw and base modified *Eucalyptus sheathiana* bark: kinetics, mechanism and equilibrium study. Process Saf Environ Prot 102:336–352

Ahluwalia SS, Goyal D (2005) Removal of heavy metals by waste tea leaves from aqueous solution. Eng Life Sci 5:158–162

Ahmadpour A, Do DD (1996) The preparation of active carbons from coal by chemical and physical activation. Carbon 34:471–479

Ahmady-Asbchin S, Andres Y, Gerente C, Le Cloirec P (2009) Natural seaweed waste as sorbent for heavy metal removal from solution. Environ Technol 30:755–762

Ajmal M, Ali Khan Rao R, Anwar S, Ahmad J, Ahmad R (2003) Adsorption studies on rice husk: removal and recovery of Cd(II) from wastewater. Bioresour Technol 86:147–149

Aksu Z (2001) Equilibrium and kinetic modelling of cadmium(II) biosorption by C. vulgaris in a batch system: effect of temperature. Sep Purif Technol 21:285–294

Al-Asheh S, Banat F, Al-Omari R, Duvnjak Z (2000) Predictions of binary sorption isotherms for the sorption of heavy metals by pine bark using single isotherm data. Chemosphere 41:659–665

Ali I, Asim M, Khan TA (2012) Low cost adsorbents for the removal of organic pollutants from wastewater. J Environ Manag 113:170–183

Amuda OS, Giwa AA, Bello IA (2007) Removal of heavy metal from industrial wastewater using modified activated coconut shell carbon. Biochem Eng J 36:174–181

Annadurai G, Juang RS, Lee DJ (2003) Adsorption of heavy metals from water using banana and orange peels. Water Sci Technol 47:185–190

Anwar J, Shafique U, Waheed-uz-Zaman SM, Dar A, Anwar S (2010) Removal of Pb(II) and Cd (II) from water by adsorption on peels of banana. Bioresour Technol 101:1752–1755

Argun ME, Dursun S, Ozdemir C, Karatas M (2007) Heavy metal adsorption by modified oak sawdust: thermodynamics and kinetics. J Hazard Mater 141:77–85

Bailey SE, Olin TJ, Bricka RM, Adrian DD (1999) A review of potentially low-cost sorbents for heavy metals. Water Res 33:2469–2479

Balköse D, Baltacioğlu H (1992) Adsorption of heavy metal cations from aqueous solutions by wool fibers. J Chem Technol Biotechnol 54:393–397

Basu M, Guha AK, Ray L (2015) Biosorptive removal of lead by lentil husk. J Environ Chem Eng 3:1088–1095

Belviso C, Cavalcante F, Fiore S (2010) Synthesis of zeolite from Italian coal fly ash: differences in crystallization temperature using seawater instead of distilled water. Waste Manag 30:839–847

Benaissa H, Benguella B (2004) Effect of anions and cations on cadmium sorption kinetics from aqueous solutions by chitin: experimental studies and modeling. Environ Pollut 130:157–163

Benguella B, Benaissa H (2002) Cadmium removal from aqueous solutions by chitin: kinetic and equilibrium studies. Water Res 36:2463–2474

Bhatnagar A, Sillanpää M (2009) Applications of chitin- and chitosan-derivatives for the detoxification of water and wastewater—a short review. Adv Colloid Interface Sci 152:26–38

Bhatnagar A, Sillanpää M (2010) Utilization of agro-industrial and municipal waste materials as potential adsorbents for water treatment—a review. Chem Eng J 157:277–296

Bhatnagar A, Vilar VJP, Botelho CMS, Boaventura RAR (2011) A review of the use of red mud as adsorbent for the removal of toxic pollutants from water and wastewater. Environ Technol 32:231–249

Bianchi A, Micheloni M, Paoletti P (1991) Thermodynamic aspects of the polyazacycloalkane complexes with cations and anions. Coor Chem Rev 110:17–113

Blöcher C, Dorda J, Mavrov V, Chmiel H, Lazaridis NK, Matis KA (2003) Hybrid flotation—membrane filtration process for the removal of heavy metal ions from wastewater. Water Res 37:4018–4026

Brown P, Atly Jefcoat I, Parrish D, Gill S, Graham E (2000) Evaluation of the adsorptive capacity of peanut hull pellets for heavy metals in solution. Adv Environ Res 4:19–29

Bulut Y, Tez Z (2007) Removal of heavy metals from aqueous solution by sawdust adsorption. J Environ Sci 19:160–166

Celis R, HermosÍn MC, Cornejo J (2000) Heavy metal adsorption by functionalized clays. Environ Sci Technol 34:4593–4599

Chen A-H, Liu S-C, Chen C-Y, Chen C-Y (2008) Comparative adsorption of Cu(II), Zn(II), and Pb (II) ions in aqueous solution on the crosslinked chitosan with epichlorohydrin. J Hazard Mater 154:184–191

Chen A, Shang C, Shao J, Lin Y, Luo S, Zhang J, Huang H, Lei M, Zeng Q (2017) Carbon disulfide-modified magnetic ion-imprinted chitosan-Fe(III): a novel adsorbent for simultaneous removal of tetracycline and cadmium. Carbohydr Polym 155:19–27

Chingombe P, Saha B, Wakeman RJ (2005) Surface modification and characterisation of a coal-based activated carbon. Carbon 43:3132–3143

Cho H, Oh D, Kim K (2005) A study on removal characteristics of heavy metals from aqueous solution by fly ash. J Hazard Mater 127:187–195

Dąbrowski A, Hubicki Z, Podkościelny P, Robens E (2004) Selective removal of the heavy metal ions from waters and industrial wastewaters by ion-exchange method. Chemosphere 56:91–106

Demiral H, Güngör C (2016) Adsorption of copper(II) from aqueous solutions on activated carbon prepared from grape bagasse. J Clean Prod 124:103–113

Demirbas A (2008) Heavy metal adsorption onto agro-based waste materials: a review. J Hazard Mater 157:220–229

Ding Y, Jing D, Gong H, Zhou L, Yang X (2012) Biosorption of aquatic cadmium(II) by unmodified rice straw. Bioresour Technol 114:20–25

Echeverría JC, Churio E, Garrido JJ (2002) Retention mechanisms of Cd on illite. Clay Clay Miner 50:614–623

El-Said AG, Badawy NA, Abdel-Aal AY, Garamon SE (2011) Optimization parameters for adsorption and desorption of Zn(II) and Se(IV) using rice husk ash: kinetics and equilibrium. Ionics (Kiel) 17:263–270

Fathima NN, Aravindhan R, Rao JR, Nair BU (2005) Solid waste removes toxic liquid waste: adsorption of chromium(VI) by iron complexed protein waste. Environ Sci Technol 39:2804–2810

Feng D, Aldrich C, Tan H (2000) Treatment of acid mine water by use of heavy metal precipitation and ion exchange. Miner Eng 13:623–642

Feng N, Guo X, Liang S, Zhu Y, Liu J (2011) Biosorption of heavy metals from aqueous solutions by chemically modified orange peel. J Hazard Mater 185:49–54

Freitas OMM, Martins RJE, Delerue-Matos CM, Boaventura RAR (2008) Removal of Cd(II), Zn (II) and Pb(II) from aqueous solutions by brown marine macro algae: kinetic modelling. J Hazard Mater 153:493–501

Fu F, Wang Q (2011) Removal of heavy metal ions from wastewaters: a review. J Environ Manag 92:407–418

Garg U, Kaur MP, Jawa GK, Sud D, Garg VK (2008) Removal of cadmium (II) from aqueous solutions by adsorption on agricultural waste biomass. J Hazard Mater 154:1149–1157

Gautam RK, Mudhoo A, Lofrano G, Chattopadhyaya MC (2014) Biomass-derived biosorbents for metal ions sequestration: adsorbent modification and activation methods and adsorbent regeneration. J Environ Chem Eng 2:239–259

Gerente C, Lee VKC, Le Cloirec P, McKay G (2007) Application of chitosan for the removal of metals from wastewaters by adsorption—mechanisms and models review. Crit Rev Env Sci Technol 37:41–127

Günay A, Arslankaya E, Tosun İ (2007) Lead removal from aqueous solution by natural and pretreated clinoptilolite: adsorption equilibrium and kinetics. J Hazard Mater 146:362–371

Guo XY, Liang S, Tian QH (2011) Removal of heavy metal ions from aqueous solutions by adsorption using modified orange peel as adsorbent. Adv Mater Res 236:237–240

Gupta VK, Ali I (2004) Removal of lead and chromium from wastewater using bagasse fly ash—a sugar industry waste. J Colloid Interface Sci 271:321–328

Gupta SS, Bhattacharyya KG (2006) Removal of Cd(II) from aqueous solution by kaolinite, montmorillonite and their poly(oxo zirconium) and tetrabutylammonium derivatives. J Hazard Mater 128:247–257

Gupta SS, Bhattacharyya KG (2008) Immobilization of Pb(II), Cd(II) and Ni(II) ions on kaolinite and montmorillonite surfaces from aqueous medium. J Environ Manag 87:46–58

Gupta VK, Rastogi A, Dwivedi MK, Mohan D (1997) Process development for the removal of zinc and cadmium from wastewater using slag—a blast furnace waste material. Sep Sci Technol 32:2883–2912

Gupta VK, Gupta M, Sharma S (2001) Process development for the removal of lead and chromium from aqueous solutions using red mud—an aluminium industry waste. Water Res 35:1125–1134

Gupta VK, Jain CK, Ali I, Sharma M, Saini VK (2003) Removal of cadmium and nickel from wastewater using bagasse fly ash—a sugar industry waste. Water Res 37:4038–4044

Gupta VK, Agarwal S, Saleh TA (2011) Synthesis and characterization of alumina-coated carbon nanotubes and their application for lead removal. J Hazard Mater 185:17–23

Gurgel LVA, Gil LF (2009) Adsorption of Cu(II), Cd(II) and Pb(II) from aqueous single metal solutions by succinylated twice-mercerized sugarcane bagasse functionalized with triethylenetetramine. Water Res 43:4479–4488
Hadi P, Barford J, McKay G (2013) Toxic heavy metal capture using a novel electronic waste-based material—mechanism, modeling and comparison. Environ Sci Technol 47:8248–8255
Hadi P, Barford J, McKay G (2014) Selective toxic metal uptake using an e-waste-based novel sorbent—single, binary and ternary systems. J Environ Chem Eng 2:332–339
Hamdy AA (2000) Biosorption of heavy metals by marine algae. Curr Microbiol 41:232–238
Han F, Zhang Z, Wang D, Yan P (2015) Hydration heat evolution and kinetics of blended cement containing steel slag at different temperatures. Thermochim Acta 605:43–51
Harja M, Buema G, Sutiman D-M, Munteanu C, Bucur D (2012) Low cost adsorbents obtained from ash for copper removal. Korean J Chem Eng 29:1735–1744
Hegazi HA (2013) Removal of heavy metals from wastewater using agricultural and industrial wastes as adsorbents. HBRC J 9:276–282
Ho YS, McKay G (1999a) Batch lead(II) removal from aqueous solution by peat. Process Saf Environ Prot 77:165–173
Ho YS, McKay G (1999b) Pseudo-second order model for sorption processes. Process Biochem 34:451–465
Hua M, Zhang S, Pan B, Zhang W, Lv L, Zhang Q (2012) Heavy metal removal from water/wastewater by nanosized metal oxides: a review. J Hazard Mater 211:317–331
Huang K, Zhu H (2013) Removal of Pb2+ from aqueous solution by adsorption on chemically modified muskmelon peel. Environ Sci Pollut Res Int 20:4424–4434
Hui KS, Chao CYH, Kot SC (2005) Removal of mixed heavy metal ions in wastewater by zeolite 4A and residual products from recycled coal fly ash. J Hazard Mater 127:89–101
Iqbal M, Saeed A, Kalim I (2009) Characterization of adsorptive capacity and investigation of mechanism of Cu^{2+}, Ni^{2+} and Zn^{2+} adsorption on mango peel waste from constituted metal solution and genuine electroplating effluent. Sep Sci Technol 44:3770–3791
Jalali R, Ghafourian H, Asef Y, Davarpanah SJ, Sepehr S (2002) Removal and recovery of lead using nonliving biomass of marine algae. J Hazard Mater 92:253–262
Joseph P (2009) Mechanisms of cadmium carcinogenesis. Toxicol Appl Pharmacol 238:272–279
Júnior OK, Gurgel LVA, de Freitas RP, Gil LF (2009) Adsorption of Cu(II), Cd(II), and Pb (II) from aqueous single metal solutions by mercerized cellulose and mercerized sugarcane bagasse chemically modified with EDTA dianhydride (EDTAD). Carbohydr Polym 77:643–650
Karnitz O Jr, Gurgel LVA, de Melo JCP, Botaro VR, Melo TMS, de Freitas Gil RP, Gil LF (2007) Adsorption of heavy metal ion from aqueous single metal solution by chemically modified sugarcane bagasse. Bioresour Technol 98:1291–1297
Karthik R, Meenakshi S (2015) Chemical modification of chitin with polypyrrole for the uptake of Pb(II) and Cd(II) ions. Int J Biol Macromol 78:157–164
Kennedy DC (1973) Treatment of effluent from manufacture of chlorinated pesticides with a synthetic, polymeric adsorbent, Amberlite XAD-4. Environ Sci Technol 7:138–141
Khan S, Cao Q, Zheng YM, Huang YZ, Zhu YG (2008) Health risks of heavy metals in contaminated soils and food crops irrigated with wastewater in Beijing, China. Environ Pollut 152:686–692
Kibe K, Takahashi M, Kameya T, Urano K (2000) Adsorption equilibriums of principal herbicides on paddy soils in Japan. Sci Total Environ 263:115–125
Krishnani KK, Meng X, Christodoulatos C, Boddu VM (2008) Biosorption mechanism of nine different heavy metals onto biomatrix from rice husk. J Hazard Mater 153:1222–1234
Langmuir I (1918) The adsorption of gases on plane surfaces of glass, mica and platinum. J Am Chem Soc 40:1361–1403
Li X, Tang Y, Xuan Z, Liu Y, Luo F (2007) Study on the preparation of orange peel cellulose adsorbents and biosorption of Cd^{2+} from aqueous solution. Sep Purif Technol 55:69–75

Li P, Lin C, Cheng H, Duan X, Lei K (2015) Contamination and health risks of soil heavy metals around a lead/zinc smelter in southwestern China. Ecotoxicol Environ Safe 113:391–399

Liang S, Guo X, Feng N, Tian Q (2010) Effective removal of heavy metals from aqueous solutions by orange peel xanthate. Trans Nonferrous Met Soc China 20:s187–s191

Liang S, Guo X, Tian Q (2011) Adsorption of Pb^{2+} and Zn^{2+} from aqueous solutions by sulfured orange peel. Desalination 275:212–216

Liu X, Wu Q (2001) PP/clay nanocomposites prepared by grafting-melt intercalation. Polymer 42:10013–10019

Liu Y, Cao Q, Luo F, Chen J (2009) Biosorption of Cd^{2+}, Cu^{2+}, Ni^{2+} and Zn^{2+} ions from aqueous solutions by pretreated biomass of brown algae. J Hazard Mater 163:931–938

Liu W, Liu Y, Tao Y, Yu Y, Jiang H, Lian H (2014) Comparative study of adsorption of Pb(II) on native garlic peel and mercerized garlic peel. Environ Sci Pollut Res Int 21:2054–2063

López E, Soto B, Arias M, Núñez A, Rubinos D, Barral MT (1998) Adsorbent properties of red mud and its use for wastewater treatment. Water Res 32:1314–1322

López-Delgado A, Pérez C, López FA (1998) Sorption of heavy metals on blast furnace sludge. Water Res 32:989–996

Maheshwari U, Mathesan B, Gupta S (2015) Efficient adsorbent for simultaneous removal of Cu (II), Zn(II) and Cr(VI): kinetic, thermodynamics and mass transfer mechanism. Process Saf Environ 98:198–210

Miraoui M, Zentar R, Abriak N-E (2012) Road material basis in dredged sediment and basic oxygen furnace steel slag. Constr Build Mater 30:309–319

Morillo E, Undabeytia T, Maqueda C, Ramos A (2000) Glyphosate adsorption on soils of different characteristics: influence of copper addition. Chemosphere 40:103–107

Naiya TK, Chowdhury P, Bhattacharya AK, Das SK (2009) Saw dust and neem bark as low-cost natural biosorbent for adsorptive removal of Zn(II) and Cd(II) ions from aqueous solutions. Chem Eng J 148:68–79

O'Connell DW, Birkinshaw C, O'Dwyer TF (2008) Heavy metal adsorbents prepared from the modification of cellulose: a review. Bioresour Technol 99:6709–6724

Ok YS, Yang JE, Zhang Y-S, Kim S-J, Chung D-Y (2007) Heavy metal adsorption by a formulated zeolite-Portland cement mixture. J Hazard Mater 147:91–96

Onundi YB, Mamun AA, Al Khatib MF, Ahmed YM (2010) Adsorption of copper, nickel and lead ions from synthetic semiconductor industrial wastewater by palm shell activated carbon. Int J Environ Sci Technol 7:751–758

Orlando US, Baes AU, Nishijima W, Okada M (2002) Preparation of chelating agents from sugarcane bagasse by microwave radiation as an alternative ecologically benign procedure. Green Chem 4:555–557

Özer A, Akkaya G, Turabik M (2006) The removal of Acid Red 274 from wastewater: combined biosorption and biocoagulation with Spirogyra rhizopus. Dyes Pigments 71:83–89

Pagnanelli F, Mainelli S, Vegliò F, Toro L (2003) Heavy metal removal by olive pomace: biosorbent characterisation and equilibrium modelling. Chem Eng Sci 58:4709–4717

Panda AK, Mishra BG, Mishra DK, Singh RK (2010) Effect of sulphuric acid treatment on the physico-chemical characteristics of kaolin clay. Colloid Surf A 363:98–104

Pandey S, Tiwari S (2015) Facile approach to synthesize chitosan based composite—characterization and cadmium(II) ion adsorption studies. Carbohydr Polym 134:646–656

Pandey RLP, Ruchi Pandey RCM, Ansari NG (2014) Removal of Cd(II) ions from simulated wastewater by HCl modified cucumis sativus peel: equilibrium and kinetic study. Air Soil Water Res 7:93–101

Paul Chen J, Lin M (2001) Equilibrium and kinetics of metal ion adsorption onto a commercial H-type granular activated carbon: experimental and modeling studies. Water Res 35:2385–2394

Pavan FA, Mazzocato AC, Jacques RA, Dias SLP (2008) Ponkan peel: a potential biosorbent for removal of Pb(II) ions from aqueous solution. Biochem Eng J 40:357–362

Pei YY, Liu JY (2012) Adsorption of Pb^{2+} in wastewater using adsorbent derived from grapefruit peel. Adv Mater Res 391:968–972

Periasamy K, Namasivayam C (1996) Removal of copper(II) by adsorption onto peanut hull carbon from water and copper plating industry wastewater. Chemosphere 32:769–789

Perić J, Trgo M, Vukojević Medvidović N (2004) Removal of zinc, copper and lead by natural zeolite—a comparison of adsorption isotherms. Water Res 38:1893–1899

Prasanna Kumar Y, King P, Prasad VSRK (2007) Adsorption of zinc from aqueous solution using marine green algae—Ulva fasciata sp. Chem Eng J 129:161–166

Purna Chandra Rao G, Satyaveni S, Ramesh A, Seshaiah K, Murthy KSN, Choudary NV (2006) Sorption of cadmium and zinc from aqueous solutions by zeolite 4A, zeolite 13X and bentonite. J Environ Manag 81:265–272

Qiu W, Zheng Y (2009) Removal of lead, copper, nickel, cobalt, and zinc from water by a cancrinite-type zeolite synthesized from fly ash. Chem Eng J 145:483–488

Rangel-Mendez JR, Monroy-Zepeda R, Leyva-Ramos E, Diaz-Flores PE, Shirai K (2009) Chitosan selectivity for removing cadmium (II), copper (II), and lead (II) from aqueous phase: pH and organic matter effect. J Hazard Mater 162:503–511

Redlich O, Peterson DL (1959) A useful adsorption isotherm. J Phys Chem 63:1024

Rinaudo M (2006) Chitin and chitosan: properties and applications. Prog Polym Sci 31:603–632

Sadrzadeh M, Mohammadi T, Ivakpour J, Kasiri N (2009) Neural network modeling of Pb^{2+} removal from wastewater using electrodialysis. Chem Eng Process 48:1371–1381

Sankararamakrishnan N, Sharma AK, Sanghi R (2007) Novel chitosan derivative for the removal of cadmium in the presence of cyanide from electroplating wastewater. J Hazard Mater 148:353–359

Santasnachok C, Kurniawan W, Hinode H (2015) The use of synthesized zeolites from power plant rice husk ash obtained from Thailand as adsorbent for cadmium contamination removal from zinc mining. J Environ Chem Eng 3:2115–2126

Sargın İ, Arslan G (2015) Chitosan/sporopollenin microcapsules: preparation, characterisation and application in heavy metal removal. Int J Biol Macromol 75:230–238

Sarı A, Tuzen M (2008) Biosorption of cadmium(II) from aqueous solution by red algae (Ceramium virgatum): equilibrium, kinetic and thermodynamic studies. J Hazard Mater 157:448–454

Sekar M, Sakthi V, Rengaraj S (2004) Kinetics and equilibrium adsorption study of lead(II) onto activated carbon prepared from coconut shell. J Colloid Interface Sci 279:307–313

Shawabkeh R, Al-Harahsheh A, Hami M, Khlaifat A (2004) Conversion of oil shale ash into zeolite for cadmium and lead removal from wastewater. Fuel 83:981–985

Sheng PX, Ting Y-P, Chen JP (2007) Biosorption of heavy metal ions (Pb, Cu, and Cd) from aqueous solutions by the marine alga sargassum sp. in single- and multiple-metal systems. Ind Eng Chem Res 46:2438–2444

Shyam R, Puri JK, Kaur H, Amutha R, Kapila A (2013) Single and binary adsorption of heavy metals on fly ash samples from aqueous solution. J Mol Liq 178:31–36

Sips R (1948) On the structure of a catalyst surface. J Chem Phys 16:490

Sprynskyy M, Buszewski B, Terzyk AP, Namieśnik J (2006) Study of the selection mechanism of heavy metal (Pb^{2+}, Cu^{2+}, Ni^{2+}, and Cd^{2+}) adsorption on clinoptilolite. J Colloid Interface Sci 304:21–28

Srivastava NK, Majumder CB (2008) Novel biofiltration methods for the treatment of heavy metals from industrial wastewater. J Hazard Mater 151:1–8

Srivastava SK, Singh AK, Sharma A (1994) Studies on the uptake of lead and zinc by lignin obtained from black liquor—a paper industry waste material. Environ Technol 15:353–361

Suzuki Y, Kametani T, Maruyama T (2005) Removal of heavy metals from aqueous solution by nonliving Ulva seaweed as biosorbent. Water Res 39:1803–1808

Tan G, Xiao D (2009) Adsorption of cadmium ion from aqueous solution by ground wheat stems. J Hazard Mater 164:1359–1363

Theocharis CR, Jacob KJ, Gray AC (1988) Enhancement of Lewis acidity in layer aluminosilicates. Treatment with acetic acid. J Chem Soc Faraday Trans 84:1509

Torab-Mostaedi M, Asadollahzadeh M, Hemmati A, Khosravi A (2013) Equilibrium, kinetic, and thermodynamic studies for biosorption of cadmium and nickel on grapefruit peel. J Taiwan Inst Chem Eng 44:295–302

Vaughan T, Seo CW, Marshall WE (2001) Removal of selected metal ions from aqueous solution using modified corncobs. Bioresour Technol 78:133–139

Vázquez G, Antorrena G, González J, Doval MD (1994) Adsorption of heavy metal ions by chemically modified *Pinus pinaster* bark. Bioresour Technol 48:251–255

Velazquez-Jimenez LH, Pavlick A, Rangel-Mendez JR (2013) Chemical characterization of raw and treated agave bagasse and its potential as adsorbent of metal cations from water. Ind Crop Prod 43:200–206

Wang Y, Qiao M, Liu Y, Zhu Y (2012) Health risk assessment of heavy metals in soils and vegetables from wastewater irrigated area, Beijing-Tianjin city cluster, China. J Environ Sci 24:690–698

Wingenfelder U, Hansen C, Furrer G, Schulin R (2005) Removal of heavy metals from mine waters by natural zeolites. Environ Sci Technol 39:4606–4613

Witek-Krowiak A, Szafran RG, Modelski S (2011) Biosorption of heavy metals from aqueous solutions onto peanut shell as a low-cost biosorbent. Desalination 265:126–134

Wu F-C, Tseng R-L, Juang R-S (2000) Comparative adsorption of metal and dye on flake- and bead-types of chitosans prepared from fishery wastes. J Hazard Mater 73:63–75

Wu F-C, Tseng R-L, Juang R-S (2001) Kinetic modeling of liquid-phase adsorption of reactive dyes and metal ions on chitosan. Water Res 35:613–618

Xu M, Hadi P, Chen G, McKay G (2014) Removal of cadmium ions from wastewater using innovative electronic waste-derived material. J Hazard Mater 273:118–123

Xu M, Hadi P, Ning C, Barford J, An KJ, McKay G (2015) Aluminosilicate-based adsorbent in equimolar and non-equimolar binary-component heavy metal removal systems. Water Sci Technol 72:2166–2178

Xue Y, Hou H, Zhu S (2009) Competitive adsorption of copper(II), cadmium(II), lead(II) and zinc (II) onto basic oxygen furnace slag. J Hazard Mater 162:391–401

Xue Y, Wu S, Zhou M (2013) Adsorption characterization of Cu(II) from aqueous solution onto basic oxygen furnace slag. Chem Eng J 231:355–364

Yu B, Zhang Y, Shukla A, Shukla SS, Dorris KL (2000) The removal of heavy metal from aqueous solutions by sawdust adsorption—removal of copper. J Hazard Mater 80:33–42

Yu B, Zhang Y, Shukla A, Shukla SS, Dorris KL (2001) The removal of heavy metals from aqueous solutions by sawdust adsorption—removal of lead and comparison of its adsorption with copper. J Hazard Mater 84:83–94

Zhu C, Luan Z, Wang Y, Shan X (2007) Removal of cadmium from aqueous solutions by adsorption on granular red mud (GRM). Sep Purif Technol 57:161–169

Chapter 6
Removal of Antibiotics from Water by Adsorption/Biosorption on Adsorbents from Different Raw Materials

José Rivera-Utrilla, Manuel Sánchez-Polo, and Raúl Ocampo-Pérez

Abstract The present chapter aimed to analyze and compare the behavior of carbon materials, both commercial (activated carbons) and prepared in our laboratories (sludge-derived materials and activated carbons from petroleum coke) with different chemical and textural characteristics in the adsorption of tetracyclines and nitroimidazoles from water. This behavior was analyzed in both static and dynamic regimes and using ultrapure water, surface water, groundwater, and urban wastewater. We also assessed the influence of the solution chemical nature (pH and ionic strength) on the adsorption of these pharmaceutical contaminants analyzing the adsorbent-adsorbate interaction types and evaluating the effectiveness of the combined use of microorganisms and activated carbon (bioadsorption) in these adsorption processes. Additionally, the mass transport mechanisms controlling the overall adsorption rate of these adsorbate-adsorbent systems were investigated in depth, and relationships between textural and chemical characteristics of these adsorbent materials with kinetic and diffusion parameters were reported.

Keywords Adsorption • Carbon materials • Tetracyclines • Nitroimidazoles

Contents

J. Rivera-Utrilla (✉) • M. Sánchez-Polo
Inorganic Chemistry Department, University of Granada, 18071, Granada, Spain
e-mail: jrivera@ugr.es; mansanch@ugr.es

R. Ocampo-Pérez
Center of Research and Postgraduate Studies, Faculty of Chemical Science, Autonomous University of San Luis Potosí, Av. Dr. M. Nava No. 6, San Luis Potosí, SLP 78210, Mexico
e-mail: ocampor@uaslp.mx

© Springer International Publishing AG 2017

A. Bonilla-Petriciolet et al. (eds.), *Adsorption Processes for Water Treatment and Purification*, DOI 10.1007/978-3-319-58136-1_6

6.1 Introduction

A constant flow of new products is generating novel contaminants with unknown short-, medium-, or long-term effects on the environment and human health that are not governed by regulations on the maximum allowable concentrations in the environment (Halling-Sørensen et al. 1998; Calamari 2002; Adler et al. 2008; Cooper et al. 2008; Rivera-Utrilla et al. 2013b). They include chemical compounds in cosmetics (creams, perfumes, makeup), domestic products (degreasants, glass cleaners, detergents), and pharmaceuticals, which are causing the greatest concern due to their very wide variety and elevated consumption (Adler et al. 2008; Cooper et al. 2008; Rivera-Utrilla et al. 2013b). For example, over 3000 different pharmaceutical substances are used in the UK (Ayscough et al. 2000), and the annual production for human consumption in the European Union is estimated to exceed 100 tons per member country (Kümmerer 2008). Antibiotics are the most heavily used medical drugs in the European Union, with an estimated annual consumption of around 10,000 tons (Kümmerer 2008). It should also be noted that these consumption figures are considerably increased by the use of many of these products in veterinary medicine. After administration, large amounts of antibiotics and their byproducts are discharged into municipal wastewater that high concentrations of antibiotics are now detected in waters intended for human consumption, reducing their quality. They generally have a low biodegradability (Al-Ahmad et al. 1999; Kümmerer et al. 2000) and high toxicity (Kümmerer 2001), and some are reported to have mutagenic and carcinogenic characteristics (Bendesky et al. 2002).

Some of the most important antibiotic groups, tetracyclines (TCs) and nitroimidazoles, have been detected in waters (Halling-Sørensen et al. 1998; Calamari 2002; Rivera-Utrilla et al. 2013b). The TCs are bacteriostatic agents that act by inhibiting bacterial protein synthesis. They show activity against a wide variety of microorganisms and are used as antibiotics in humans and animals (Mathers et al. 2011; Gao et al. 2012b). Due to their low cost, TCs are also used as a food additive to enhance the growth rate of animals, adding a further pathway for their entry into the environment alongside emissions from the manufacture and formulation of the compounds and the disposal of unused or expired products. In surface waters, TCs were detected at concentrations ranging from 0.11 to 4.20 μg/L (Lin et al. 2009), while concentrations in an effluent of a wastewater treatment plant ranged from 46 to 1300 ng/L for tetracycline, 270 to 970 ng/L for chlortetracycline, and 240 ng/L for oxytetracycline (Ternes et al. 2002; Yang et al. 2005; Batt et al. 2007; Stackelberg et al. 2007; Lin et al. 2009; Gao et al. 2012b). On the other hand, nitroimidazole antibiotics are widely used to treat infections caused by anaerobic and protozoan bacteria (e.g., *Trichomonas vaginalis* and Giardia lamblia) in humans and animals and are added to chow for fish and fowl (Tally et al. 1981; Lau et al. 1992; Lindberg et al. 2004), leading to their accumulation in animals, fish farm waters, and, especially, meat industry effluents (Kümmerer 2001). Nitroimidazoles have also been detected in waters at concentrations of 0.1–90.2 μg/L (Lindberg et al. 2004). Importantly, the presence of traces of TCs and nitroimidazoles in the environment can lead to the appearance of microorganisms that are resistant to these antibiotics and to which humans may be exposed via drinking water (Stackelberg et al. 2007).

Nowadays, studies have demonstrated that conventional treatment plants, mainly based on the use of microorganisms, have proven inadequate to effectively remove antibiotics from water, largely due to their complex molecular structure (Kümmerer et al. 2000; Carballa et al. 2004; Rivera-Utrilla et al. 2013b). Thus, the US Environmental Protection Agency recommended the adsorption on activated carbon as the best available technology for removing nonbiodegradable toxic organic compounds from drinking water and industrial wastewater (USEPA 1991).

Both granular (GAC) and powdered (PAC) activated carbons have been widely used for the adsorption of organic micropollutants in solution due to their chemical and textural properties (Radovic et al. 2001; Moreno-Castilla 2004; Dias et al. 2007; Rivera-Utrilla et al. 2011; Beita-Sandí et al. 2016). The capacity of activated carbon to adsorb pharmaceutical-related pollutants has attracted research interest (Snyder et al. 2007; Choi et al. 2008; Simazaki et al. 2008; Yu et al. 2008; Calisto et al. 2015; Zhu et al. 2015; Vidal et al. 2015). Besides, an important advantage of using activated carbon to remove pharmaceuticals is that toxic or pharmacologically active products are not generated.

Snyder et al. (2007) assessed the mechanisms underlying the adsorption of various pharmaceuticals and hormones on GAC and PAC and obtained removal percentages of around 90% for most of the pharmaceuticals studied. Optimal performances were obtained for acetaminophen (73–84%), carbamazepine (74–86%), triclosan (90–96%), and fluoxetine (91%), but the removal percentage did not exceed 50% for naproxen, diclofenac, gemfibrozil, sulfamethoxazole, and ibuprofen, among other drugs. They observed that the effectiveness of activated carbon was markedly reduced in the presence of natural organic matter (NOM), which competes for the active sites on the carbon, blocking its porosity.

Various authors have studied the adsorption of tetracyclines using adsorbents other than carbon, including apatites (Misra 1991), clays, and soils (Figueroa et al. 2004; Kulshrestha et al. 2004; Jones et al. 2005; Başakçilardan-Kabakci et al. 2007; Turku et al. 2007; Gu and Karthikeyan 2008; Parolo et al. 2008; Chang et al. 2012). Thus, Jones et al. (2005) evidenced that the iron oxide content, cation exchange capacity, and soil texture have high influence on the sorption of TCs on soils with organic carbon content between 0% and 4%. Parolo et al. (2008) investigated the removal of tetracycline on montmorillonite as a function of pH and ionic strength. The results revealed that tetracycline can intercalate into the interlayer space of montmorillonite. Additionally, at pH 4, the highest tetracycline removal is obtained because, at this pH, tetracycline is in cationic form favoring the cation exchange. Finally, the presence of sodium ions reduces the adsorption capacity of montmorillonite.

TCs can be also removed from aqueous solutions by adsorption on different carbon materials such as $MnFe_2O_4$/activated carbon composites (Shao et al. 2012), graphene oxide (Gao et al. 2012a), and multiwalled carbon nanotubes (Zhang et al. 2011). However, fewer data are available on activated carbon as TC adsorbent. Choi et al. (2008) found activated carbon columns to be highly effective for the adsorption of seven TCs in aqueous medium, obtaining percentage removal values of around 90% that varied according to the type of TC and characteristics of the water, especially the concentration of NOM. They reported a higher adsorption capacity of activated carbon for TCs than for sulfonamide, despite the greater hydrophobicity of the latter.

In reference to nitroimidazole adsorption, Carrales-Alvarado et al. (2014) investigated the removal of MNZ on novel carbon materials with different chemical and textural characteristics. It was found that the chemical and textural characteristics of carbon adsorbents play a key role in the adsorption of MNZ in aqueous solution. Carbon materials with a large surface area and low content of carboxylic groups are preferable for the removal of nitroimidazoles from aqueous solution. Carbon materials can be effectively applied to remove MNZ from wastewater, because the adsorption of electrolytes in the water cooperates rather than competes with its adsorption. Additionally, the adsorption of MNZ on carbon materials is reversible, allowing the exhausted adsorbents to be regenerated by contacting them with water. Adsorption kinetic data published by Ahmed and Theydan (2013) showed that the adsorption rate of metronidazole (MNZ) on activated carbon from an agricultural waste follows a pseudo-second-order model, and estimation of the heat of adsorption revealed that the adsorption of MNZ was endothermic.

In order to further deepen in the interactions established in the adsorption of tetracyclines and nitroimidazole drugs on different adsorbent materials, the present chapter aimed to outline results obtained during the development of a wider project to analyze and compare the behavior of carbon materials, both commercial (activated carbons) and prepared in our laboratories (sludge-derived materials and activated carbons from petroleum coke) with different chemical and textural natures in the adsorption of tetracyclines and nitroimidazoles from water. This behavior was analyzed in both static and dynamic regime and using ultrapure water, surface water, groundwater, and urban wastewater. We also assessed the influence of solution chemical nature (pH and ionic strength) on the adsorption of these compounds analyzing the adsorbent-adsorbate interaction types and evaluated the combined use of microorganisms and activated carbon (biosorption) in these adsorption processes. Besides, the design of an adsorption system to treat wastewater also requires data on the adsorbate concentration decay curves of the adsorbate/adsorbent system and on the mechanisms controlling the adsorption kinetics. Therefore, other objectives of this project were (i) to apply diffusional and kinetic models to explain the overall adsorption rate of both tetracyclines and nitroimidazoles on the adsorbents selected, (ii) to investigate the mass transport mechanism controlling the overall adsorption rate, and (iii) to analyze the relationship of textural and chemical characteristics of these adsorbent materials with kinetic and diffusion parameters.

The first part of this chapter is concerned with the chemical and textural characterization of the adsorbents to remove TCs and nitroimidazoles from water. The preparation methods of both sludge-derived materials and activated carbons from petroleum coke will be also analyzed. The particular objective of this part was to optimize adsorbent material preparation by means of a statistical experimental planning method, obtaining materials derived from sludge and petroleum coke by chemical activation with NaOH and KOH, respectively, at high temperatures and establishing the properties required for the removal of contaminants from water. Special attention was paid to the effect of binders on the surface and adsorbent characteristics of these new materials. Some of the results obtained in this wider project have already been published (Sánchez-Polo and Rivera-Utrilla 2006; Rivera-Utrilla et al. 2009; Méndez-Díaz et al. 2010; Gómez-Pacheco et al. 2012; Ocampo-Pérez et al. 2012; Ocampo-Pérez et al. 2013; Rivera-Utrilla et al. 2013a; Ocampo-Pérez et al. 2015). This chapter summarizes these findings.

6.2 Adsorbent Materials

The adsorbents used in this study were commercial activated carbons, sludge-derived materials, and activated carbons from petroleum coke. All adsorbents were characterized by determining their surface area, pore volume accessible to water, pore size distribution, oxygen surface groups, and pH of the point of zero charge (pH_{PZC}). Sludge-derived materials were also characterized by X-ray fluorescence and chemical and elemental analyses.

6.2.1 Commercial Activated Carbons

Sorbo (S) and Merck (M) commercial activated carbons, with a particle diameter ranging between 0.6 and 1 mm were used in this study. The experimental methods followed to characterize the adsorbents were described in detail elsewhere (Sánchez-Polo and Rivera-Utrilla 2003; Rivera-Utrilla and Sánchez-Polo 2004; Bautista-Toledo et al. 2008; Gómez-Pacheco et al. 2012). Table 6.1 depicts the textural characteristics of activated carbons used in this study: both of them have a large surface area (>1200 m^2/g) and a highly developed microporosity. Micropore

Table 6.1 Textural characteristics of commercial activated carbons

Activated carbon	S_{BET}[a] (m^2/g)	$W_o(N_2)$[b] (cm^3/g)	$W_o(CO_2)$[c] (cm^3/g)	$L_o(N_2)$[d] (nm)	$L_o(CO_2)$[e] (nm)	S_{ext}[f] (m^2/g)	V_2[g] (cm^3/g)	V_3[h] (cm^3/g)
S	1225	0.39	0.29	1.02	0.70	46.90	0.04	0.48
M	1301	0.42	0.29	1.69	0.70	41.90	0.10	0.28

[a]Surface area determined from N_2 adsorption isotherms at 77 K
[b,c]Volumes of micropores determined by N_2 and CO_2 adsorption, respectively
[d,e]Mean widths of micropores determined with the Dubinin equation
[f]External surface area determined by mercury porosimetry
[g]Volume of pores with diameter of 6.6–50 nm determined by mercury porosimetry
[h]Volume of pores with diameter >50 nm determined by mercury porosimetry

Table 6.2 Chemical characteristics of the commercial activated carbons

Activated carbon	Carboxylic groups (meq/g)	Carbonyl groups (meq/g)	Acidic groups (meq/g)	Basic groups (meq/g)	pH_{PZC}
S	0.00	0.15	0.45	1.08	9.0
M	0.04	0.84	0.40	0.44	7.7

volumes, deduced from N_2 adsorption, were considerably higher than those from CO_2 adsorption (Table 6.1), indicating a very heterogeneous micropore distribution in the activated carbon. Because CO_2 is only adsorbed in smaller size micropores (ultramicropores), whereas N_2 is adsorbed on the surface of all micropores (Garrido et al. 1987; Rodriguez-Reinoso and Linares-Solano 1989), N_2 adsorption data yield the total micropore volume, $W_0(N_2)$. Thus, the mean micropore size (L_0) was higher when determined by N_2 versus CO_2 adsorption (Table 6.1).

Table 6.2 lists the chemical characteristics of both activated carbons, showing that they are predominantly basic, with pH of point of zero charge values of 7.7 (carbon M) and 9.0 (carbon S).

6.2.2 *Sludge-Derived Materials*

Environmental concerns have been raised about the management of the sludge remaining after the primary and secondary treatment of urban wastewaters (Werther and Ogada 1999). It is a semisolid slurry with a content of little economic value and can have a major impact on the environment. Sludge, which is a subproduct of most wastewater treatment processes, is considered dangerous toxic waste and is generally used as fertilizer (Campbell 2000; Oleszkiewicz and Mavinic 2001). Sludge has conventionally been removed to monofills or sanitary landfills, but alternative options have been proposed. One of these is to incorporate sludge into the soil after its stabilization, e.g., by composting techniques; taking advantage of its high content in organic matter, phosphorus, nitrogen, and potassium, among others; and enabling its utilization in the recovery of eroded land (Passuello et al. 2012). In this investigation, treatment plant sludge was used to prepare materials with the appropriate chemical and textural properties for the adsorption of organic and inorganic compounds.

6.2.2.1 Preparation of Adsorbent Materials from Sludge

The adsorbent materials were obtained from treatment plant sludge supplied by *Aguas y Servicios de la Costa Tropical de Granada (Spain)*. Sludge was transformed into adsorbent materials by thermal pyrolysis and chemical activation. It was chemically activated with sodium hydroxide by sludge impregnation, mixing the precursor with a solution of NaOH and binding agent (humic matter, phenolic resins, or clayey soil). After impregnation, the sample was left to dry under infrared lamps for approximately 12 h. The proportion of NaOH ranged from 5 to 100 g/100 g of sample and the proportion of binder from 0 to 20 g/100 g. After drying, the sample underwent pyrolysis in a model RO 10/100 Heraeus tubular oven equipped with Jumo-Digimat temperature programmer under controlled N_2 atmosphere (99.998%) with flow of 5 L/min, from 300 °C to 700 °C (ramps of 10 °C/min), maintaining the maximum temperature for periods ranging from 30 min to 3 h. The yield obtained in the preparation of these adsorbent materials was around 30%.

The binders or agglutinants selected were humic acid (Sigma Aldrich), clayey soil, and phenolic resins (both Ismael Quesada Chemical products). Humic acid was selected for its high carbon content and its utilization in briquetting processes, clayey soil for its binding capacity and low cost, and phenolic resins due to reports on their ability to bind different types of particles (Cavdar et al. 2008; Benk 2010; Correa et al. 2010; Wang et al. 2011). Table 6.3 lists the designations of the prepared samples.

6.2.2.2 Optimization of Sludge Activation Process

Optimization of Sludge Activation Process Without Binder (Linear Model)

Multivariate analyses were performed to optimize the sludge activation process, using a mathematical and statistical methodology that permits optimal planning of the experiment sequence, minimizing the cost of the experiment and the influence of experimental error (Hunter and Hunter 1978). The MODDE 7.0 statistical program was used for the experimental design and to optimize the preparation of adsorbent materials from treatment plant sludge.

The experimental conditions were varied in order to determine the optimal NaOH dose, pyrolysis temperature, and pyrolysis time for obtaining adsorbent materials with maximum tetracycline adsorption capacity. This effect was visualized by using a linear scanning model to study the response surface, conducting 11 experiments and varying the following factors: (A) amount of NaOH (from 5 to

Table 6.3 Designations assigned to the adsorbent samples prepared from biological treatment plant sludge (Gómez-Pacheco et al. 2012)

Experimental conditions						
Sample name	T (°C)	Type of binder	Binder (g)	NaOH (g)	Pyrolysis time (h)	Response q_e (mg/L)
CL	700	–	–	–	3.0	–
C1′	300	–	–	5.0	0.5	–
C2′	300	–	–	100.0	0.5	–
C3′	700	–	–	5.0	0.5	100
C4′	700	–	–	100.0	0.5	300
C5′	300	–	–	5.0	3.0	–
C6′	300	–	–	100.0	3.0	10
C7′	700	–	–	5.0	3.0	400
C8′	700	–	–	100.0	3.0	560
C9′	500	–	–	52.5	1.8	200
C10′	500	–	–	52.5	1.8	210
C11′	500	–	–	52.5	1.8	205
C2	700	–	–	50.0	3.0	–
C3	700	Humic ac.	10.0	25.0	3.0	418.3
C4	700	Humic ac.	10.0	50.0	3.0	466.6
C5	700	Humic ac.	20.0	25.0	3.0	406.8
C6	700	Humic ac.	0.0	25.0	3.0	407.5
C8	550	Humic ac.	10.5	30.0	2.0	352.4
C9	700	Humic ac.	20.0	50.0	3.0	415.9
C10	400	Humic ac.	20.0	50.0	3.0	–
C11	400	Humic ac.	20.0	50.0	1.0	–
C12	700	Humic ac.	20.0	10.0	3.0	242.9
C13	700	Humic ac.	1.0	10.0	1.0	278.4
C15	700	Humic ac.	20.0	10.0	1.0	219.7
C16	700	Humic ac.	20.0	50.0	1.0	380.5
C19	700	Humic ac.	1.0	50.0	1.0	460.2
C20	400	Humic ac.	1.0	10.0	1.0	–
C21	400	Humic ac.	20.0	10.0	1.0	–
C22	700	Humic ac.	1.0	10.0	3.0	278.8
C23	550	Humic ac.	10.5	30.0	2.0	349.0
C24	550	Humic ac.	6.5	30.0	2.0	360.0
C25	700	Humic ac.	1.0	50.0	3.0	512.9
CR1	700	Resin 1	20.0	25.0	3.0	–
CR2	700	Resin 2	20.0	25.0	3.0	–
CAR	700	Clay	20.0	25.0	3.0	–
CH	700	Humic ac.	20.0	25.0	3.0	–

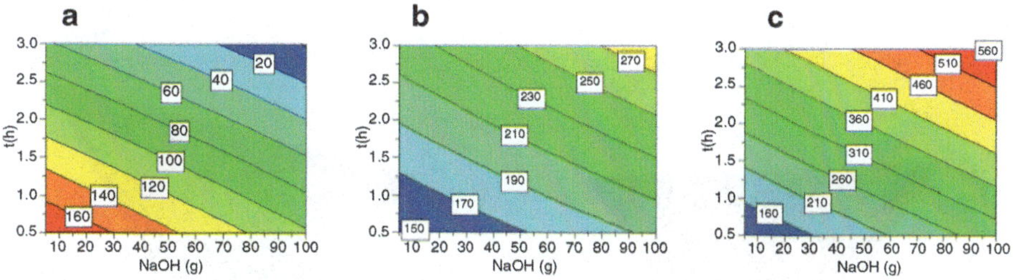

Fig. 6.1 Response surface values obtained using linear planning at pyrolysis temperatures of (**a**) 300 °C, (**b**) 500 °C, and (**c**) 700 °C (Gómez-Pacheco et al. 2012)

100 g), (B) pyrolysis temperature (from 300 °C to 700 °C), and (C) pyrolysis time (from 30 min to 3 h). The response considered was the capacity of the carbon to adsorb tetracycline (q_e). The adsorption capacity was obtained by using a tetracycline initial concentration of 700 mg/L, 0.1 g of adsorbent and 0.1 L of solution volume.

For the study of factor A (amount of NaOH added), experiments were conducted with fixed values of B (300 °C) and C (0.5 h) but two different values of A (5 and 100 g), allowing any variations in the response to be attributed to factor A (Samples C1′ and C2′ in Table 6.3). The same procedure was carried out for the other two factors. Figure 6.1 depicts the response surface obtained using linear planning to prepare the adsorbents from sludge. The maximum adsorption capacity of these adsorbents (in red) was obtained with pyrolysis for 3 h at a constant temperature of 700 °C, reaching a tetracycline adsorption capacity of 560 mg/g. As shown in Fig. 6.1a, a low adsorption capacity was found in samples pyrolyzed at a temperature of 300 °C, observing a decrease in adsorption capacity with longer pyrolysis time or higher NaOH concentration; the opposite behaviors to those observed at a pyrolysis temperature of 500 °C or 700 °C.

Optimization of Sludge Activation Process with Binder (Orthogonal Model)

Preparation of adsorbent materials by sludge chemical activation requires the addition of a binder to facilitate inter-particle union and obtain materials with suitable mechanical properties for application in water treatments. We investigated the effect of the binder on the adsorption capacity of these materials using orthogonal planning (MODDE 7.0 program) to study response surface values (Hunter and Hunter 1978). Figure 6.2 shows the response surface obtained to prepare these materials, maintaining a constant amount of NaOH (25 g) for different pyrolysis times (1, 2, and 3 h, respectively).

Figure 6.2 shows that the optimal values (in red) for the adsorption capacity of the activated carbon were 500 °C and 20 g of binder or 700 °C and 2 g of binder, achieving a tetracycline adsorption capacity of 470 mg/g. The results depicted in Fig. 6.2b show that the adsorption capacity of these materials increases with higher

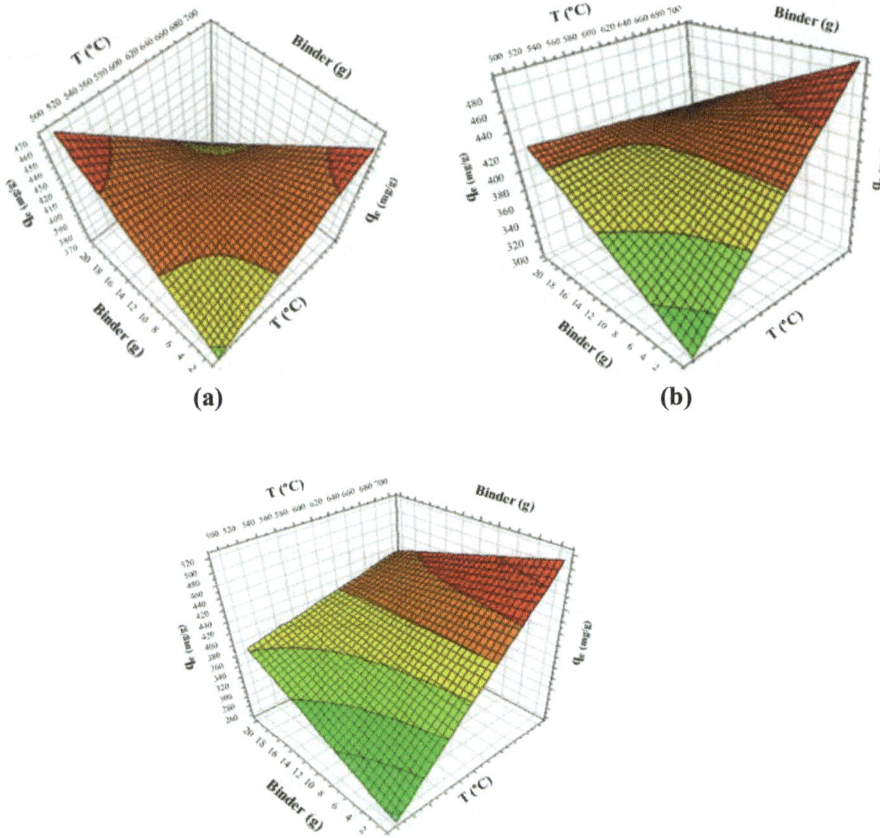

Fig. 6.2 Response surface by orthogonal planning, maintaining a constant amount of NaOH (25 g): (**a**) 1 h of pyrolysis, (**b**) 2 h of pyrolysis, (**c**) 3 h of pyrolysis. V = 100 mL, C_{A0} = 700 mg/L, m = 0.1 g (Gómez-Pacheco et al. 2012)

pyrolysis temperatures and lower amounts of binder, reaching values close to 500 mg/g. The absence of binder in the sample may increase the external surface area available for the adsorbate.

The results in Fig. 6.2c show that the optimal surface increases with higher pyrolysis temperature, and the model predicts that a good adsorption is even maintained with larger amounts of binder. These findings indicate that samples pyrolyzed at high temperatures for 3 h have good adsorbent properties, reaching values above 400 mg/g, even with a large amount of binder.

6.2.2.3 Characterization of Sludge-Derived Adsorbent Materials

Table 6.4 lists the results of the chemical analysis of the baseline sludge; it had a high content of organic matter (64.0%) and total nitrogen (7.8%) and metals, mainly zinc, copper, lead, and nickel. The textural and chemical properties of the sludge depend of the urban sludge considered. Sludge characterization is usually linked to the municipal and industrial activity of the city.

Textural Characterization of Adsorbents with Humic Acid as Binding Agent

Table 6.5 lists the values for surface area (S_{BET}) and mean micropore width (L_0) determined by N_2 adsorption at 77 K of the sludge-derived adsorbent materials containing humic acid as binder. The general behaviors shown in Table 6.5 are in agreement with previous reports (Méndez et al. 2005; Kang et al. 2006; Ros et al. 2006) on the use of chemical activation to prepare adsorbents from different raw materials:

Table 6.4 Chemical analysis of the dehydrated sludge (Gómez-Pacheco et al. 2012)

Component	% (by weight)	Component	μg/g
Dry matter	27.0	Cadmium	1.0
Organic matter	64.0	Chrome	2.7
pH	7.7	Copper	270.0
Total nitrogen	7.8	Lead	75.0
Phosphorus	3.8	Zinc	544.0
Calcium	0.3	Nickel	16.5
		Mercury	1.0

Table 6.5 Surface area and mean micropore width of the adsorbent materials prepared with humic acid as binder (Gómez-Pacheco et al. 2012)

Sample	Temperature (°C)	Activation time (h)	Binder (g)	NaOH (g)	S_{BET} (m^2/g)	$L_0(N_2)$ (nm)
C25	700	3	1.0	50	105	1.76
C3	700	3	10.0	25	175	2.25
C22	700	3	1.0	10	94	1.74
C9	700	3	20.0	50	124	1.90
C5	700	3	20.0	25	164	1.13
C12	700	3	20.0	10	104	1.28
C8	550	2	10.5	30	59	2.29
C23	550	2	10.5	30	61	2.20
C24	550	2	6.5	30	97	2.15
C16	700	1	20.0	50	87	1.75
C15	700	1	20.0	10	85	1.83
C19	700	1	1.0	50	87	1.79
C13	700	1	1.0	10	103	1.28

(i) The surface area of these materials was very low, with values ranging from 164 to 59 m^2/g, indicating the low effectiveness of the chemical activation of sludge in comparison to that of other adsorbent materials from sludge, with surface areas of 1000 m^2/g (Ros et al. 2006) or 380 m^2/g (Wang et al. 2008b).
(ii) Regardless of the sample in question, the surface area was enlarged with higher temperature or longer activation time.
(iii) Regardless of the amount of binder added, the surface area value was always higher for samples prepared with 25 g NaOH (sludge: NaOH ratio of 100:25 by weight).

Results in Table 6.5 show the optimal values for activating and binding agents to prepare adsorbent materials from sludge, considering the surface area value as response. As an example, Fig. 6.3 depicts the results obtained by applying the statistical optimization model at a pyrolysis temperature of 700 °C and a residence time of 3 h. As can be observed, the surface area of the materials increases with the amount of NaOH and, especially, with the amount of binder. The optimal experimental conditions are 3 h of pyrolysis at 700 °C with 30 g NaOH and 20 g of binder. As observed in Fig. 6.3, the surface area of these materials is not strongly influenced by the amount of NaOH and is most influenced by the amount of binder.

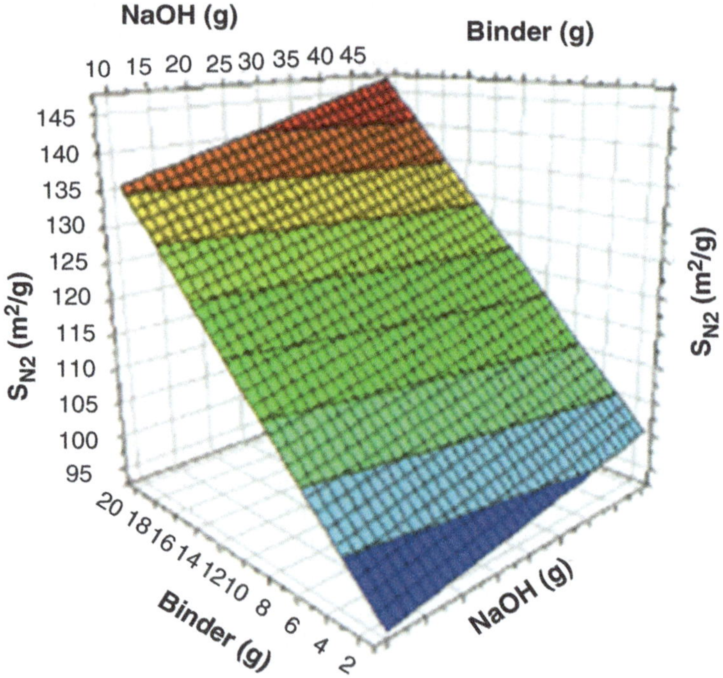

Fig. 6.3 Surface obtained with pyrolysis at 700 °C for 3 h, applying the statistical model to optimize adsorbent material preparation (Gómez-Pacheco et al. 2012)

Table 6.6 Textural characteristics of sludge-derived adsorbents with humic acid as binder (Gómez-Pacheco et al. 2012)

Sample	$W_0(N_2)$ (cm^3/g)	$W_0(CO_2)$ (cm^3/g)	$L_0(N_2)$ (nm)	$L_0(CO_2)$ (nm)	$W_0(N_2)/W_0(CO_2)$
C2	0.06	0.05	1.97	0.50	1.20
C3	0.06	0.02	2.25	1.25	3.00
C5	0.06	0.05	1.13	0.47	1.20
C8	0.03	0.04	2.29	0.50	0.75
C9	0.05	0.06	1.90	0.48	0.83
C12	0.04	0.06	1.28	0.50	0.67
C13	0.04	0.05	1.24	0.49	0.80
C15	0.04	0.06	1.83	0.52	0.67
C16	0.03	0.04	1.75	0.48	0.75
C19	0.04	0.03	1.79	0.49	1.33
C22	0.04	0.05	1.74	0.49	0.80
C23	0.03	0.04	2.20	0.48	0.75
C24	0.03	0.04	2.15	0.49	0.75
C25	0.05	0.03	1.76	0.52	1.66

Table 6.6 lists the results of the textural analysis from N_2 and CO_2 adsorption isotherms, showing that the mean micropore width determined with CO_2 (all ~0.5 nm except for sample C3) is lower than that determined with N_2 (> 1.13 mm for all samples). This is due to the fact that CO_2 is only adsorbed in smaller size micropores (ultramicropores), whereas N_2 is adsorbed on the surface of greater size micropores (Garrido et al. 1987; Rodríguez-Reinoso and Linares-Solano 1989). In the majority of samples, the CO_2-determined micropore volume was larger than the N_2-determined micropore volume, indicating a very narrow microporosity that is not totally accessible to N_2 molecules under these experimental adsorption conditions.

Influence of Binding Agent on Properties of the Adsorbent Materials

After optimizing the preparation of these materials from sludge for their tetracycline adsorption capacity and surface area (response variables), we studied the influence of binder type on their properties. Samples were prepared using humic acid (CH), clayey soil (CAR), or phenolic resins (CR1 or CR2) as binder. Table 6.7 shows the textural characteristics of these samples and of a sample prepared without NaOH or binder (CL) and one prepared without binder (C2).

The results in Table 6.7 show that the surface area was increased by the activation, as discussed above. They reveal that the textural properties of these materials were not substantially affected by the addition of phenolic resins as binding agents, with samples C2 (reference), CH, CR1 and CR2 having similar textural features. However, we highlight the similar surface areas of sample CAR

Table 6.7 Textural characteristics of adsorbent materials prepared with different binders (Gómez-Pacheco et al. 2012)

Samples	S_{BET} (m^2/g)	$W_0(N_2)$ (cm^3/g)	$W_0(CO_2)$ (cm^3/g)	$L_0(N_2)$ (nm)	$L_0(CO_2)$ (nm)	$W_0(N_2)/W_0(CO_2)$
CL	47	0.02	0.03	1.18	0.99	0.67
C2	139	0.06	0.05	1.97	0.50	1.20
CH	163	0.06	0.05	1.13	0.47	1.20
CAR	62	0.03	0.02	1.24	1.35	1.50
CR1	147	0.06	0.03	1.13	0.83	2.00
CR2	152	0.07	0.03	1.16	0.95	2.33

Table 6.8 Adsorbent characteristics obtained by mercury porosimetry (Gómez-Pacheco et al. 2012)

Sample	S_{ext} (m^2/g)	V_2[a] (cm^3/g)	V_3[b] (cm^3/g)	ρ_a[c] (g/cm^3)
CL	31.39	0.07	0.19	0.16
C2	134.02	0.29	1.54	0.09
CH	44.65	0.12	0.24	0.18
CAR	106.77	0.20	0.78	0.13
CR1	163.21	0.38	0.45	0.10
CR2	102.98	0.27	0.66	0.10

[a]V_2 Volume of pores with diameter of 6.6–50 nm determined by mercury porosimetry
[b]V_3 Volume of pores with diameter >50 nm determined by mercury porosimetry
[c]ρ_a Apparent density determined by mercury porosimetry

and the sample CL, which was not NaOH-activated, indicating that the surface area was reduced by the presence of clayey soil in the sample. In general, samples had a larger micropore volume by N_2 than by CO_2 determination, reflecting a heterogeneous distribution of microporosity in these materials.

Table 6.8 exhibits the pore volumes of the adsorbent materials obtained by mercury porosimetry. Except for the development of mesoporosity in sample CR1, the mesoporosity (V_2) and macroporosity (V_3) of all samples decreased with the presence of the binder, especially with humic acid (sample CH). Table 6.9 shows the elemental analysis of the samples; all had a low C content that was higher with the presence of binders; it was only 5.76% in the sample without binder and around 18% in the samples with binder, with the exception of CAR. X-ray fluorescence results in Table 6.10 show SiO_2 and CaO to be the predominant inorganic compounds in these samples. The following components were identified in the X-ray diffraction diagrams: calcium pyrophosphate, β–$Ca_2P_2O_7$(peak at 30.8°2θ); calcium hydroxyapatite, $Ca_5(PO_4)_3OH$ (peak at 31.6° 2θ); goethite, α-FeOOH (peak at 21°2θ); and hematites, α-Fe_2O_3 (peak at 33° 2θ). The remaining oxides in Table 6.10 were not observed in the diagrams, possibly due to their lack of crystallinity. Table 6.9 compiles some chemical characteristics of the adsorbent material samples, which were predominantly of basic nature, with pH_{PZC} values ranging from 8.7 (CAR) to 10.3 (C2).

Table 6.9 Elemental analysis of the samples (dry basis) (Gómez-Pacheco et al. 2012)

Sample	C (%)	H (%)	N (%)	(O + remaining elements)$_{dif}$ (%)	pH_{PZC}
CL	23.09	1.05	1.87	73.99	9.6
C2	5.76	0.70	0.41	93.13	10.3
CH	18.25	1.06	1.29	79.40	9.4
CAR	5.13	1.05	0.00	93.82	8.7
CR1	17.75	1.11	0.62	80.52	8.9
CR2	17.80	1.11	1.06	80.03	8.9

6.2.3 *Activated Carbons from Petroleum Coke*

Petroleum coke is a dark solid composed mainly of carbon, produced by the thermal decomposition and polymerization of heavy liquid hydrocarbon derived from crude oil. The coke produced from distillation residues tends to form the sponge coke, while coke produced from a cracking residue forms the premium coke. The sponge type, due to its practically amorphous structure, has little commercial value and may be an economical and environmental problem. Petroleum coke is a residue of the petrochemical industry, which generates around 4 tons of carbon for every 100 tons of crude oil refined. Because of its high concentration of heavy metals (Ni, V, Fe), this residue cannot be used in any productive process. However, due to its high carbon content, it is an excellent raw material for the production of activated carbon.

6.2.3.1 Preparation of Activated Carbons by Chemical Activation of Coke

A series of activated carbons were prepared from calcinated petroleum coke supplied by REPSOL-YPF (Alicante, Spain). This coke will be designated as C, and its particle size ranged from 0.5 to 0.8 mm.

KOH was used as activating agent for the petroleum coke activation. Different activated cokes were prepared, with a KOH/coke mass ratio ranging from 1 to 4. The samples will be designated with the letter C followed by the KOH/coke mass ratio used in the activation process.

The procedure described by Otowa et al. (1993) was used for the coke activation. Briefly, 25 g of petroleum coke was mixed with 100 mL of a KOH solution at the appropriate concentration for each sample. This mixture was first heated at 60 °C for 48 h and was then treated in N_2 atmosphere (flow 300 mL/min) at 400 °C for 2 h. Finally, it was carbonized by heating the mixture at 700 °C for 1 h under an N_2 flow of 300 mL/min. In all heat treatments, the oven heating rate was 10 °C/min. Finally, the samples were washed with deionized water to constant conductivity.

Table 6.10 Adsorbent characteristics obtained by X-ray fluorescence (Gómez-Pacheco et al. 2012)

Sample	SiO_2 (%)	Al_2O_3 (%)	Fe_2O_3 (%)	MnO (%)	MgO (%)	CaO (%)	Na_2O (%)	K_2O (%)	TiO_2 (%)	P_2O_5 (%)	LOI (%)
C2	23.75	11.65	9.64	0.06	4.13	21.81	3.42	0.15	1.00	11.29	12.41
CH	21.24	10.47	11.22	0.06	3.54	18.10	5.96	0.11	0.94	9.78	17.68
CAR	27.15	13.39	11.00	0.10	5.35	11.46	4.37	0.25	1.03	11.53	13.61
CR2	23.44	9.55	11.26	0.08	4.53	16.59	2.00	0.16	0.97	8.24	22.74

LOI loss on ignition

6.2.3.2 Characterization of Activated Carbons from Coke

Elemental analysis of the petroleum coke used was done with a Fisons Instruments 1108 CHNS analyzer and showed a composition of C = 87.7 ± 0.1%, H = 4.2 ± 0.1%, N = 0.4 ± 0.1%, S = 6.4 ± 0.1%, and O = 1.3 ± 0.1%.

Table 6.11 shows the results of the textural characterization of the activated carbons prepared with different KOH/coke mass ratios and of both original and demineralized cokes. These results indicate that the activation process considerably developed the porosity in all of the samples studied, increasing the volume of micro-, meso- (V_2), and macropores (V_3). Thus, the surface area of the original coke markedly increased after the activation, with sample C-1 showing the highest value. Moreover, it was observed that the micropore volume reduced with an increase in the amount of KOH added to the coke; the K_2O and K generated during the activation blocked the entrance to the pores, hampering the diffusion of CO_2 into a fraction of the micropores, as discussed below.

The value of S_{BET} was always higher than the value of micropore surface area (S_{mic}) regardless of the activated coke sample considered (Table 6.11). According to these results, a large fraction of the surface of these four samples corresponded to meso- and macropores, which are determined by N_2 at 196 °C but not by CO_2 at 0 °C (Garrido et al. 1987; Rodriguez-Reinoso and Linares-Solano 1989). Thus, the proportion of this fraction ranged from 41% in sample C-4 to 5% in sample C-1. It was also observed that the volume of macropores (V_3) was slightly larger when the amount of KOH in the samples was increased, whereas the volume of micro- and mesopores reduced with increases in the amount of KOH.

Several authors have studied the mechanism by which KOH activates carbonaceous materials (Marsh et al. 1984; Otowa et al. 1997). Thus, Marsh et al. (1984) showed that the oxygen of the alkali can remove cross-linking and stabilizing carbon atoms in crystallites. K metal obtained at the reaction temperatures may intercalate and force apart the separate lamellae of the crystallites. The microporosity of activated carbon in the new structure is created by the removal of potassium salts by washing and the removal of carbon atoms from the internal volume of the carbon by activation reaction. The results presented in Table 6.11 indicate that W_0 and S_{BET} values decreased with an increase in the KOH/coke

Table 6.11 Textural characterization of the original and activated cokes (Sánchez-Polo and Rivera-Utrilla 2006)

Sample	KOH/ coke	S_{BET} (m^2/g)	S_{mic}[a] (m^2/g)	$W_0(N_2)$ (cm^3/g)	V_2 (cm^3/g)	V_3 (cm^3/g)
C	0	<30	<30	0.02 ± 0.01	Nil	0.011 ± 0.003
C-1	1	1619 ± 30	1539 ± 30	0.55 ± 0.01	0.063 ± 0.003	0.132 ± 0.003
C-2	2	1261 ± 30	1165 ± 30	0.41 ± 0.01	0.061 ± 0.003	0.154 ± 0.003
C-3	3	1021 ± 30	716 ± 30	0.25 ± 0.01	0.058 ± 0.003	0.176 ± 0.003
C-4	4	970 ± 30	569 ± 30	0.20 ± 0.01	0.051 ± 0.003	0.263 ± 0.003

[a]Micropore surface area obtained from CO_2 adsorption isotherm at 273.15 K

ratios, which may be due to a blockage of the pores by remains of the K and K_2O that were generated during the activation process and not completely removed by the washing treatment. As reported above, the macroporosity of the coke increases with a higher KOH/coke ratio in the activation process. According to Otowa et al. (1993), elevated temperatures and high KOH/coke ratios produce large pores in the carbon structure due to the presence of KOH-derived K_2O, which expands the carbon atomic layers. K_2O also acts as catalyst of the carbon gasification process, so that its generation during petroleum coke activation potentiates the development of meso- and macroporosity on the carbon surface. Moreover, at temperatures above 700 °C, a considerable amount of K is formed by the reduction of K_2O with carbon (Otowa et al. 1993). As a result of the consumption of the inner carbon atoms, pores are formed in the structure.

Regarding the surface chemistry of the samples, it was observed that activation of the coke modified their chemical nature (Table 6.12). Thus, whereas the original coke was a mildly acid material ($pH_{PZC} = 6.5$), the pH_{PZC} of the KOH-activated coke ranged from 8.4 for sample C-1 to 9.7 for sample C-4. This change is largely due to the generation of surface basic groups during the activation process which increases with the KOH/coke ratio (Table 6.12). Nevertheless, the pH_{PZC} values, as with the above textural characteristics, might be slightly influenced by the presence of K and K_2O in the activated coke samples. In order to know it, the ash content of the activated coke samples was determined. The results obtained, 0.40%, 0.42%, 0.46%, and 0.52% for the samples C-1, C-2, C-3, and C-4, respectively, indicated a slight increase in the ash content as the KOH/coke ratio was increased. Therefore, as mentioned above, the presence of these species of K can influence, in part, on the basicity of the activated coke samples.

The functional groups generated on the carbon surface during activation were studied by means of X-ray photoemission spectroscopy (Table 6.13). The results

Table 6.12 pH_{PZC} of the activated cokes (Sánchez-Polo and Rivera-Utrilla 2006)

Sample	pH_{PZC}
C	6.5 ± 0.1
C-1	8.4 ± 0.1
C-2	8.8 ± 0.1
C-3	9.3 ± 0.1
C-4	9.7 ± 0.1

Table 6.13 Results of the deconvolution of the XPS O1s spectrum of the activated coke samples (Sánchez-Polo and Rivera-Utrilla 2006)

Sample	(- C = O) (%) 530.7 ± 0.2 eV	(C - OH, C – O - C) (%) 532.1 ± 0.2 eV	(- COOH) (%) 533.3 ± 0.2 eV	H_2O and/or O_2 (%) 535.3 ± 0.2 eV
C-1	76 ± 1	14 ± 1	6 ± 1	4 ± 1
C-2	80 ± 1	8 ± 1	8 ± 1	4 ± 1
C-3	84 ± 1	8 ± 1	3 ± 1	5 ± 1
C-4	90 ± 1	4 ± 1	2 ± 1	4 ± 1

showed a percentage of surface oxygen of around 10% in all the activated coke samples, much higher value than the percentage detected by elemental analysis in the original coke (% O = 1.3), which would confirm the creation of oxygen groups on the coke during its KOH activation.

In order to determine the surface functional groups generated by the activation process, the spectrum of the O1s region was studied in the activated coke samples following the method described elsewhere (Moreno-Castilla et al. 2003b). The results obtained are shown in Table 6.13. In the four samples, the oxygen percentage corresponding to the –COOH group, which is the main responsible group of the surface acidity, is very low. This fact may be due to the low thermal stability of this group (242–367 °C), which may be desorbed in the form of CO, CO_2, and H_2O during the activation process (Zielke et al. 1996); these results justify the pH_{PZC} values found for the activated coke samples. As the KOH/coke ratio increased, the concentration of –C = O on the activated coke sample also increased. These results explain, in part, the enhancement in the surface basicity as the KOH/coke used in the activation process increased.

6.3 Kinetic Study of the Adsorption of Tetracyclines and Nitroimidazoles on Sludge-Derived Materials and Activated Carbons

6.3.1 Tetracyclines and Nitroimidazoles Characterization

6.3.1.1 Tetracyclines

The tetracyclines (TCs) studied in this chapter were tetracycline (TC), oxytetracycline (OTC), and chlortetracycline (CTC). The chemical structures of the TCs are depicted in Fig. 6.4a, and their physicochemical properties are summarized in Table 6.14. Additionally, the speciation diagram of TC as a function of solution pH is shown in Fig. 6.4b, as an example.

6.3.1.2 Nitroimidazoles

Nitroimidazoles used in the present study were dimetridazole (DMZ), metronidazole (MNZ), ronidazole (RNZ), and tinidazole (TNZ). Figure 6.5 depicts the molecular structure and speciation diagrams of these compounds. Dimensions of the nitroimidazole molecules were determined with the computer software Advanced Chemistry Development (ACD/Labs) Software v8.14. Octanol-water partition coefficients (K_{OW}) were obtained from ChemIDplus Advanced database. Table 6.15 lists the main physicochemical properties of nitroimidazoles.

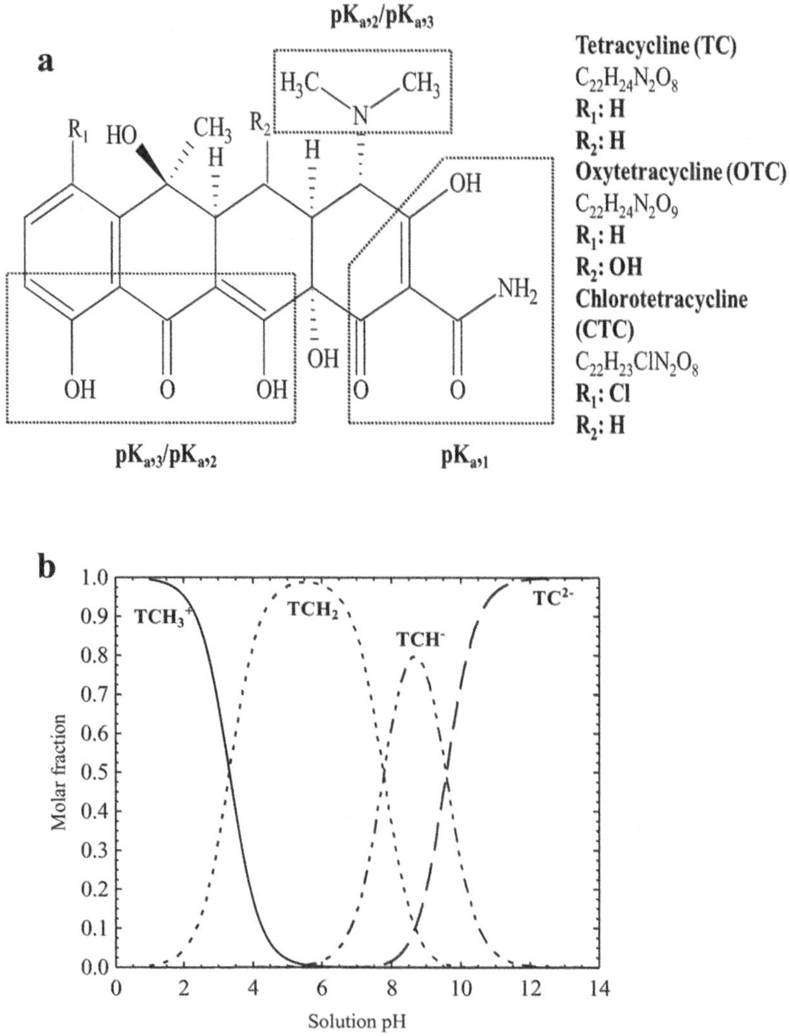

Fig. 6.4 Molecular structure (**a**) and speciation diagram (**b**) of tetracyclines in aqueous solution

6.3.2 *Kinetic and Diffusional Models*

In kinetic models, it is usually assumed that the overall adsorption rate is exclusively controlled by the adsorption rate of the solute on the adsorbent surface, and the intraparticle diffusion and external mass transport can be ignored. It is also considered that the adsorption rate of a solute on the surface can be represented in the same manner as the rate of a chemical reaction. As stated in Chap. 3, adsorption kinetics are commonly interpreted with first- and second-order kinetic models. On the other hand, diffusional models assume the adsorption process to occur by means

Table 6.14 Physicochemical properties of TCs

TCs	Molecular volume[a] (A^3)	Cross-sectional area[a] (A^2)	Water solubility[b] (mg/L)	Log K_{ow}[b]	pK_a[b]
TC	403.72	396.90	22	−1.30	3.32 7.78 9.58
OTC	413.13	407.49	17	−0.90	3.22 7.46 8.94
CTC	416.90	410.50	4.2	−0.62	3.33 7.55 9.33

[a]Determined by applying a Monte Carlo algorithm with the Spartan08 program
[b]Determined by means of the Advanced Chemistry Development (ACD/Labs) Software v8.14 program. pK_a corresponding to the successive deprotonation reactions

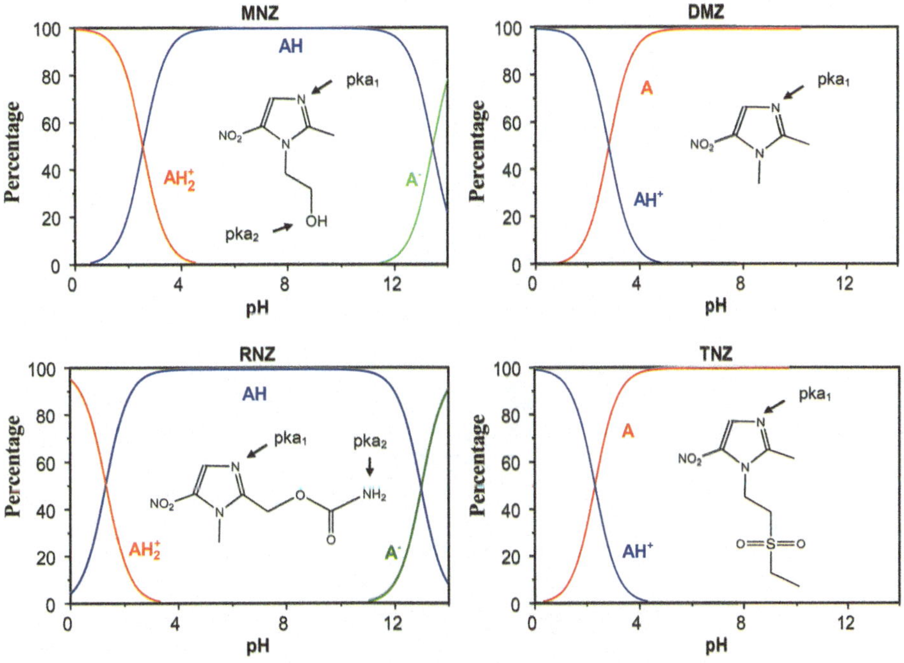

Fig. 6.5 Chemical structure and speciation diagram of nitroimidazoles

of three well-differentiated steps: (i) external mass transfer, where the adsorbate is transferred from the bulk of the solution to the adsorbent external surface; (ii) intraparticle diffusion, where the adsorbate molecule diffuses through the pores from the external surface of the adsorbent particle to the adsorption sites; and iii) adsorption rate of the compound on an active site (Leyva-Ramos and Geankoplis 1985). In this last step, the adsorption equilibrium between solute in

Table 6.15 Properties of nitroimidazoles

Nitroimidazole	Molecular weight (g/mol)	Transversal section (A^2)	$D_{AB} \times 10^6$ (cm^2/s)	Volume (A^3)	Solubility in water (mol/L)	pK_{ow}	pK_{a1}	pK_{a2}
MNZ	171.15	101	9.63	186	0.041	0.02	2.58	14.44
DMZ	141.13	88	8.48	157	0.062	−0.31	2.81	–
TNZ	247.27	138	8.25	258	0.008	0.35	2.30	–
RNZ	200.15	113	6.72	206	0.015	0.38	1.32	12.99

solution and solute on adsorbent is considered to be instantaneous (Ocampo-Perez et al. 2010); therefore the amount of solute adsorbed on the pore surface is at equilibrium with the solute concentration in the solution. Under these conditions, external transfer and/or intraparticle diffusion will be largely responsible for the global adsorption rate. In the following sections, the most important kinetic models and diffusional models used to predict the concentration decay curves for several adsorbate-adsorbent systems will be discussed in detail. Note that details of adsorption kinetic models have been also covered in Chap. 3 of this book.

6.3.2.1 Pseudo First-Order Kinetic Model

Lagergren advanced the first-order kinetic model to predict the adsorption rate of oxalic and malonic acids onto charcoal (Ho and Mckay 1998). The pseudo first-order kinetic model has been extensively applied to interpret the adsorption rate of solutes on different adsorbents (Srivastava et al. 2006). The pseudo first-order kinetic model can be mathematically represented by the following equation:

$$\frac{dq_t}{dt} = k_1(q_e - q_t) \tag{6.1}$$

where k_1 is the rate constant of the pseudo first-order kinetic model (1/h), t is the time (min), and q_e and q_t are the mass of adsorbate adsorbed at the equilibrium and at time t (mg/g), respectively.

This equation can be integrated using the initial condition $q_t = 0$ when $t = 0$, and the resulting equation is:

$$q_t = q_e\left(1 - e^{-k_1 t}\right) \tag{6.2}$$

The above equation can also be expressed in terms of C_A and C_{Ae} by using the mass balance equation at time t and equilibrium as follows:

$$q_t = \frac{V(C_{A0} - C_A)}{m} \tag{6.3}$$

$$q_e = \frac{V(C_{A0} - C_{Ae})}{m} \tag{6.4}$$

where V is the volume of the solution (L); C_{A0}, C_A, and C_{Ae} are the initial, at a time t and the equilibrium concentration of adsorbate in aqueous solution (mg/L); and m is the mass of adsorbent (g). These mathematical relationships can be substituted into Eq. (6.2), and the following equation can be obtained:

$$C_A = C_{A0} - \left(\frac{m}{V}\right) q_e \left(1 - e^{-k_1 t}\right) \tag{6.5}$$

6.3.2.2 Pseudo Second-Order Kinetic Model

The pseudo-second-order kinetic model can be represented by the following differential equation (Blanchard et al. 1984):

$$\frac{dq_t}{dt} = k_2 (q_e - q_t)^2 \tag{6.6}$$

where k_2 is the rate constant of the pseudo-second-order kinetic model (g/mg/h).

Integrating Eq. (6.6) and using the initial condition $q_t = 0$ when $t = 0$, the following equation can be obtained:

$$q = \frac{q_e^2 k_2 t}{1 + q_e k_2 t} \tag{6.7}$$

This equation can be also formulated in terms of C_A and C_{A0}, yielding the following equation:

$$C_A = C_{A0} - \left(\frac{m}{V}\right) \frac{q_e^2 k_2 t}{1 + q_e k_2 t} \tag{6.8}$$

6.3.2.3 Intraparticle Diffusion Model

A functional relationship common to most treatments of intraparticle diffusion is that uptake varies almost proportionately with the half-power of time ($t^{0.5}$), rather than t. Good linearization of the data is observed for the initial phase of the reaction in accordance with expected behavior if intraparticle diffusion is the rate-limiting step (Ho and Mckay 1998; Ip et al. 2010). The intraparticle diffusion equation is the following:

$$q_t = k_i t^{0.5} \tag{6.9}$$

where k_i is the intraparticle diffusion rate constant (mg/g^{-1}h$^{-0.5}$).

The k_i values are calculated from the slope of the straight line of the respective plots. The plot of q_t versus $t^{0.5}$ may present multi-linearity, which indicates that two or more rate controlling steps occur in the adsorption process.

6.3.2.4 Surface and Pore Volume Diffusion Model

The diffusional model is based on the following assumptions: (i) the intraparticle diffusion occurs by pore volume diffusion (Fick diffusion) and surface diffusion, (ii) the rate of adsorption on an active site is instantaneous, and (iii) the adsorbent particles are spherical. The model equations and initial and boundary conditions are the following (Leyva-Ramos and Geankoplis 1994; Ocampo-Perez et al. 2011):

$$V\frac{dC_A}{dt} = -mSk_L\left(C_A - C_{Ar}|_{r=R_p}\right) \tag{6.10}$$

$$t = 0 \qquad C_A = C_{A0} \tag{6.11}$$

$$\varepsilon_p\frac{\partial C_{Ar}}{\partial t} + \rho_p\frac{\partial q}{\partial t} = \frac{1}{r^2}\frac{\partial}{\partial r}\left[r^2\left(D_{ep}\frac{\partial C_{Ar}}{\partial r} + D_s\rho_p\frac{\partial q}{\partial r}\right)\right] \tag{6.12}$$

$$C_{Ar} = 0 \qquad t = 0 \qquad 0 \leq r \leq R_p \tag{6.13}$$

$$\left.\frac{\partial C_{Ar}}{\partial r}\right|_{r=0} = 0 \tag{6.14}$$

$$D_{ep}\left.\frac{\partial C_{Ar}}{\partial r}\right|_{r=R} + D_s\rho_p\frac{\partial q}{\partial r} = k_L\left(C_A - C_{Ar}|_{r=R_p}\right) \tag{6.15}$$

where S is the external surface area determined from $S = 3/(R_p \times \rho_p)$, R_p is the radius of the particle (cm), k_L is the external mass transfer coefficient in liquid phase (cm/s), R is the distance in radial direction of the particle (cm), $C_{Ar}\ /_{r=R}$ is the concentration of adsorbate at the external surface of the particle at $r = R$ (mg/L), ε_p is void fraction of particles, C_{Ar} is the concentration of adsorbate within the particle at distance r (mg/L), ρ_p the density of adsorbent particles (g/mL), D_{ep} is the effective pore volume diffusion coefficient (cm^2/s), D_s is the surface diffusion coefficient (cm^2/s), and q is the adsorption capacity (mg/g).

The Eqs. (6.10), (6.11), (6.12), (6.13), (6.14), and (6.15) represent the pore volume and surface diffusional model (PVSDM). The parameters k_L, D_S, and D_{ep} correspond to external mass transport, surface diffusion, and pore volume diffusion mechanisms, respectively. The PVSDM model can be simplified by considering that the intraparticle diffusion mechanism may be exclusively due to either pore volume diffusion (PVDM) ($D_{ep} \neq 0$, $D_S = 0$) or surface diffusion (SDM) ($D_{ep} = 0$, $D_S \neq 0$).

If the adsorption rate on an active site is considered to be instantaneous, there is local equilibrium between the adsorbate concentration in the solution within the pore of the adsorbent and the mass of adsorbate adsorbed on the surface of the pore. This equilibrium relationship between C_{Ar} and q_m is represented by the adsorption isotherm:

$$q_m = f(C_{Ar}) \tag{6.16}$$

6.3.3 Results and Discussion

6.3.3.1 Kinetic Study of Tetracycline Adsorption on Sludge-Derived Adsorbents

Adsorption kinetics were obtained by adding 0.1 g adsorbent material in Erlenmeyer flasks that contained 100 mL of a 700 mg/L TC solution. The flasks were maintained in thermostatic bath at 298 K and under agitation for 10 days, and samples were periodically extracted to determine the amount of TC adsorbed as a function of time. Solution pH ranged between 7 and 8 in all cases. Figure 6.6 depicts the adsorption kinetics of TC on selected sludge-derived adsorbents.

In order to quantify the rate of TC adsorption on the adsorbents and to identify the chemical and textural properties involved in this process, adsorption rate constants (k_1, k_2) were determined by fitting the experimental adsorption kinetics data to pseudo first-order and second-order kinetic models, respectively. Table 6.16 shows the results obtained and also the values of the amounts of TC adsorbed at equilibrium and the corresponding determination coefficients. A comparison between q_e values (theoretically calculated) and the corresponding q_e values (experimental) shows that the pseudo-first-order model fits the experimental data for all TC-adsorbents systems better than does the pseudo-second-order model.

The adsorption rate constants (k_1) (Table 6.16) were related to some of the characteristics of the corresponding adsorbents (Tables 6.7, 6.8, and 6.9). In general, no clear relationship was observed between the k_1 values and most of the textural parameters of the adsorbents, except for a tendency to higher k_1 values with increased pore volume, V_2, and macropore volume, V_3 (see Fig. 6.7a, b, respectively).

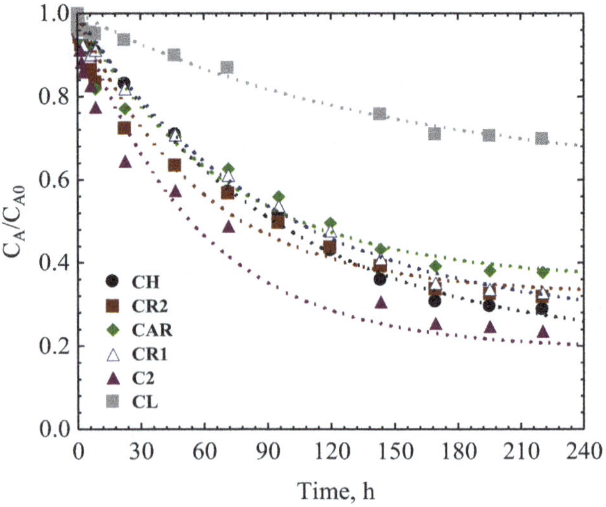

Fig. 6.6 TC adsorption kinetics on the adsorbents. pH ≈ 7, $[TC]_o = 700$ mg/L, T = 298 K. The *lines* represent the predictions of the first-order kinetic model (Ocampo-Pérez et al. 2012)

Table 6.16 Adsorption rate constants, determination coefficients, and amounts of TC adsorbed at equilibrium obtained experimentally and according to pseudo-first-order, pseudo-second-order, and intraparticle diffusion models (Ocampo-Pérez et al. 2012)

	Pseudo-first order				Pseudo-second order			Intraparticle diffusion	
Adsorbent	q_e (exp.) (mg/g)	$k_1 \times 10^2$ (1/h)	R^2	q_e (predicted) (mg/g)	$k_2 \times 10^5$ (L (mg/h)	R^2	q_e (predicted) (mg/g)	k_i (mg/(g $h^{0.5}$))	R^2
CL	210	0.68 ± 0.24	0.978	278 ± 55	1.11 ± 0.79	0.978	424 ± 110	10.82	0.952
C2	532	2.60 ± 0.079	0.955	565 ± 43	12.1 ± 3.8	0.977	537 ± 31	45.33	0.992
CH	497	1.04 ± 0.095	0.997	565 ± 23	2.16 ± 0.51	0.993	638 ± 42	31.12	0.977
CAR	434	1.42 ± 0.022	0.988	450 ± 26	2.81 ± 0.76	0.991	555 ± 40	31.50	0.982
CR1	469	1.08 ± 0.012	0.995	521 ± 27	3.89 ± 1.25	0.986	518 ± 40	29.93	0.980
CR2	476	1.73 ± 0.026	0.988	472 ± 23	4.37 ± 1.03	0.992	543 ± 29	37.50	0.991

TC adsorption rate constants also increased with lower carboxylic group content of the adsorbents (Fig. 6.7c), but no relationship was observed with the other chemical parameters of the adsorbents, such as the pH_{PZC} or basic group content. According to these findings, carboxylic groups in these materials reduce the adsorption rate because, at the working pH of 7–8, repulsive interactions are established between the negative charge of these groups and TC molecules, which are negatively charged as a consequence of deprotonation (see Fig. 6.4b).

The mechanism of adsorption is generally considered to involve (i) mass transfer of adsorbate from the bulk phase to particle surface, (ii) adsorption on surface site, and (iii) intraparticle diffusion of the adsorbate molecules to an adsorption site by a pore diffusion and/or surface diffusion mechanism. Step (ii) is often assumed to be extremely rapid; therefore the adsorption of large molecules, with long contact times to equilibrium, is always considered to be via diffusion controlled by external film resistance and/or internal diffusion mass transport or intraparticle diffusion.

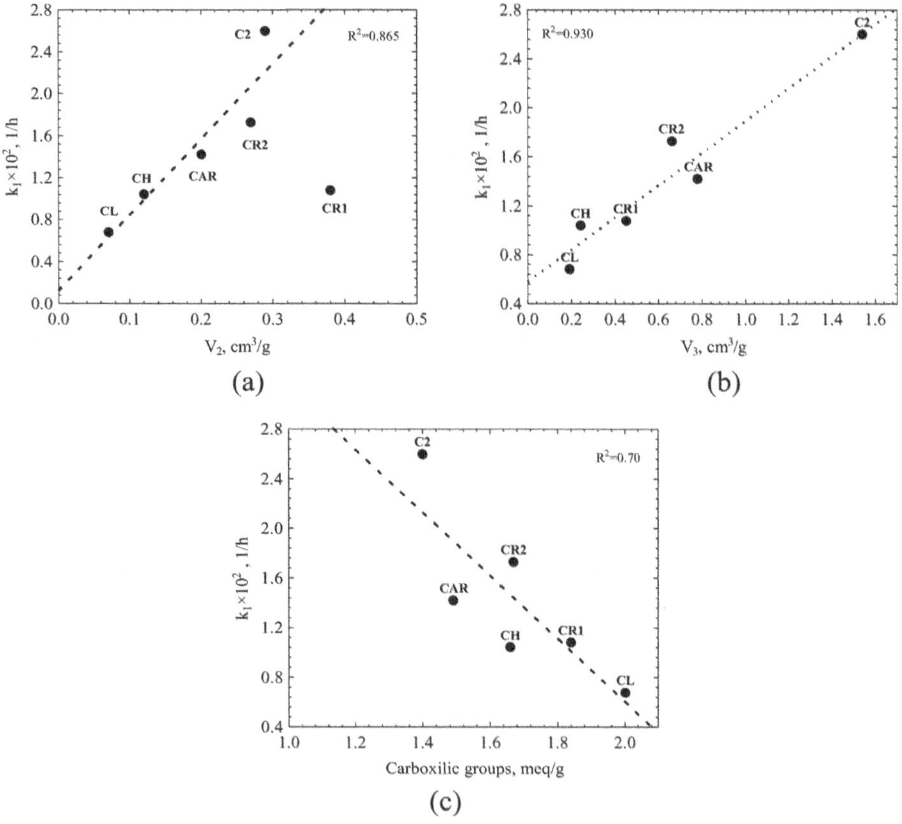

Fig. 6.7 Variation of k_1 with textural and chemical parameters of adsorbents (Ocampo-Pérez et al. 2012)

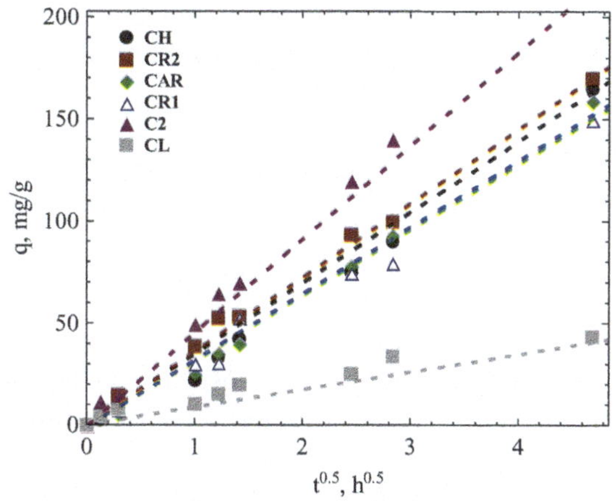

Fig. 6.8 Root time plot for the adsorption of tetracycline on the adsorbents. pH ≈ 7, $[TC]_o = 700$ mg/L, T = 298 K (Ocampo-Pérez et al. 2012)

A classical approach to analyze whether an adsorption process is controlled by intraparticle diffusion is to plot the amount adsorbed versus the square root of time, $t^{0.5}$ for a short contact time, i.e., $q/q_e < 0.3$. If the plot is linear and passes through the origin, it indicates that intraparticle diffusion is controlling the rate of adsorption. Thus, Fig. 6.8 shows the amount of tetracycline adsorbed as a function of $t^{0.5}$ for the sludge-derived adsorbents. It can be seen that intraparticle diffusion model satisfactorily fitted the experimental data, obtaining a linear section that passes through the origin in a short time, indicating that tetracycline adsorption on these adsorbents is governed by intraparticle diffusion. The values of the intraparticle diffusion rate constant (k_i) are given in Table 6.16 together with determination coefficient values (R^2). The k_i values ranged from 10.82 mg/(g $h^{0.5}$) for sample CL to 45.33 mg/(g $h^{0.5}$) for sample C2.

Surface diffusion, i.e., the movement of the adsorbate through the solid surface, is influenced by the distribution of the solid and fluid phases, and the surface concentration gradients are the main driving force. The surface diffusion model (SDM) is the most widely used to interpret concentration decay curves of aromatic compounds on adsorbents (Traegner and Suidan 1989; Ganguly and Goswami 1996); it assumes that the intraparticle diffusion is exclusively due to surface diffusion and that there is no pore volume diffusion. For its part, pore volume diffusion refers to the movement of the adsorbate due to concentration gradients in the fluid phase (i.e., molecular mechanisms) and is affected by the geometry of pore. Hence, the pore volume diffusion model (PVDM) ignores surface diffusion and assumes that pore volume diffusion is the sole mechanism of intraparticle diffusion (Traegner and Suidan 1989).

The TC concentration decay curves were predicted with SDM and PVDM models. Values of the external mass transfer coefficient (k_L) and the effective diffusion coefficients (D_s and D_{ep}) were required to resolve both models. Experimental k_L values were estimated from Eq. (6.17), while D_s and D_{ep} were obtained by matching the numerical solution of the models to the experimental data. The optimal values of D_s and D_{ep} were the values that best fitted the experimental data by minimizing the objective function given by Eq. (6.18).

$$\left[\frac{d\left(\frac{C_A}{C_{A0}}\right)}{dt}\right]_{t=0} = \frac{-mSk_L}{V} \tag{6.17}$$

$$\sum_{1}^{N}\left(\phi_{\text{exp}} - \phi_{\text{pred}}\right)^2 = \text{Minimum} \tag{6.18}$$

where ϕ_{exp} and ϕ_{pred} are the experimental dimensionless concentration of adsorbate in the solution and predicted with the diffusion models, respectively, and N is number of experimental data.

Figure 6.9 depicts the TC concentration decay curves on all adsorbents and the predictions of SDM (Fig. 6.9a) and PVDM (Fig. 6.9b) models. It can be observed that both models satisfactorily fitted the experimental data, which were better interpreted by the PVDM model than by the SDM model. It should be noted that the D_s values ranged from 2.43×10^{-10} to 9.7×10^{-10} cm^2/s, whereas the D_{ep} values ranged from 0.85×10^{-7} to 2.91×10^{-7} cm^2/s. On average, D_{ep} values were around 300-fold higher than D_s values, which suggests that TC diffuses within the pores by pore volume diffusion rather than by surface diffusion.

In order to elucidate the mechanism of intraparticle diffusion that governs the diffusion of tetracycline on these adsorbent materials, the experimental data were interpreted with the PVSDM model. This model considers that surface diffusion and pore volume diffusion are both important in the overall adsorption rate. Fig. 6.10a depicts, as an example, the experimental data of TC adsorption rate on sample CR2 and the prediction of the PVSDM model, showing a satisfactory fit of the experimental data. The optimal values of D_{ep} and D_s were 1.73×10^{-7} and 9.7×10^{-11} cm^2/s, respectively. The D_{ep} value obtained with the PVDSM model is very similar to that obtained with the PVDM model ($D_{ep} = 1.94 \times 10^{-7}$ cm^2/s), whereas the D_s value is lower than that obtained with the SDM model.

The relative contribution of pore volume diffusion to the overall intraparticle diffusion was estimated from these results by using the following equation:

$$\frac{N_{AP}}{N_{AS} + N_{AP}} = \frac{D_{ep}\frac{\partial C_{Ar}}{\partial r}}{D_s\rho_p\frac{\partial q}{\partial r} + D_{ep}\frac{\partial C_{Ar}}{\partial r}} \tag{6.19}$$

where N_{AP} and N_{AS} are the mass transport due to the pore volume diffusion and surface diffusion, respectively.

Figure 6.10b shows the relative contribution of pore volume diffusion as a function of time at different dimensionless radial positions $\xi(r/R)$ in sample CR2, as an example; the pore volume contribution always represented more than 84% of the total intraparticle diffusion regardless of the radial position and time. Similar results were obtained for the other adsorbents. These results confirm that

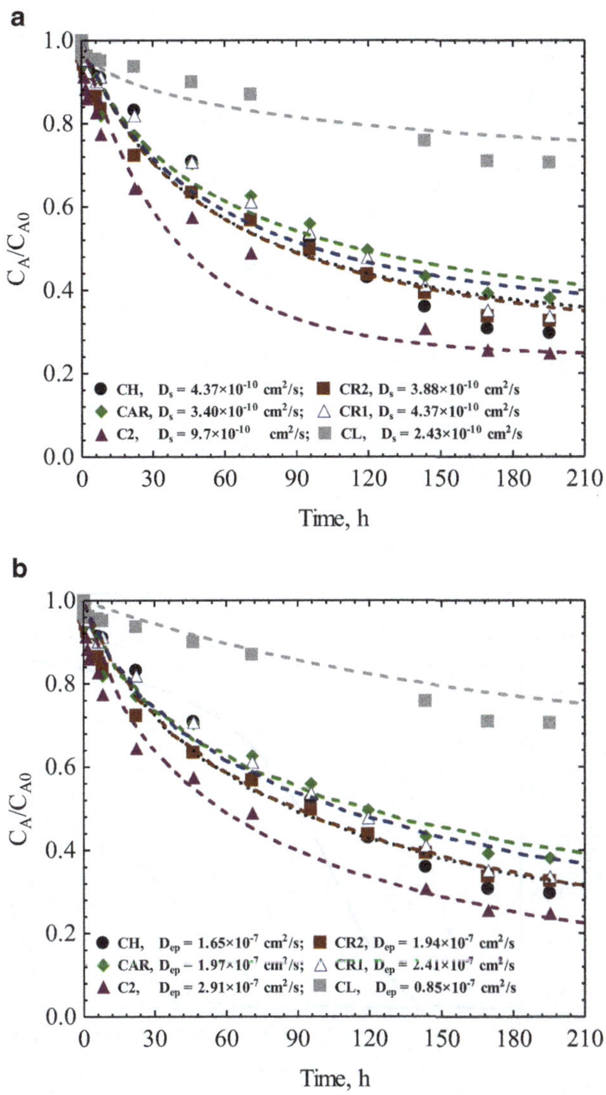

Fig. 6.9 TC adsorption kinetics on the adsorbents. The *lines* represent the prediction of (**a**) SDM model and (**b**) PVDM model. pH = 7, $[TC]_o$ = 700 mg/L, T = 298 K (Ocampo-Pérez et al. 2012)

tetracycline diffusion on adsorbents derived from sewage sludge is governed by pore volume diffusion.

TC molecular diffusion within the adsorbent pores can be drastically affected by the physical properties of the adsorbent, such as the porosity and tortuosity. The porosity of a material is directly related to the total pore volume, while the tortuosity is a function of the pore shape and size distribution. In Fig. 6.11, the D_{ep} values are

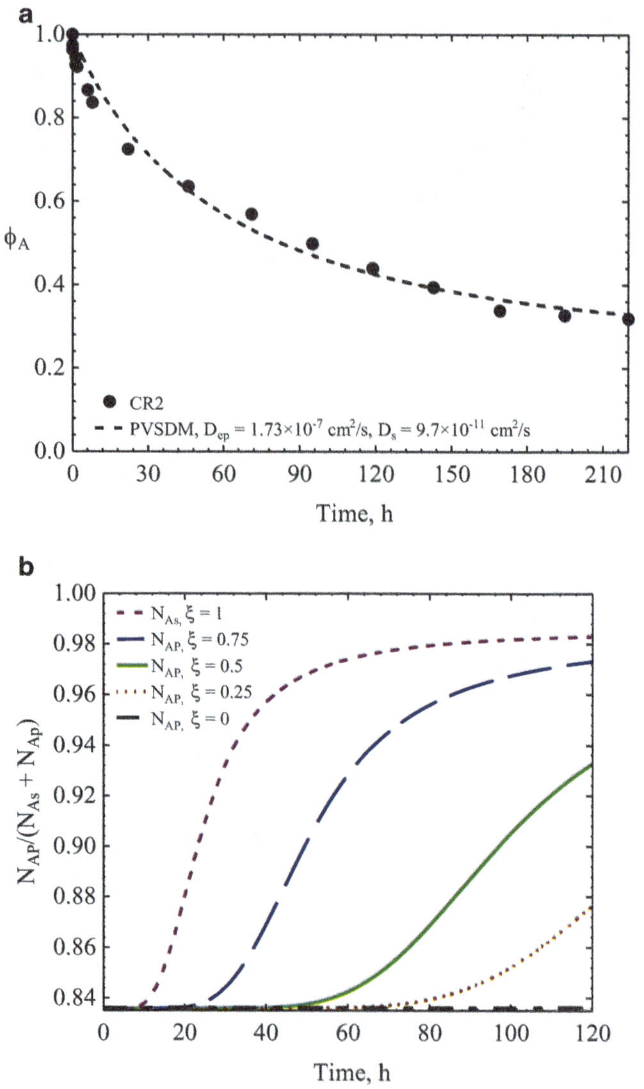

Fig. 6.10 (**a**) TC adsorption kinetics on sample CR2. (**b**) Contribution of pore volume diffusion to the total intraparticle diffusion at different radial positions during TC adsorption on sample CR2 (Ocampo-Pérez et al. 2012)

plotted against the main textural characteristics of the adsorbents (V_2, V_3, $W_0(N_2)$, and S_{ext}). This figure shows a tendency for D_{ep} values to increase with higher V_2, V_3, $W_0(N_2)$, and S_{ext} values. These results indicate that the TC diffusion is directly related to its accessibility to the interior pores. Thus, higher effective pore volume

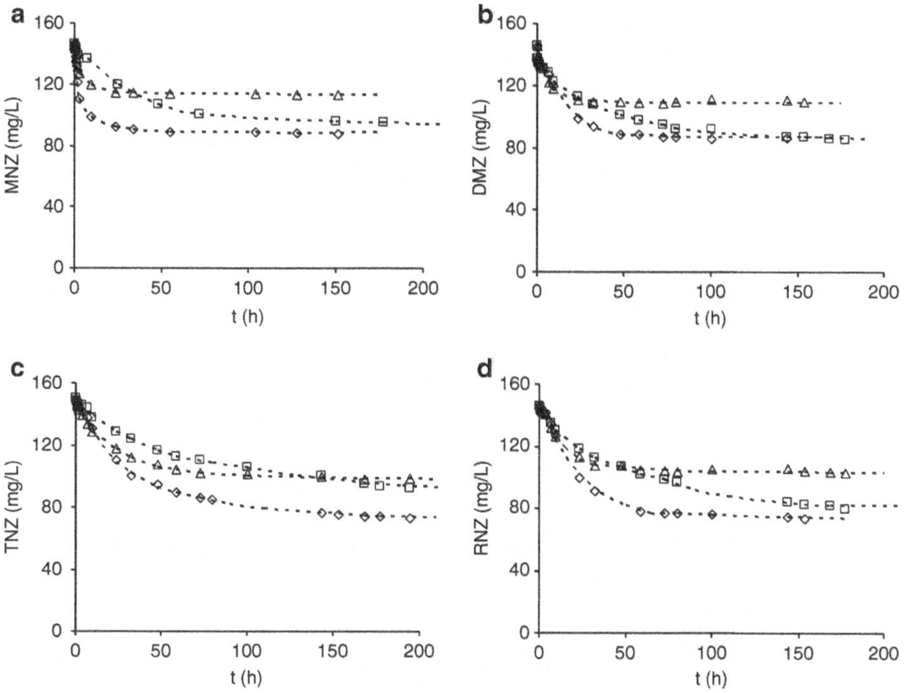

Fig. 6.14 Adsorption kinetics of nitroimidazoles on activated carbons. pH = 7, T = 298 K, [activated carbon] = 0.2 g/L. (◇) S, (Δ) M, (□) C-2. (**a**) $[DMZ]_0$ = 150 mg/L, (**b**) $[MNZ]_0$ = 150 mg/L, (**c**) $[RNZ]_0$ = 150 mg/L, (**d**) $[TNZ]_0$ = 150 mg/L. The *lines* represent the predictions of the pseudo-first-order kinetic model (Méndez-Díaz et al. 2010)

relationship was observed between the pseudo-first- or pseudo-second-order rate constants and the textural properties of the carbons, except that the rate constants k_1 or k_2 are augmented, increasing the micropore volume ($W_0(N_2)$) and mean micropore size ($L_0(N_2)$) of the activated carbon. The relationships between k_1 and $W_0(N_2)$ or $L_0(N_2)$ are depicted in Fig. 6.16a, b for the four nitroimidazoles.

The results plotted in these figures appear to indicate that the adsorption rate k_1 is largely dependent upon the volume and mean micropore size. Nevertheless, meso- and macroporosity do not appear to have a major effect on the adsorption rate in these systems. It is expected that the porous structure does not affect the rate constants of the kinetic models since in these models the diffusion through the pores (intraparticle diffusion) has been neglected, and the surface adsorption in the micropores is the controlling step (Poling et al. 2001; Leyva-Ramos et al. 2009).

The nitroimidazole adsorption rate constant k_1 was also related to surface chemical characteristics of the activated carbons. The kinetic constants (k_1) increased, reducing the oxygen percentage of the activated carbon (Fig. 6.17) and diminishing the content of carbonylic (Fig. 6.18) and phenolic groups (Fig. 6.19). However, the rate constants appeared to be independent upon the other chemical properties of carbon such as the pH_{PZC} or concentration of basic sites. Thus, the

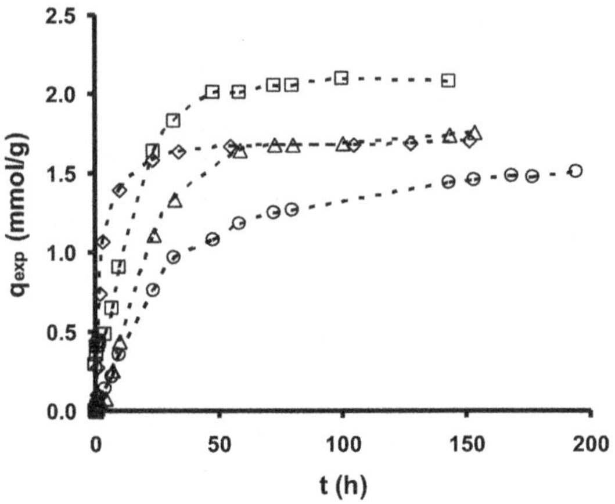

Fig. 6.15 Adsorption kinetics of DMZ, MNZ, RNZ, and TNZ on activated carbon S. pH = 7, $[nitroimidazole]_0 = 150$ mg/L, T = 298 K, [activated carbon] = 0.2 g/L. (□), DMZ; (Δ), MNZ; (◇), RNZ; (○), TNZ. (- - -) The *lines* represent the predictions with the pseudo-first-order- (DMZ, MNZ, and RNZ) and pseudo-second-order (TNZ) kinetic models (Méndez-Díaz et al. 2010)

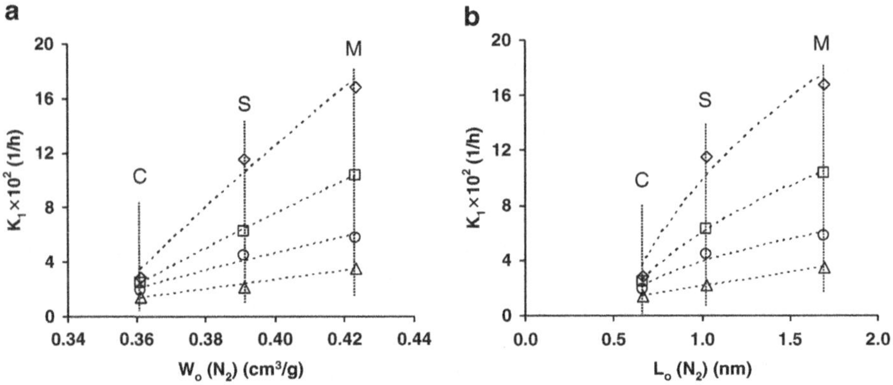

Fig. 6.16 Adsorption rate constants of (◇), MNZ; (□), DMZ; (○), RNZ; (△), TNZ as a function of (**a**) $W_0(N_2)$ and (**b**) $L_0(N_2)$. pH = 7, $[NMZ]_0 = 150$ mg/L, T = 298 K, [activated carbon] = 0.2 g/L (Méndez-Díaz et al. 2010)

adsorption rate constants increased by lessening the oxygen content and carbonylic and phenolic sites in the carbons. This behavior was common to all nitroimidazoles.

The relationships presented in Figs. 6.17, 6.18, and 6.19 indicate that the adsorption rate k_1 is directly related to the carbon hydrophobicity. The carbon hydrophobicity increased, diminishing the oxygen content, and there is a decrease in the competition between water and nitroimidazole molecules for the active adsorption sites of the carbon surface, increasing the adsorption rate of nitroimidazole. Thus, the nitroimidazole adsorption rate constant k_1 can be related

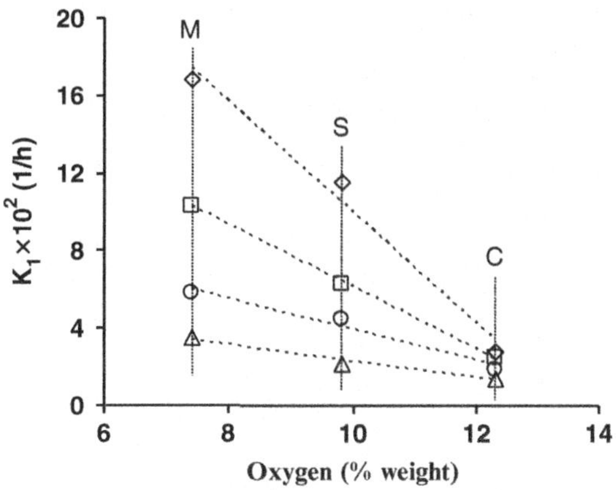

Fig. 6.17 Effect of the activated carbon oxygen content on the nitroimidazole adsorption rate constant. pH = 7, [nitroimidazole]$_0$ = 150 mg/L, T = 298 K, [activated carbon] = 0.2 g/L. (○), DMZ; (Δ), MNZ; (◇), RNZ; (□), TNZ (Méndez-Díaz et al. 2010)

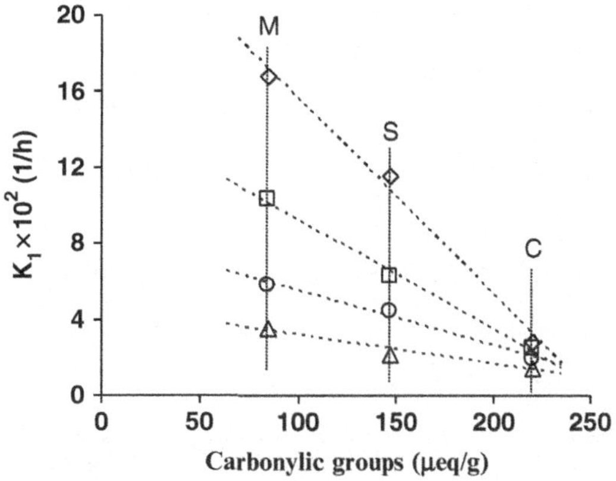

Fig. 6.18 Effect of the carbonylic group content of activated carbons on the nitroimidazole adsorption rate constant. pH = 7, [nitroimidazole]$_0$ = 150 mg/L, T 298 K, [activated carbon] = 0.2 g/L. (○), DMZ; (Δ), MNZ; (◇), RNZ; (□), TNZ (Méndez-Díaz et al. 2010)

to the relative carbon hydrophobicity as illustrated in Fig. 6.20, and the rate constant k_1 increases with the carbon hydrophobicity.

The adsorption kinetics of nitroimidazoles on activated carbons was also studied by interpreting the experimental kinetic data with the PVDM model. An initial guess of the D_{ep} was calculated using Eq. (6.20) and, assuming a tortuosity factor of $\tau_p = 3.50$, recommended for activated carbons (Leyva-Ramos and Geankoplis 1994; Ocampo-Perez et al. 2011).

The concentration decay data predicted with the diffusional model, as well as the experimental data for the adsorption of DMZ on activated carbon S, are depicted in Fig. 6.21. It can be seen that the diffusional model interpreted satisfactorily well the experimental concentration decay, and the value of τ_p was 3.74 which was very close to $\tau_p = 3.50$, recommended by Leyva-Ramos and Geankoplis (1994).

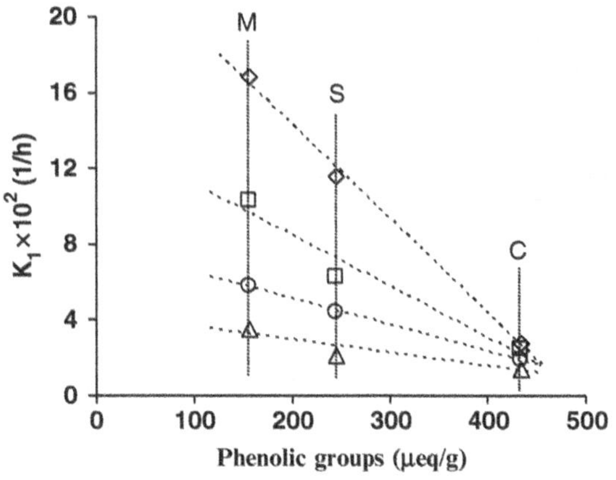

Fig. 6.19 Effect of the phenolic group content of activated carbons on the nitroimidazole adsorption rate constant. pH = 7, [nitroimidazole]$_0$ = 150 mg/L, T = 298 K, [activated carbon] = 0.2 g/L. (○), DMZ; (Δ), MNZ; (◇), RNZ; (□), TNZ (Méndez-Díaz et al. 2010)

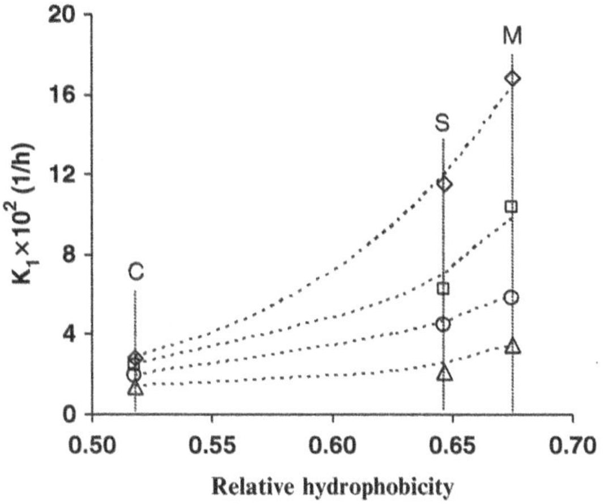

Fig. 6.20 Effect of the activated carbon hydrophobicity on the nitroimidazole adsorption rate constant. pH = 7, [nitroimidazole]$_0$ = 150 mg/L, T = 298 K, [activated carbon] = 0.2 g/L. (○), DMZ; (Δ), MNZ; (◇), RNZ; (□), TNZ (Méndez-Díaz et al. 2010)

The values of τ_p for RNZ and TNZ varied from 3.84 to 4.91, and these values are within the range of values reported for activated carbons (Leyva-Ramos and Geankoplis 1994; Ocampo-Perez et al. 2011). In general, the diffusional model interpreted well the experimental concentration decay data for the adsorption of RNZ and TNZ on the three carbons. To illustrate the satisfactory fitting obtained with the diffusional model, the predicted and experimental concentration decay data for the adsorption of RNZ on carbon S are plotted in Fig. 6.22. As illustrated in this figure, the diffusional model with τ_p = 4.33 predicted reasonably well the experimental concentration decay curve. In this case the diffusional model with τ_p = 3.50 overpredicted the experimental concentration decay data.

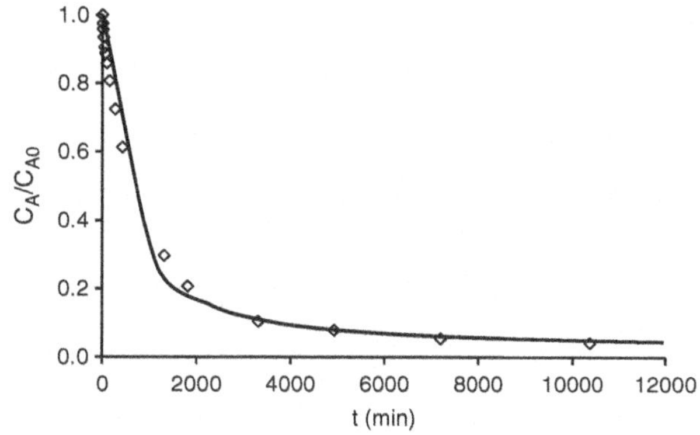

Fig. 6.21 Adsorption kinetics of DMZ on activated carbon S. pH = 7, $[DMZ]_0 = 150$ mg/L, T = 298 K, [activated carbon] = 1 g/L. (◇), experimental data; (—), PVDM model prediction (Méndez-Díaz et al. 2010)

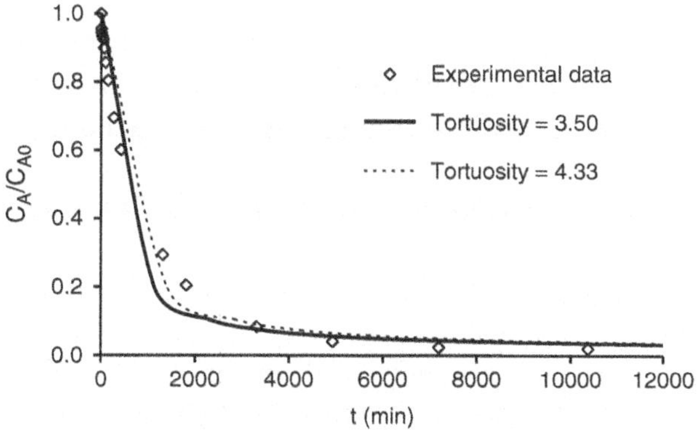

Fig. 6.22 Adsorption kinetics of RNZ on activated carbon S. pH = 7, $[RNZ]_0 = 150$ mg/L, T = 298 K, [activated carbon] = 1 g/L. (◇), experimental data; (—), PVDM model prediction for $\tau_p = 3.5$; (- -), model prediction for $\tau_p = 4.33$ (Méndez-Díaz et al. 2010)

The diffusional model predicted reasonably well the experimental concentration decay data of MNZ during adsorption on the three carbons. As an example, the concentration decay predicted with the diffusional model and the experimental concentration decay for the adsorption of MNZ on carbon S are displayed in Fig. 6.23.

However, the values of τ_p for MNZ range from 1.47 to 2.17. These values are rather smaller than those for the other nitroimidazoles. The concentration decay curve predicted with the diffusional model and $\tau_p = 3.50$ is also graphed in Fig. 6.23, and it can be noticed that the diffusional model with $\tau_p = 3.50$ underpredicted the experimental concentration decay data. This behavior clearly indicates that the tortuosity factor for the adsorption rate of MNZ on carbon S is smaller than $\tau_p = 3.50$.

The low values of τ_p for MNZ can be explained considering that surface diffusion is possibly taking place in the case of the intraparticle diffusion of MNZ. In other words, intraparticle diffusion of MNZ is due to pore volume diffusion as well as surface diffusion. The intraparticle diffusion would be faster when surface diffusion was occurring simultaneously with pore volume diffusion. If surface diffusion was not included in the model such as in this case, this would result in greater values of D_{ep} and smaller values of τ_p.

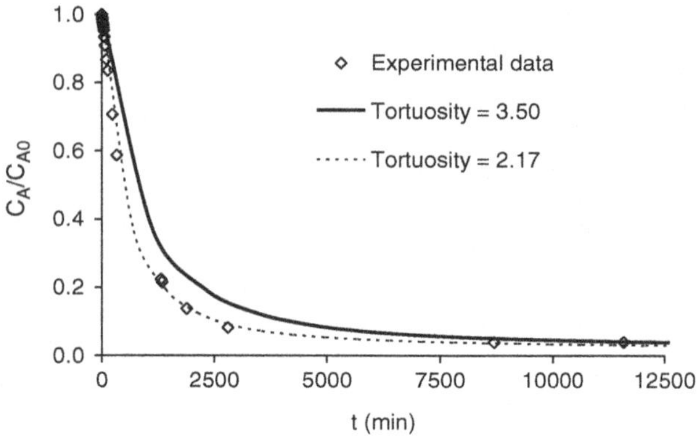

Fig. 6.23 Adsorption kinetics of MNZ on activated carbon S. pH = 7, $[MNZ]_0 = 150$ mg/L, T = 298 K, [activated carbon] = 1 g/L. (◇), experimental data; (—), PVDM model prediction for $\tau_p = 3.5$; (- -), model prediction for $\tau_p = 2.17$ (Méndez-Díaz et al. 2010)

6.4 Adsorption/Biosorption Equilibrium Isotherms of Tetracyclines and Nitroimidazoles on Sludge-Derived Materials and Activated Carbons

6.4.1 Nitroimidazole Adsorption Processes

Both the mechanisms involved in the process of nitroimidazole adsorption and the adsorption capacity of activated carbons were determined by obtaining the corresponding adsorption isotherms. Figure 6.24 depicts, as an example, the normalized adsorption isotherms of the four nitroimidazoles for activated carbon S. It shows the millimoles of nitroimidazole adsorbed per gram of carbon (q_e) versus the equilibrium concentrations normalized for the solubility of each nitroimidazole in water at 298 K (C_{eq}/S_{BET}). Isotherms show the L form of the Giles classification (Giles et al. 1974a, b; Moreno-Castilla 2004), suggesting that the aromatic rings of nitroimidazole molecules are adsorbed in parallel to the carbon surface and that there is no major competition between nitroimidazoles and water molecules for the active adsorption centers on the carbon.

Langmuir (Eq. (6.21)) and Freundlich (Eq. (6.22)) models were applied to the experimental adsorption isotherm data. As stated in Chap. 2, they are the most widely used models for describing this type of process and are represented mathematically as:

$$q_e = \frac{K q_m C_{eq}}{1 + K C_{eq}} \tag{6.21}$$

$$q_e = K_f C_{eq}^{1/n} \tag{6.22}$$

where q_m is the adsorption capacity (mg/g), C_{eq} is the concentration of adsorbate at equilibrium (mg/L), K is the Langmuir constant, K_f is the Freundlich's affinity parameter, and n is the exponential Freundlich's coefficient.

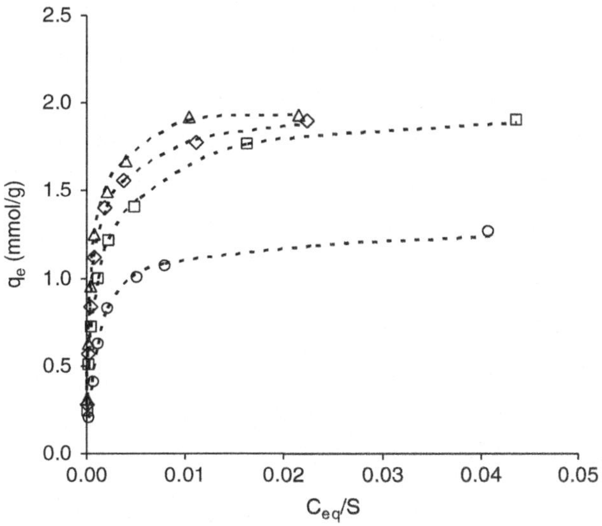

Fig. 6.24 Adsorption isotherms of nitroimidazoles on carbon S. (◇), MNZ; (△), DMZ; (○), TNZ; (□), RNZ. pH = 7, T = 298 K (Rivera-Utrilla et al. 2009)

The determination coefficients obtained were >0.99 for all systems with the Langmuir equation (Table 6.18) but ranged from 0.842 for the TNZ-carbon S system to 0.987 for DMZ-carbon M when the Freundlich equation was applied (Table 6.19). Although both models served to explain our results, the Langmuir model fitted better the experimental data.

The adsorption capacity was very elevated for all carbons and nitroimidazoles (Table 6.18), with q_m values ranging from 1.04 mmol/g for TNZ-carbon C-2 to 2.04 mmol/g for DMZ-carbon C-2. In general, relative affinity values, Kq_m, were also higher than those usually reported in aromatic compound adsorption (Radovic et al. 2001), indicating the high chemical affinity of nitroimidazoles for carbon. Interestingly, the constant K value in Langmuir's equation (related to adsorption energy) increased in the order DMZ < MNZ < RNZ < TNZ, which may be related to the solubility of these nitroimidazoles in water, with an increase in their adsorption energy as solubility decreases (Table 6.15). The corresponding values of the Freundlich exponential coefficient, $1/n$, were low, ranging from 0.214 to 0.295 (Table 6.19), which also indicates strong adsorbent-adsorbate interactions (Freundlich 1926).

The adsorption capacity of carbon for the different nitroimidazoles was expressed per unit of carbon surface area (q'_m) to enable comparisons among them (Table 6.18). Except in the case of TNZ, the adsorption capacity of carbon increased in the order M < S < C-2, which may be related to the oxygen content of carbon, which increased in the same direction. Thus, in the case of carbon C-2 and S, this oxygen is mainly forming phenolic groups, which are electronic activators of the aromatic rings of carbon graphene planes; this favors the adsorption of aromatic compounds like nitroimidazoles, which can be adsorbed by dispersion interactions of π electrons of their aromatic rings with π electrons of

Table 6.18 Parameters obtained by applying Langmuir's equation to the adsorption isotherms of nitroimidazoles on activated carbons S, M, and C-2 (Rivera-Utrilla et al. 2009)

	q_m (mmol/g)			$q'_m\ 10^3$ (mmol/m^2)			K (L/mmol)			Kq_m (L/g)		
Nitroimidazole	S	*M*	*C*-2	S	*M*	*C*-2	S	*M*	*C*-2	S	*M*	*C*-2
MNZ	1.92	1.25	1.68	1.56	0.96	1.98	52.11	9.44	23.50	100.00	11.82	39.37
DMZ	1.99	1.32	2.04	1.62	1.01	2.40	43.84	9.38	23.29	87.24	12.39	47.62
TNZ	1.37	1.56	1.04	1.12	1.20	1.23	75.79	94.24	56.45	104.17	147.06	58.82
RNZ	1.97	1.82	1.89	1.61	1.40	2.23	52.85	25.49	24.57	104.11	46.51	46.51

Table 6.19 Parameters obtained by applying Freundlich's equation to the adsorption isotherms of nitroimidazoles on activated carbons S, M, and C-2 (Rivera-Utrilla et al. 2009)

	S			M			C-2		
Nitroimidazole	1/n	K_f (L/g)	R^2	1/n	K_f (L/g)	R^2	1/n	K_f (L/g)	R^2
MNZ	0.247	403.27	0.925	0.268	171.67	0.968	0.282	248.42	0.932
DMZ	0.235	408.97	0.901	0.261	191.20	0.987	0.260	345.46	0.893
TNZ	0.295	268.47	0.842	0.214	439.44	0.915	0.280	200.26	0.885
RNZ	0.223	479.40	0.935	0.262	317.03	0.967	0.240	361.80	0.928

the carbon graphene planes (Radovic et al. 2001; Moreno-Castilla 2004). Furthermore, the presence of oxygen in the carbon favors the establishment of hydrogen bonds between nitroimidazoles and the carbon surface. These bonds may be responsible for the strong adsorbent-adsorbate interactions detected in these systems and commented above.

Comparison of the adsorption capacity of carbon for each nitroimidazole (q'_m) shows that, with the exception of carbon M, q'_m decreased in the order DMZ > RNZ > MNZ > TNZ. These results show some relationship with the size of the nitroimidazole molecules and hence with their accessibility to the carbon porosity. Thus, TNZ has the largest molecule size and was the least adsorbed, whereas DMZ has the smallest molecule size and therefore highest accessibility to the carbon surface and showed the highest adsorption. Besides the accessibility of nitroimidazoles to the carbon surface, the electronic density of their aromatic ring also increases their adsorption, since it enhances the abovementioned π-π adsorbate-adsorbent dispersion interactions. Based on the chemical composition of these nitroimidazoles and the electronic activating/deactivating power of the groups they contain, the electronic density of aromatic rings decreases in the order DMZ > RNZ ≈ MNZ > TNZ, the same order found for the increase in adsorption capacity of these nitroimidazoles. These results demonstrate that the adsorption process is mainly determined by the adsorbent-adsorbate dispersion interactions described above.

6.4.2 *Tetracyclines Adsorption Isotherms*

Table 6.20 exhibits the data obtained by applying Langmuir and Freundlich equations to the adsorption isotherms of TCs on activated carbons M and S. In all systems, they yielded determination coefficients close to unity. Although both models could be used to explain the results, in general, the Langmuir model best fits the experimental data.

The results in Table 6.20 indicate that, regardless of the carbon and TC considered, the adsorption capacity of the activated carbons was very high, with q_m values ranging from 471.1 mg/g for the TC-carbon M system to 65.1 mg/g for the CTC-S system. In general, the values of relative affinity (Kq_m) were relatively high, indicating the high chemical affinity of the TCs for the carbons. The corresponding values of Freundlich's exponential coefficient ($1/n$) were low, ranging from 0.10 to 0.52, indicating strong adsorbent-adsorbate interactions. All of these results demonstrate the high effectiveness of these commercial activated carbons for TC removal from water.

As shown in Table 6.20, the capacity of carbon M to adsorb the three TCs was higher than that of carbon S, and its relative affinity with these adsorbates was also considerably superior. For both carbons, the capacity to adsorb these TCs increased in the order CTC < OTC < TC.

Radovic et al. (2001) in their extensive review of the literature on organic compound adsorption on carbons concluded that the adsorption mechanism of these processes remains controversial. They reported that both electrostatic and dispersive adsorbent-adsorbate interactions are involved in the adsorption of aromatic compounds. Coughlin and Ezra (1968) suggested that the adsorption of aromatic compounds on activated carbon is based on the establishment of dispersion interactions between π electrons of the organic compound aromatic ring and π electrons of the activated carbon graphene planes (π-π interactions). Furthermore, Leon and Leon et al. (1992) found that basic carbons ($pH_{PZC} > 7$) with a low oxygen percentage are characterized by a high content of electron-rich sites in their graphene planes and a low concentration of surface electron-attracting oxygen groups, which could enhance the adsorption of aromatic compounds according to the mechanism proposed by Coughlin and Ezra (1968).

Given that TC molecules have neutral charge at the study pH (4–5), Figure 6.4b, electrostatic interactions between positively charged activated carbon surface and TC molecules are weak, and the adsorption mechanism must be largely governed by dispersive adsorbent-adsorbate interactions (Rivera-Utrilla and Sánchez-Polo 2002). Consequently, the greater adsorption capacity of carbon M is partly due to its lower oxygen content and hence higher hydrophobicity. The chemical composition of the TCs and the electron activating/deactivating power of their functional groups means that the electron density of the aromatic ring decreases in the order TC > OTC > CTC, the same order as found for their capacity for adsorption. These results demonstrate that the adsorption process is mainly governed by

Table 6.20 Results obtained from applying Langmuir and Freundlich equations to the TC adsorption isotherms (Rivera-Utrilla et al. 2013a)

		Langmuir			Freundlich		
TCs	Carbon	R^2	q_m (mg/g)	Kq_m (L/g)	R^2	1/n	K (mg/g) (L/mg)
TC	S	0.9998	375.4	52.0	0.9939	0.17	178
	M	0.9999	471.1	142.0	0.9851	0.10	263
OTC	S	0.9911	252.6	6.1	0.9912	0.45	19
	M	0.9991	413.2	15.0	0.9782	0.23	100
CTC	S	0.9535	65.1	4.2	0.9939	0.52	3
	M	0.9984	309.9	10.0	0.9893	0.21	86

dispersive interactions. In addition, the establishment of hydrogen bonds between the phenolic groups of TCs and the oxygenated groups of carbon would also contribute to the high adsorption of these pollutants.

A further aspect to be considered in adsorption is the accessibility of the adsorbate to the activated carbon surface, which is determined by the adsorbate molecule size and carbon pore dimensions. Thus, the higher adsorption capacity of carbon M is partly due to its larger mean micropore size and mesopore volume (Table 6.1), giving TCs greater access to its pores and presenting a larger surface fraction that is effective in adsorption in comparison to carbon S. As stated above, the adsorption capacity of carbons for these TCs shows that it decreases in the order TC > OTC > CTC. These results are in part related to the size of TC molecules (Table 6.14) and their consequent accessibility to the carbon porosity. Thus, the tetracycline with largest molecule size (CTC) shows the lowest adsorption and the one with the smallest size (TC) the highest.

Table 6.21 shows that sludge-derived adsorbents had a higher capacity to adsorb TC in comparison to the commercial activated carbons used in the present study (Table 6.20) and other adsorbent reported in the literature (Pils and Laird 2007; Wang et al. 2010; Zou et al. 2012; Chang et al. 2012; Zhou et al. 2012; Liu et al. 2012, 2013; Huang et al. 2013; Li et al. 2013; Liao et al. 2013; Lin et al. 2013), reaching a value of 672.0 mg/g for the C2 sample. Taking into account the small surface area and low C content of these materials, their high TC adsorption capacity can only be explained by the formation of complexes between TC and the metal elements present in these samples. The formation of chelate-type coordination compounds between TC and metal elements has been extensively studied in the literature (Figueroa et al. 2004; Jones et al. 2005; Başakçilardan-Kabakci et al. 2007). Thus, a characteristic chemical property of TCs is their capacity to form insoluble chelates at neutral pH with certain metal ions, such as Fe^{2+}, Al^{3+}, Ca^{2+}, and Mg^{2+} (Van der Bijl and Pitigoi-Aron 1995; Carson and Breslyn 1996; Chakrawarti 2001; Wang et al. 2008a). Therefore, the presence of these metal cations in our adsorbents (Table 6.10) would explain their high adsorption capacity.

6.4.3 Influence of Operational Variables

6.4.3.1 Influence of Solution pH

Figure 6.25 depicts, as an example, the influence of solution pH on TC adsorption on activated carbon M, showing a pH range from 2 to 7 at which the amount adsorbed remains constant, with a reduction in adsorption at higher pH values. This behavior was observed for all three TCs studied, maintaining the adsorption order TC > OTC > CTC observed for the adsorption isotherms. A similar behavior was detected for carbon S, except that adsorption was reduced from pH values of around 9.

Table 6.21 Results of applying Langmuir equation to tetracycline adsorption isotherms of adsorbents obtained from sludge and others from literature (Rivera-Utrilla et al. 2013a)

	Langmuir			
Adsorbent	R^2	q_m (mg/g)	Kq_m (L/g)	References
C2	0.9988	672.0	10.96	In this work
CR2	0.9898	593.4	7.64	In this work
CH	0.9966	553.3	6.23	In this work
CAR	0.9979	512.1	4.70	In this work
HCl-modified Zeolite	0.9998	20.4	7.29	Zou et al. (2012)
Magnetic resin (Q100)[a]	0.9292	429.7	13.42	Zhou et al. (2012)
Alkali biochar	0.9570	58.82	0.882	Liu et al. (2012)
MCM-41 impregnated with zeolite[b]	0.9811	419.3	3.04	Liu et al. (2013)
Graphene oxide functionalized magnetic particles	0.9920	39.1	17.6	Lin et al. (2013)
Anaerobic granular sludge[a]	0.9910	4.61	–	Li et al. (2013)
Bamboo charcoal	0.9211	22.7	0.454	Liao et al. (2013)
Activated carbon fiber modified by microwave	0.9818	312	13.54	Huang et al. (2013)
Illite	0.988	32	–	Chang et al. (2012)
Clays, humic substances, and clay-humic complexes[a]	–	12	–	Pils and Laird (2007)
The red soil (RS, UdicFerrosols)	–	12	–	Wang et al. (2010)

[a]T = 45 °C
[b]T = 50 °C

The behavior depicted in Fig. 6.25 is due to the progressive ionization of the surface oxygenated groups of the activated carbon with higher solution pH, which produces an increase in its negative surface charge density and a consequent reduction in adsorption. This is because repulsive electrostatic interactions are established between the carbon surface, negatively charged at pH values above its pH_{PZC}, and the TC molecules, negatively charged at pH values above their pK_{a2} values. These results indicate that, depending on the solution pH, electrostatic adsorbent-adsorbate interactions also play an important role in these adsorption processes.

Regarding nitroimidazoles, Fig. 6.26 shows, as an example, the influence of solution pH on nitroimidazole adsorption of activated carbon M. The solution pH did not significantly affect the process of adsorption of nitroimidazoles on carbon M between pH 4 and 11, but all four nitroimidazoles showed a slight decrease at pH 2. According to their pK_{a1} values (Table 6.15 and Fig. 6.5), this is because of the protonation of nitroimidazoles at pH values close to 2. As a result, repulsive electrostatic interactions are established between the carbon surface, positively

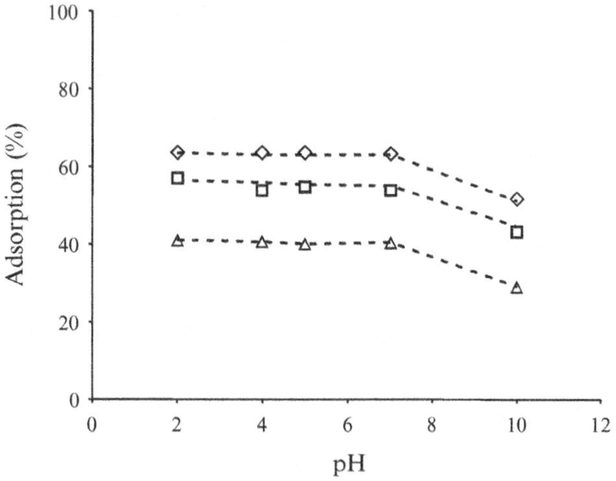

Fig. 6.25 Amount of TCs adsorbed on M carbon, at equilibrium-8 days, as a function of solution pH. (◇), TC; (□), OTC; (△), CTC. T = 298 K. [M] = 1 g/L. $[TCs]_0$ = 700 mg/L (Rivera-Utrilla et al. 2013a)

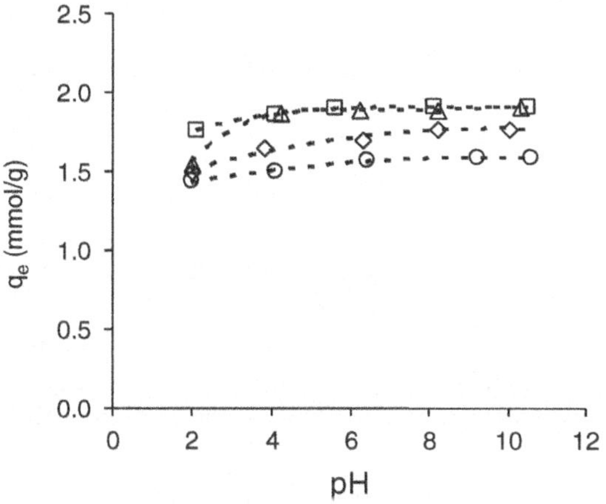

Fig. 6.26 Influence of pH in nitroimidazole adsorption process on carbon M. (◇), MNZ; (△), DMZ; (○), TNZ; (□), RNZ. T = 298 K. [M] = 1 g/L. $[\text{nitroimidazole}]_o$ = 600 mg/L (Rivera-Utrilla et al. 2009)

charged at pH 2 ($pH_{solution} < pH_{PZC}$) and the nitroimidazoles, positively charged at pH values close to 2. Similar results were observed for the other carbon samples.

Results depicted in Fig. 6.26 verify that electrostatic interactions do not play a major role in the adsorption of nitroimidazoles on activated carbon surface between pH 4 and 11, since nitroimidazole molecules are neutral under these conditions, and surface oxygenated groups of the activated carbon progressively ionize with higher solution pH. Therefore, the increase in negative surface charge does not produce a reduction in the adsorption of the contaminants. These data confirm that non-electrostatic interactions are largely responsible for the adsorption of these compounds on activated carbon, as reported in the above section.

6.4.3.2 Influence of Solution Ionic Strength

Figure 6.27 depicts, as an example, the results for TC adsorption on carbon M in the presence of increasing NaCl concentrations. Ionic strength can affect the adsorption process on activated carbons. According to the results obtained by López-Ramón et al. (2003), the presence of electrolytes in solution can modify the strength of adsorbate-adsorbent electrostatic interactions. These interactions, either attractive or repulsive, can be increased or reduced by varying the solution ion strength.

Figure 6.27 shows that, with an increase in NaCl concentration, in the medium from 0.0 to 0.1 molar, the adsorption percentage is reduced by around 20%, which remains constant at higher NaCl concentrations. These results can be explained by considering that electrostatic interactions between the carbon surface and adsorbate are attractive at the study pH (7), because the carbon surface is positively charged but the TC molecules are partially ionized (Figure 6.4b). Therefore, adsorption capacity is decreased with greater ion strength due to a possible screening effect of the carbon surface charge produced by the added NaCl, i.e., by placing Cl^- and Na^+ ions between carbon surface and TC molecules, reducing attractive electrostatic interactions between carbon surface and the fraction of ionized TC. A similar behavior was found for the rest of tetracyclines (Rivera-Utrilla et al. 2013a). In the case of nitroimidazole adsorption, the effect of solution ionic strength was less marked (Rivera-Utrilla et al. 2009).

6.4.3.3 Influence of the Presence of Microorganisms

The presence of microorganisms in natural water and wastewater may exert a considerable influence on the effectiveness of activated carbon because microorganisms can be adsorbed on the carbon during water treatment, giving rise to the

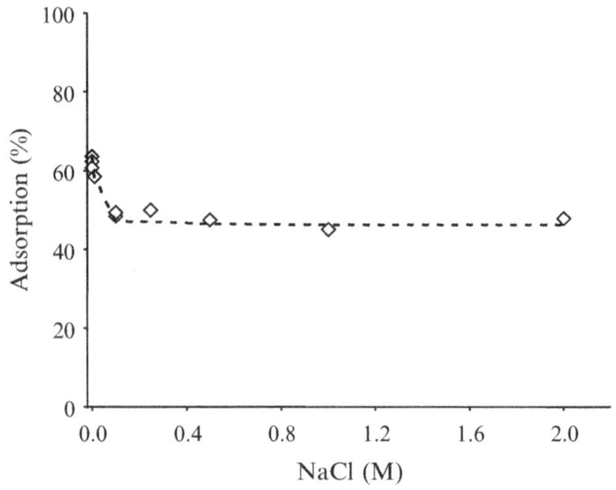

Fig. 6.27 Adsorption of TC on carbon M as a function of solution ionic strength. pH = 7, T = 298 K. [M] = 1 g/L. $[TC]_0$ = 700 mg/L (Rivera-Utrilla et al. 2013a)

formation of bacteria colonies and the consequent modification of the chemical and textural characteristics of the carbon surface (Rivera-Utrilla et al. 2001, 2003; Moreno-Castilla et al. 2003a; Bautista-Toledo et al. 2008).

This aspect was analyzed by obtaining TC and CTC adsorption isotherms on carbon M in the presence of bacteria from secondary effluents of a wastewater treatment plant. Figure 6.28 depicts the adsorption/biosorption isotherms of TC and CTC on carbon M in the presence and the absence of these bacteria. Table 6.22 shows the results of applying Langmuir's equation to these isotherms.

The effect of the adsorption of bacteria on the chemical and textural properties of activated carbon was previously reported by our research group (Rivera-Utrilla et al. 2001, 2003), finding (i) a decrease in the surface area due to pore blocking and (ii) a reduction in the pH_{PZC}, which increases the negative surface charge density of the activated carbon. Moreover, given that the external walls of bacteria are formed by phospholipids (Nikaido and Vaara 1985), their adsorption on activated carbon (Fig. 6.29) increases the hydrophobicity of the carbon surface.

The results obtained (Table 6.22) show that the presence of microorganisms during the adsorption of both TCs decreases the adsorption capacity of the activated carbon and reduces the adsorbate-adsorbent relative affinity values (Kq_m) by around 80% and 30% for TC and CTC, respectively. This behavior may be due to two factors: (i) an increase in electrostatic repulsions between the ionized TC fraction and the activated carbon surface, which are both negatively charged at the study pH, and (ii) the appearance of exopolymers released by bacteria outside the cell in the presence of the TC (Fig. 6.29c). These exopolymers are mainly formed by lipids and carbohydrates (Beech et al. 1999; Sheng et al. 2005) which form a biofilm (see Fig. 6.29) that reduces the activated carbon surface area accessible to the TC molecules and hence the amount adsorbed.

In the case of nitroimidazoles, Fig. 6.30 depicts the adsorption/biosorption isotherms of DMZ and RNZ on carbon S in the presence and absence of microorganisms. Application of Langmuir's equation to these isotherms (Table 6.23) showed, unlike what was found in the case of TCs, that the presence of microorganisms during the adsorption of these nitroimidazoles increases the adsorption capacity of carbon S by 22% and decreased the adsorbate-adsorbent relative affinity values (Kq_m) by around 60%. These results suggest that the presence of bacteria produces a change in the main interactions responsible for the adsorption and may therefore affect the adsorption mechanism. Nitroimidazole biodegradation kinetics by the bacteria under study were investigated before obtaining the biosorption isotherms, observing no biodegradation of these compounds under the experimental conditions used. Thus, results shown in Fig. 6.30 and Table 6.23 may be explained by the increase in the hydrophobicity of the activated carbon surface, which is known to considerably favor the adsorption process. All these results indicate that the effect of the presence of microorganisms on the adsorption process depends on the adsorbate-adsorbent system.

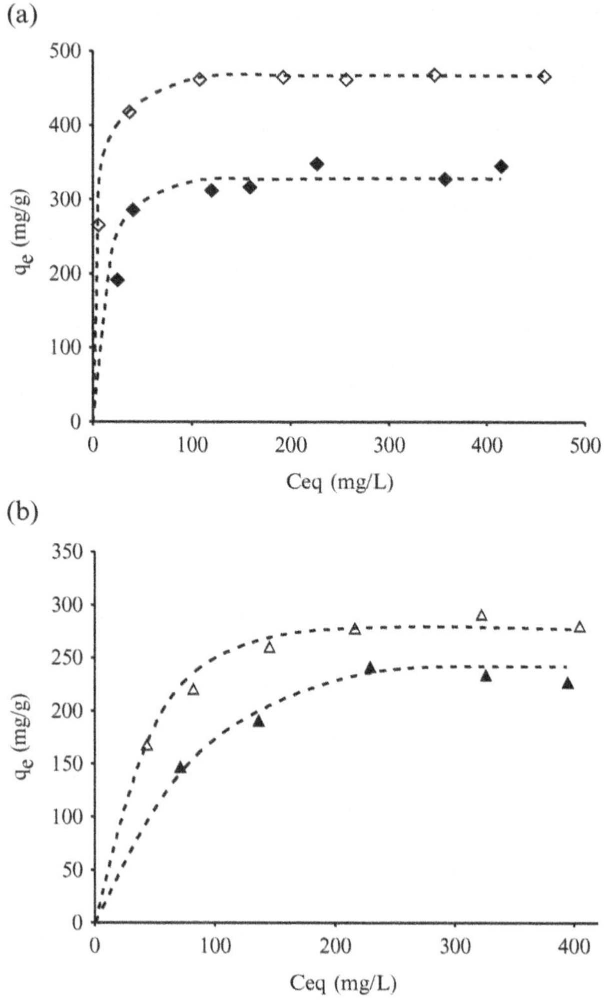

Fig. 6.28 Adsorption isotherms of TC (**a**) and CTC (**b**) on carbon M in the presence (♦, ▲) and absence (◇, △) of bacteria. pH = 7, T = 298 K. [M] = 1 g/L (Rivera-Utrilla et al. 2013a)

Table 6.22 Results of applying Langmuir's equation on TC and CTC adsorption isotherms of M carbon in the presence and absence of microorganisms (Rivera-Utrilla et al. 2013a)

	With bacteria		Without bacteria	
Tetracycline	q_m (mg/g)	Kq_m (L/g)	q_m (mg/g)	Kq_m (L/g)
TC	353.33	24.36	471.18	142.17
OTC	261.42	6.15	309.99	10.22

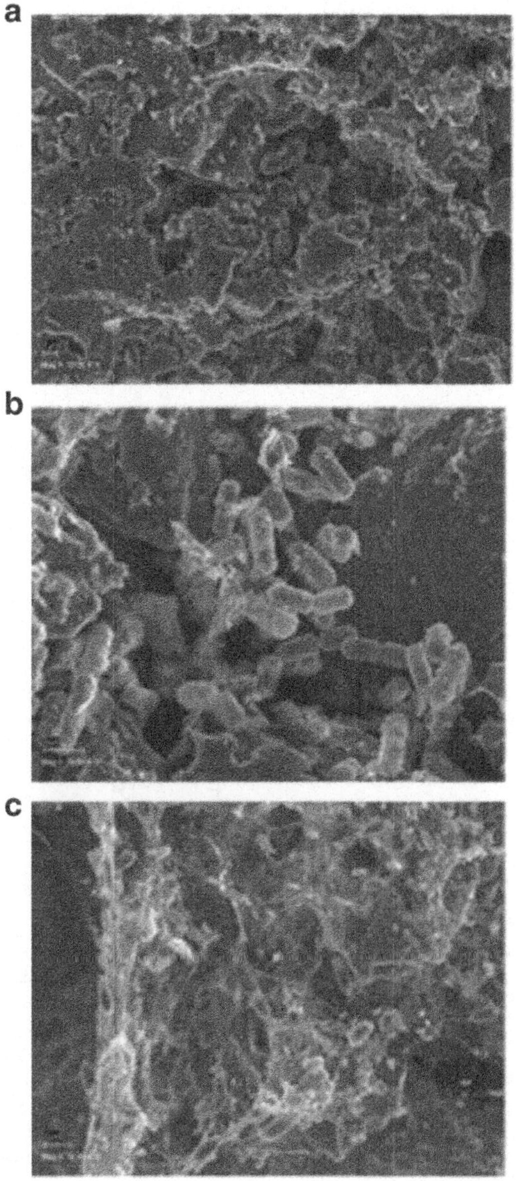

Fig. 6.29 SEM images of carbon M before and after TC biosorption. (**a**) Carbon M, (**b**) carbon M + bacteria, (**c**) carbon M + bacteria + TC (Rivera-Utrilla et al. 2013a)

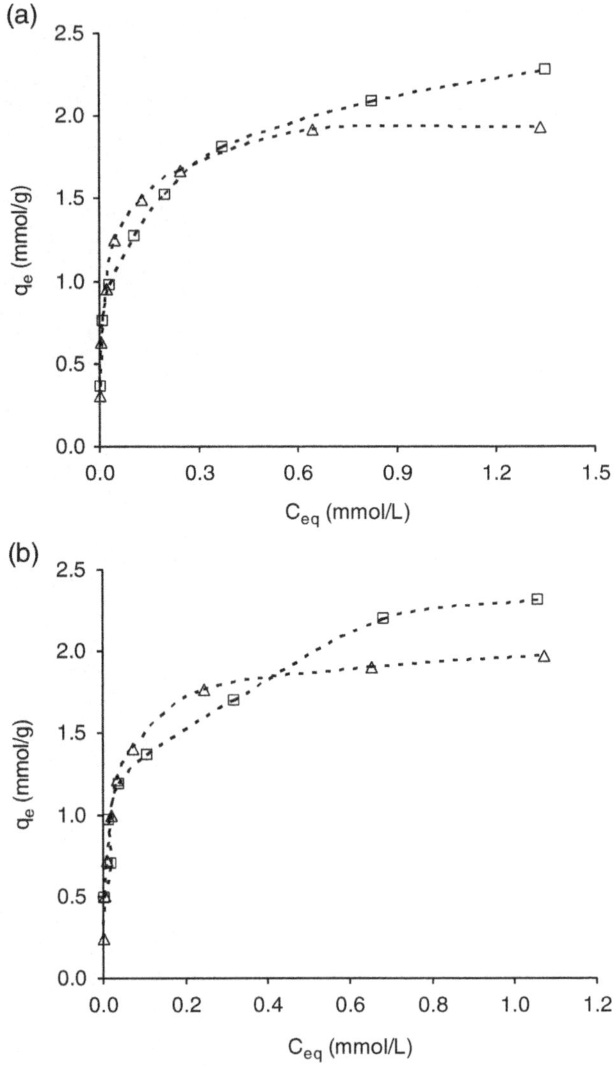

Fig. 6.30 Adsorption isotherms of DMZ (**a**) and RNZ (**b**) in the presence (□) and absence (△) of microorganisms on carbon S. pH = 7, T = 298 K (Rivera-Utrilla et al. 2009)

Table 6.23 Results obtained by applying Langmuir's equation to the adsorption isotherms of DMZ and RNZ on carbon S in the presence and absence of microorganisms (Rivera-Utrilla et al. 2009)

DMZ				RNZ			
With bacteria		Without bacteria		With bacteria		Without bacteria	
q_m (mmol/g)	Kq_m (L/g)	q_m (mmol/g)	Kq_m (L/g)	q_m (mmol/g)	Kq_m (L/g)	q_m (mmol/g)	Kq_m (L/g)
2.44	31.06	1.99	87.24	2.42	44.25	1.97	104.11

6.5 Adsorption of Tetracyclines and Nitroimidazoles on Sludge-Derived Materials and Activated Carbons in Dynamic Regime. Determination of the Breakthrough Curves and Characteristics of the Adsorbent Columns

TC and TNZ adsorption was also studied in dynamic regime, constantly passing a TC (20 mg/L) or TNZ (100 mg/L) solution through adsorbent columns (2 g, height 7 cm, diameter 1 cm, packing density 0.36 g/cm^3) at a flow of 1.5 mL/min and obtaining both the bed breakthrough curves up to 0.9 breakthrough and the bed characteristics. Fig. 6.31 shows the column breakthrough curves for TC adsorption on CR2, CH, CAR, and M samples. The adsorbent sample obtained from sludge in the presence of a phenolic resin (CR2) allows a larger volume of TC-contaminated water to be treated in comparison to the samples obtained in the presence of humic acid (CH) or clayey soil (CAR). Nevertheless, the water volume treated by the commercial activated carbon M sample is around threefold higher than the volume treated by sample CR2.

Based on the column breakthrough curves, a previously reported method (Zogorski and Faust 1977; Ferro-García et al. 1990) was used to determine the column characteristics listed in Table 6.24, showing that the TC amount adsorbed at the breakthrough point ($q_{0.02}$) was very low, especially in sludge-derived samples. Regardless of the adsorbent sample considered, the amount adsorbed at column breakthrough point ($q_{0.02}$) was lower than the adsorption capacity observed in static regime (Tables 6.20 and 6.21). These results indicate that adsorption is much less effective in dynamic than static regime due to problems of TC diffusion into the adsorbent pores and the shorter contact time between adsorbate and adsorbent.

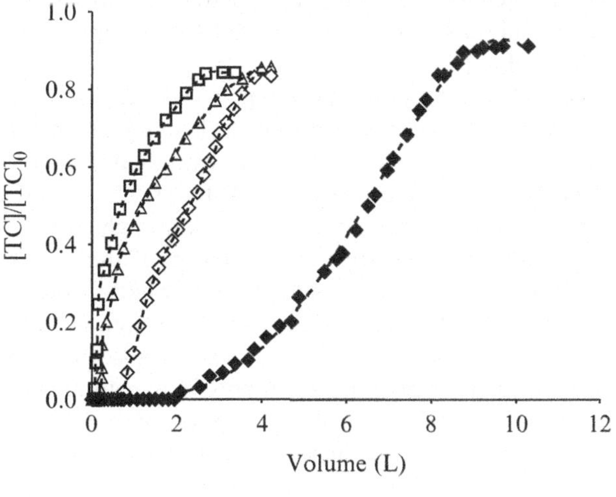

Fig. 6.31 Breakthrough curves for TC adsorption on sludge-derived adsorbents and activated carbon M. (♦), M; (◇), CR2; (Δ), CAR; (□), CH. T = 25 °C, pH = 7, $[TC]_0$ = 20 mg/L (Rivera-Utrilla et al. 2013a)

Table 6.24 Characteristics of the TC adsorption columns (Rivera-Utrilla et al. 2013a)

Adsorbent	$q_{0.02}$ (mg/g)	$q_{0.80}$ (mg/g)	$V_{0.02}$ (L)	Φ	H_{MTZ} (cm)	Gu (%)
CH	0.7	21	0.07	0.32	18.59	19.93
CR2	7.5	28	0.75	0.54	8.01	39.24
CAR	2.0	29	0.21	0.43	12.80	20.55
M	21.0	60	2.10	0.65	6.53	44.97

$X_{0.02}$ is the amount of TC adsorbed at column breakthrough point. $X_{0.80}$ is the amount adsorbed on the column when TC concentration in the effluent is 80% of the initial concentration. $V_{0.02}$ is the volume treated at column breakthrough point. Φ is the fractional capacity of the mass transference zone. H_{MTZ} is the height of the mass transference zone. Gu is the grade of utility of the column

Table 6.25 Characteristics of carbon M columns in TC adsorption as a function of the type of water used (Rivera-Utrilla et al. 2013a)

Water	$q_{0.02}$ (mg/g)	$q_{0.80}$ (mg/g)	$V_{0.02}$ (L)	Φ	H_{MTZ} (cm)	Gu (%)
Ultrapure	21	60	2.10	0.65	6.53	44.97
Surface	5	49	0.50	0.45	10.87	22.03
Groundwater	15	51	1.50	0.50	8.26	37.46
Wastewater	2	26	0.20	0.41	11.91	19.43

The amount of TC adsorbed at the column breakthrough point was higher for the adsorbent bound with phenolic resin than for the other samples obtained with different binding agents. Moreover, the height of the mass transference zone of this CR2 column was smaller and its utility grade (Gu) higher, indicating that it is the most effective sludge sample column to remove TC from water. Nevertheless, these sludge-derived materials are not appropriate for column adsorption processes, mainly because of their low compaction capacity. Hence, according to these results, these materials can mainly be used for adsorption in static regime (conventional batch technique).

TC adsorption in dynamic regime was also studied using columns of activated carbon M and water with different chemical compositions, in an attempt to reproduce as far as possible the real hypothetic situation by using surface water, groundwater, and wastewater. Table 6.25 shows the column characteristics obtained from the corresponding breakthrough curves. We highlight the differences in column characteristics as a function of the type of water used. As shown in Table 6.25, $V_{0.02}$, $q_{0.02}$, and the Gu values were lower in surface water and, especially, wastewater than in ultrapure water, and the height of the mass transference zone was greater. This is because the adsorbent porosity is partially blocked by adsorption of the organic matter present in natural waters, reducing the surface area available for TC molecule adsorption and markedly decreasing the treatment effectiveness. This reduction in effectiveness is accentuated by competition between the organic matter and the TC molecules for adsorbent surface adsorption sites, in agreement with previous reports (Carter et al. 1992; Kilduff and Wigton 1999).

Figure 6.32 depicts the breakthrough curves for TNZ adsorption on columns of activated carbon M in different types of water. In all types of water, the amount adsorbed at the breakthrough point of the column ($q_{0.02}$) was lower than the amount adsorbed in static regime. As mentioned above, this less effective adsorption in dynamic versus static regime can be attributed to problems of TNZ diffusion to the interior of carbon pores. The differences in the values of the column characteristics as a function of the chemical composition of the different waters were analyzed. Unlike the results obtained for tetracycline, $V_{0.02}$ and $q_{0.02}$ values and the degree of utility of the column (Gu) were higher in surface water and especially in groundwater than in ultrapure water (Table 6.26), and the height of the mass transference zone was lower in the former waters, indicating an increase in the effectiveness of the treatment. Two effects may explain these findings: (i) lower solubility of nitroimidazoles in surface and groundwater as a consequence of their higher alkalinity and salinity, since nonpolar organic compounds are known to markedly decrease their solubility in the presence of salts (West and Harwell 1992; Flaming et al. 2003), explaining the increase in $V_{0.02}$, $q_{0.02}$, and Gu, and (ii) the presence of Ca^{2+} ions may modify the adsorption of organic molecules on activated carbon,

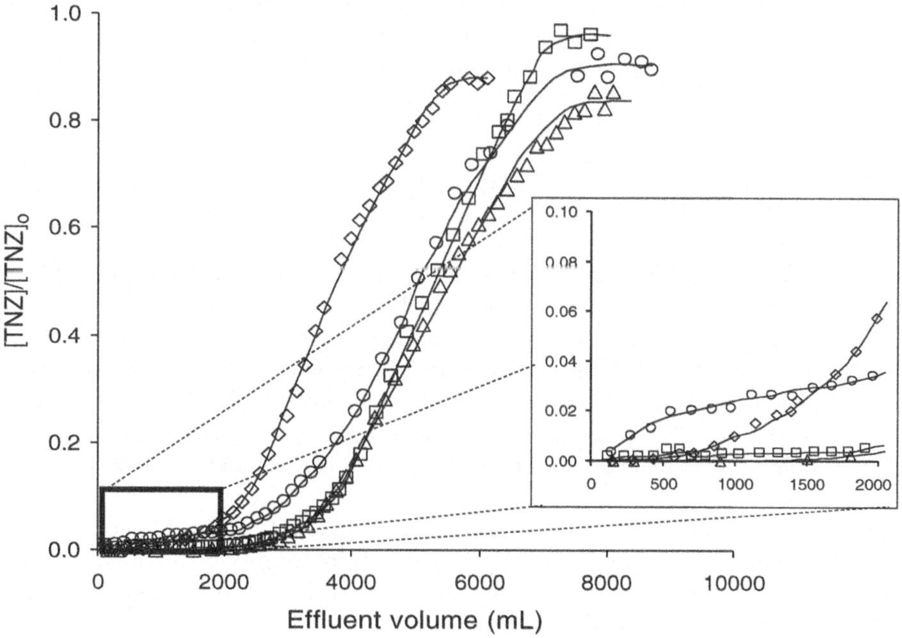

Fig. 6.32 Breakthrough curves of TNZ on carbon M as a function of the type of water used. (◇), ultrapure water; (□), surface; (△), groundwater; (○), wastewater. T = 298 K. [nitroimidazole]$_o$ = 100 mg/L (Rivera-Utrilla et al. 2009)

Table 6.26 Characteristics of carbon M columns in the adsorption of TNZ as a function of the type of water used (Rivera-Utrilla et al. 2009)

Water	$q_{0.02}$ (mmol/g)	$V_{0.02}$ (mL)	Φ	H_{MTZ} (cm)	Gu (%)
Ultrapure	0.35	1377	0.636	6.044	41.14
Surface	0.64	2693	0.660	4.341	56.13
Groundwater	0.73	2870	0.591	4.697	57.50
Waste	0.17	700	0.694	7.670	15.77

since this ion can react as a co-adsorbate by creating a bridge between the structure of activated carbon and adsorbed molecules (Lafrance and Mazet 1989; Cannon et al. 1994).

Markedly lower $V_{0.02}$, $q_{0.02}$, and Gu values and a higher H_{MTZ} value were obtained for TNZ adsorption in wastewaters, because adsorption of dissolved organic matter on carbon surface would reduce the surface area available for the adsorption of TNZ molecules. The effectiveness of treatment would be reduced by the competition between dissolved organic matter and TNZ molecules for active sites on the activated carbon, as previously reported (Carter et al. 1992; Kilduff and Wigton 1999). Similar values for the fractional capacity of the mass transference zone (Φ) were observed in all water samples studied (Table 6.26).

The results presented in Table 6.26 and Fig. 6.32 are very interesting from an application point of view. Thus, according to the results obtained, nitroimidazoles could be efficiently removed from surface and groundwater by adsorption on activated carbon; however, due to the high concentration of organic carbon dissolved in wastewater, activated carbon adsorption would not be the best technological alternative to remove nitroimidazoles from wastewater.

6.6 Conclusions

Sludge chemical activation considerably increases the adsorption capacity of sludge for tetracyclines. The addition of a binding agent (humic acid, phenolic resin, or clayey soil) before chemical activation of the material slightly reduces this adsorption capacity but improves its granulometry and allows the production of large granules, enhancing their technological applicability and, therefore, their commercialization. The capacity of all sludge adsorbents to adsorb tetracyclines is much higher than that of a commercial activated carbon (Merck) widely used in water treatment.

Petroleum coke was chemically activated with KOH. The activation process considerably develops the micro-, meso-, and macroporosity of the raw material. The surface chemical nature of the original coke was also modified by the activation process, increasing its surface basicity.

Experimental data for TC concentration decay curves on the treatment sludge-derived adsorbents are adequately interpreted by the kinetic and diffusion models applied. The pseudo first-order kinetic model provides the best interpretation of TC adsorption kinetics on all adsorbents under study, and their rate constants vary in a linear manner with their macro- and mesopore volumes and decrease with the carboxylic group content. The intraparticle diffusion model demonstrated that the TC adsorption rate on all adsorbents is controlled by intraparticle diffusion. Pore volume diffusion represents more than 80% of the total intraparticle diffusion, indicating that surface diffusion does not play a major role in TC diffusion on these adsorbents.

The concentration decay data of the TCs during the adsorption was interpreted quite well using the PVSDM model, and it was demonstrated that the pore volume diffusion and surface diffusion are important in the adsorption rate of the TC_S. The contribution of surface diffusion is directly related to the adsorption capacity of the activated carbons. In general, the higher the adsorption capacity, the greater is the contribution of surface diffusion.

The adsorption rates of the nitroimidazoles DMZ, MNZ, RNZ, or TNZ on activated carbons were better interpreted by a pseudo-first-order model or a pseudo-second-order model depending on the adsorbate-adsorbent system studied. The adsorption rate of the nitroimidazoles decreases in the order MNZ > DMZ > RNZ > TNZ. Therefore, in the case of MNZ, molecular size does not appear to be a determining factor in the adsorption rate. The adsorption rate of nitroimidazoles is related to the decrease in oxygen percentage of activated carbons and, therefore, their increase in hydrophobicity. Thus, hydrophobic interactions appear to favor the kinetics of the adsorption process. A diffusional model was applied that combines external mass transfer and intraparticle diffusion, achieving an adequate fit to the experimental data in the majority of the systems studied.

The TC adsorption capacity of the sludge-derived materials is very high (512–672 mg/g) and greater than that of the commercial activated carbons studied, attributable to the strong tendency of TCs to form complex ions with some of the metal ions in these materials.

The commercial activated carbons studied have a high TC adsorption capacity (65–471 mg/g). At the study pH (pH 4–5), the capacity of carbons to adsorb TCs is directly related to the density of delocalized π electrons in both the graphene layers of the carbon and the TC aromatic ring.

The solution pH and the presence of electrolytes had a major influence on TC adsorption on the commercial activated carbons, indicating that electrostatic adsorbent-adsorbate interactions also play an important role in TC adsorption at pH values that produce TC deprotonation. However, in the case of nitroimidazoles, the pH of the medium and the concentration of the electrolyte present do not have a major effect on the adsorption of these compounds on activated carbon, indicating that adsorbent-adsorbate electrostatic interactions do not play an important role in these adsorption processes.

The presence of bacteria in solution reduced TC adsorption/biosorption on the commercial activated carbons, weakening interactions between the adsorbate and the biofilm formed on the carbon surface. These results are explained by the formation of exopolymers released by bacteria in the presence of

TC. Nitroimidazoles are not degraded by the microorganisms used in the biological treatment stage of a wastewater treatment plant. However, the presence of these microorganisms during the adsorption of these compounds increases their adsorption/biosorption on the activated carbon, although interactions between adsorbate and carbon surface are weakened.

TC adsorption on both sludge-derived materials and activated carbons was markedly lower in dynamic versus static regime, attributable to problems of TC diffusion into the adsorbent pores and the shorter contact time between adsorbent and adsorbate in the case of dynamic regime. The extend of TC adsorption was markedly lower in natural waters (surface water, groundwater, and wastewater) than in ultrapure water, which may be explained by the reduced adsorbent surface area available for TC adsorption due to the adsorption of dissolved organic matter in these type of water. Nevertheless, in the case of nitroimidazoles, results obtained in dynamic regime show that the adsorption capacity of the activated carbon was markedly higher in surface and groundwater than in ultrapure water and urban wastewaters.

References

Adler NE, Koschorreck J, Rechenberg B (2008) Environmental impact assessment and control of pharmaceuticals: the role of environmental agencies. Water Sci Technol 57:91–97

Ahmed MJ, Theydan SK (2013) Microporous activated carbon from Siris seed pods by microwave-induced KOH activation for metronidazole adsorption. J Anal Appl Pyrol 99:101–109

Al-Ahmad A, Daschner FD, Kümmerer K (1999) Biodegradability of cefotiam, ciprofloxacin, meropenem, penicillin G, and sulfamethoxazole and inhibition of waste water bacteria. Arch Environ Con Tox 37:158–163

Ayscough NJ, Fawell J, Franklin G (2000) Review of human pharmaceuticals in the environment. Environment Agency, Bristol

Başakçilardan-Kabakci S, Thompson A, Cartmell E, Le Corre K (2007) Adsorption and precipitation of tetracycline with struvite. Water Environ Res 79:2551–2556

Batt AL, Kim S, Aga DS (2007) Comparison of the occurrence of antibiotics in four full-scale wastewater treatment plants with varying designs and operations. Chemosphere 68:428–435

Bautista-Toledo MI, Méndez-Díaz JD, Sánchez-Polo M, Rivera-Utrilla J, Ferro-García MA (2008) Adsorption of sodium dodecylbenzenesulfonate on activated carbons: effects of solution chemistry and presence of bacteria. J Colloid Interf Sci 317:11–17

Beech I, Hanjagsit L, Kalaji M, Neal AL, Zinkevich V (1999) Chemical and structural characterization of exopolymers produced by Pseudomonas sp NCIMB 2021 in continuous culture. Microbiology 145:1491–1497

Beita-Sandí W, Ersan MS, Uzun H, Karanfil T (2016) Removal of N-nitrosodimethylamine precursors with powdered activated carbon adsorption. Water Res 88:711–718

Bendesky A, Menéndez D, Ostrosky-Wegman P (2002) Is metronidazole carcinogenic? Mutat Res-Rev Mutat 511:133–144

Benk A (2010) Utilisation of the binders prepared from coal tar pitch and phenolic resins for the production metallurgical quality briquettes from coke breeze and the study of their high temperature carbonization behaviour. Fuel Process Technol 91:1152–1161

Blanchard G, Maunaye M, Martin G (1984) Removal of heavy metals from waters by means of natural zeolites. Water Res 18:1501–1507

Calamari D (2002) Assessment of persistent and bioaccumulating chemicals in the aquatic environment. Toxicology 181:183–186

Calisto V, Ferreira CI, Oliveira JA, Otero M, Esteves VI (2015) Adsorptive removal of pharmaceuticals from water by commercial and waste-based carbons. J Environ Manag 152:83–90

Campbell HW (2000) Sludge management–future issues and trends. Water Sci Technol 41:1–8

Cannon FS, Snoeyink VL, Lee RG, Dagois G (1994) Reaction mechanism of calcium-catalyzed thermal regeneration of spent granular activated carbón. Carbon 32:1285–1301

Carballa M, Omil F, Lema JM, Llompart M, Garcia-Jares C, Rodriguez I, Gómez M, Ternes TM (2004) Behavior of pharmaceuticals, cosmetics and hormones in a sewage treatment plant. Water Res 38:2918–2926

Carrales-Alvarado DH, Ocampo-Pérez R, Leyva-Ramos R, Rivera-Utrilla J (2014) Removal of the antibiotic metronidazole by adsorption on various carbon materials from aqueous phase. J Colloid Interf Sci 436:276–285

Carson MC, Breslyn W (1996) Simultaneous determination of multiple tetracycline residues in milk by metal chelate affinity chromatography: collaborative study. J AOAC Int 79:29–42

Carter MC, Weber WJ Jr, Olmstead KP (1992) Effects of background dissolved organic matter on TCE adsorption by GAC. J Am Water Works Ass 84:81–91

Cavdar AD, Kalaycioglu H, Hiziroglu S (2008) Some of the properties of oriented strandboard manufactured using kraft lignin phenolic resin. J Mater Process Tech 202:559–563

Chakrawarti PB (2001) Chelation and antibiotic activity. J Ind Chem Soc 78:273–279

Chang P, Li Z, Jean J, Jiang W, Wang C, Lin K (2012) Adsorption of tetracycline on 2:1 layered non-swelling clay mineral illite. Appl Clay Sci 67–68:158–163

Choi KJ, Kim SG, Kim S (2008) Removal of tetracycline and sulfonamide classes of antibiotic compound by powdered activated carbon. Environ Technol 29:333–342

Cooper ER, Siewicki TC, Phillips K (2008) Preliminary risk assessment database and risk ranking of pharmaceuticals in the environment. Sci Total Environ 398:26–33

Correa M, Laza JM, Vilas JL, Bilbao E, Rodríguez M, León LM (2010) Reutilization of thermostable polyester wastes by means of agglomeration with phenolic resins. Waste Manag 30:2305–2311

Coughlin RW, Ezra FS (1968) Role of surface acidity in the adsorption of organic pollutants on the surface of carbon. Environ Sci Technol 2:291–297

Dias JM, Alvim-Ferraz MCM, Almeida MF, Rivera-Utrilla J, Sánchez-Polo M (2007) Waste materials for activated carbon preparation and its use in aqueous-phase treatment: a review. J Environ Manag 85:833–846

Ferro-García MA, Carrasco-Marín F, Rivera-Utrilla J, Utrera-Hidalgo E, Moreno-Castilla C (1990) The use of activated carbon columns for the removal of ortho-phosphate ions from aqueous solutions. Carbon 28:91–95

Figueroa RA, Leonard A, Mackay AA (2004) Modeling tetracycline antibiotic sorption to clays. Environ Sci Technol 38:476–483

Flaming JE, Knox RC, Sabatini DA, Kibbey TC (2003) Surfactant effects on residual water and oil saturations in porous media. Vadose Zone J 2:168–176

Freundlich H (1926) Colloid and capillary chemistry. Metheun, London

Ganguly SK, Goswami AN (1996) Surface diffusion kinetics in the adsorption of acetic acid on activated carbon. Separ Sci Technol 31:1267–1278

Gao Y, Li Y, Zhang L, Huang H, Hu J, Shah SM, Su X (2012a) Adsorption and removal of tetracycline antibiotics from aqueous solution by graphene oxide. J Colloid Interf Sci 368:540–546

Gao P, Mao D, Luo Y, Wang L, Xu B, Xu L (2012b) Occurrence of sulfonamide and tetracycline-resistant bacteria and resistance genes in aquaculture environment. Water Res 46:2355–2364

Garrido J, Linares-Solano A, Martin-Martinez JM, Molina-Sabio M, Rodriguez-Reinoso F, Torregrosa R (1987) Use of nitrogen vs. carbon dioxide in the characterization of activated carbons. Langmuir 3:76–81

Giles CH, Smith D, Huitson A (1974a) A general treatment and classification of the solute adsorption isotherm I Theoretical. J Colloid Interf Sci 47:755–765

Giles CH, D'Silva AP, Easton IA (1974b) A general treatment and classification of the solute adsorption isotherm part II Experimental interpretation. J Colloid Interf Sci 47:766–778

Gómez-Pacheco CV, Rivera-Utrilla J, Sánchez-Polo M, López-Peñalver JJ (2012) Optimization of the preparation process of biological sludge adsorbents for application in water treatment. J Hazard Mater 217:76–84

Gu C, Karthikeyan KG (2008) Sorption of the antibiotic tetracycline to humic-mineral complexes. J Environ Qual 37:704–711

Halling-Sørensen B, Nielsen SN, Lanzky PF, Ingerslev F, Lützhøft HH, Jørgensen SE (1998) Occurrence, fate and effects of pharmaceutical substances in the environment-A review. Chemosphere 36:357–393

Ho YS, Mckay G (1998) Kinetic models for the sorption of dye from aqueous solution by wood. Process Saf Environ 76:183–191

Huang L, Shi C, Zhang B, Niu S, Gao B (2013) Characterization of activated carbon fiber by microwave heating and the adsorption of tetracycline antibiotics. Separ Sci Technol 48:1356–1363

Hunter WG, Hunter JS (1978) Statistics for experimenters: an introduction to design, data analysis, and model building. Wiley, New York, p 319

Ip AW, Barford JP, McKay G (2010) A comparative study on the kinetics and mechanisms of removal of Reactive Black 5 by adsorption onto activated carbons and bone char. Chem Eng J 157:434–442

Jones AD, Bruland GL, Agrawal SG, Vasudevan D (2005) Factors influencing the sorption of oxytetracycline to soils. Environ Toxicol Chem 24:761–770

Kang HY, Park SS, Rim YS (2006) Preparation of activated carbon from paper mill sludge by KOH-activation. Korean J Chem Eng 23:948–953

Kilduff JE, Wigton A (1999) Sorption of TCE by humic-preloaded activated carbon: incorporating size- exclusion and pore blockage phenomena in a competitive adsorption model. Environ Sci Technol 33:250–256

Kulshrestha P, Giese RF Jr, Aga DS (2004) Investigating the molecular interactions of oxytetracycline in clay and organic matter: insights on factors affecting its mobility in soil. Environ Sci Technol 38:4097–4105

Kümmerer K (2001) Drugs in the environment: emission of drugs, diagnostic aids and disinfectants into wastewater by hospitals in relation to other sources–a review. Chemosphere 45:957–969

Kümmerer K (ed) (2008) Pharmaceuticals in the environment: sources, fate, effects and risks. Springer Science Business Media, Berlin/Heidelberg, pp 73–93

Kümmerer K, Al-Ahmad A, Mersch-Sundermann V (2000) Biodegradability of some antibiotics, elimination of the genotoxicity and affection of wastewater bacteria in a simple test. Chemosphere 40:701–710

Lafrance P, Mazet M (1989) Adsorption of humic substances in the presence of sodium salts. J Am Water Works Ass 81:155–162

Lau AH, Lam NP, Piscitelli SC, Wilkes L, Danziger LH (1992) Clinical pharmacokinetics of metronidazole and other nitroimidazole anti-infectives. Clin Pharmacokinet 23:328–364

Leon y Leon CA, Solar JM, Calemma V, Radovic LR (1992) Evidence for the protonation of basal plane sites on carbon. Carbon 30:797–811

Leyva-Ramos R, Geankoplis CJ (1985) Model simulation and analysis of surface diffusion in liquids in porous solids. Chem Eng Sci 40:799–807

Leyva-Ramos R, Geankoplis CJ (1994) Diffusion in liquid-filled pores of activated carbon I Pore volume diffusion. Can J Chem Eng 72:262–271

Leyva-Ramos R, Rivera-Utrilla J, Medellín-Castillo NA, Sánchez-Polo M (2009) Kinetic modelling of naphthalenesulphonic acid adsorption from aqueous solution onto untreated and ozonated activated carbons. Adsorpt Sci Technol 27:395–411

Li K, Ji F, Liu Y, Tong Z, Zhan X, Hu Z (2013) Adsorption removal of tetracycline from aqueous solution by anaerobic granular sludge: equilibrium and kinetic studies. Water Sci Technol 67:1490–1496

Liao P, Zhan Z, Dai J, Wu X, Zhang W, Wang K, Yuan S (2013) Adsorption of tetracycline and chloramphenicol in aqueous solutions by bamboo charcoal: a batch and fixed-bed column study. Chem Eng J 228:496–505

Lin AYC, Yu TH, Lateef SK (2009) Removal of pharmaceuticals in secondary wastewater treatment processes in Taiwan. J Hazard Mater 167:1163–1169

Lin Y, Xu S, Li J (2013) Fast and highly efficient tetracyclines removal from environmental waters by graphene oxide functionalized magnetic particles. Chem Eng J 225:679–685

Lindberg R, Jarnheimer PÅ, Olsen B, Johansson M, Tysklind M (2004) Determination of antibiotic substances in hospital sewage water using solid phase extraction and liquid chromatography/mass spectrometry and group analogue internal standards. Chemosphere 57:1479–1488

Liu P, Liu W, Jiang H, Chen J, Li W, Yu H (2012) Modification of bio-char derived from fast pyrolysis of biomass and its application in removal of tetracycline from aqueous solution. Bioresour Technol 121:235–240

Liu M, Hou L, Yu S, Xi B, Zhao Y, Xia X (2013) MCM-41 impregnated with A zeolite precursor: synthesis, characterization and tetracycline antibiotics removal from aqueous solution. Chem Eng J 223:678–687

López-Ramón V, Moreno-Castilla C, Rivera-Utrilla J, Radovic LR (2003) Ionic strength effects in aqueous phase adsorption of metal ions on activated carbons. Carbon 41:2020–2022

Marsh H, Yan DS, O'Grady TM, Wennerberg A (1984) Formation of active carbons from cokes using potassium hydroxide. Carbon 22:603–611

Mathers JJ, Flick SC, Cox LA (2011) Longer-duration uses of tetracyclines and penicillins in US food-producing animals: indications and microbiologic effects. Environ Int 37:991–1004

Méndez A, Gascó G, Freitas MMA, Siebielec G, Stuczynski T, Figueiredo JL (2005) Preparation of carbon-based adsorbents from pyrolysis and air activation of sewage sludges. Chem Eng J 108:169–177

Méndez-Díaz JD, Prados-Joya G, Rivera-Utrilla J, Leyva-Ramos R, Sánchez-Polo M, Ferro-García MA, Medellín-Castillo NA (2010) Kinetic study of the adsorption of nitroimidazole antibiotics on activated carbons in aqueous phase. J Colloid Interf Sci 345:481–490

Misra DN (1991) Adsorption and orientation of tetracycline on hydroxyapatite. Calcified Tissue Int 48:362–367

Moreno-Castilla C (2004) Adsorption of organic molecules from aqueous solutions on carbon materials. Carbon 42:83–94

Moreno-Castilla C, Bautista-Toledo I, Ferro-García MA, Rivera-Utrilla J (2003a) Influence of support surface properties on activity of bacteria immobilised on activated carbons for water denitrification. Carbon 41:1743–1749

Moreno-Castilla C, Perez-Cadenas AF, Maldonado-Hodar FJ, Carrasco-Marın F, Fierro JLG (2003b) Influence of carbon–oxygen surface complexes on the surface acidity of tungsten oxide catalysts supported on activated carbons. Carbon 41:1157–1167

Nikaido H, Vaara M (1985) Molecular basis of bacterial outer membrane permeability. Microbiol Rev 49:1–32

Ocampo-Perez R, Leyva-Ramos R, Alonso-Davila P, Rivera-Utrilla J, Sanchez-Polo M (2010) Modeling adsorption rate of pyridine onto granular activated carbon. Chem Eng J 165:133–141

Ocampo-Perez R, Leyva-Ramos R, Mendoza-Barron J, Guerrero-Coronado RM (2011) Adsorption rate of phenol from aqueous solution onto organobentonite: surface diffusion and kinetic models. J Colloid Interf Sci 364:195–204

Ocampo-Pérez R, Rivera-Utrilla J, Gómez-Pacheco C, Sánchez-Polo M, López-Peñalver JJ (2012) Kinetic study of tetracycline adsorption on sludge-derived adsorbents in aqueous phase. Chem Eng J 213:88–96

Ocampo-Pérez R, Orellana-Garcia F, Sánchez-Polo M, Rivera-Utrilla J, Velo-Gala I, López-Ramón MV, Alvarez-Merino MA (2013) Nitroimidazoles adsorption on activated carbon cloth from aqueous solution. J Colloid Interf Sci 401:116–124

Ocampo-Pérez R, Leyva-Ramos R, Rivera-Utrilla J, Flores-Cano JV, Sánchez-Polo M (2015) Modeling adsorption rate of tetracyclines on activated carbons from aqueous phase. Chem Eng Res Des 104:579–588

Oleszkiewicz JA, Mavinic DS (2001) Wastewater biosolids: an overview of processing, treatment, and management. Can J Civil Eng 28:102–114

Otowa T, Tanibata R, Itoh M (1993) Production and adsorption characteristics of MAXSORB: high-surface-area active carbon. Gas Sep Purif 7:241–245

Otowa T, Nojima Y, Miyazaki T (1997) Development of KOH activated high surface area carbon and its application to drinking water purification. Carbon 35:1315–1319

Parolo ME, Savini MC, Vallés JM, Baschini MT, Avena MJ (2008) Tetracycline adsorption on montmorillonite: pH and ionic strength effects. Appl Clay Sci 40:179–186

Passuello A, Cadiach O, Perez Y, Schuhmacher M (2012) A spatial multicriteria decision making tool to define the best agricultural areas for sewage sludge amendment. Environ Int 38:1–9

Pils JRV, Laird DA (2007) Sorption of tetracycline and chlortetracycline on K- and Ca-saturated soil clays, humic substances, and clay humic complexes. Environ Sci Technol 41:1928–1933

Poling BE, Prausnitz JM, John Paul OC, Reid RC (2001) The properties of gases and liquids, vol 5. McGraw-Hill, New York

Radovic LR, Moreno-Castilla C, Rivera-Utrilla J (2001) Carbon materials as adsorbents in aqueous solutions. Chem Phys Carbon 27:227–405

Rivera-Utrilla J, Sánchez-Polo M (2002) The role of dispersive and electrostatic interactions in aqueous phase adsorption of naphthalenesulphonic acids on ozone-treated activated carbons. Carbon 40:2685–2269

Rivera-Utrilla J, Sánchez-Polo M (2004) Ozonation of naphthalenesulphonic acid in the aqueous phase in the presence of basic activated carbons. Langmuir 20(21):9217–9222

Rivera-Utrilla J, Bautista-Toledo I, Ferro-García MA, Moreno-Castilla C (2001) Activated carbon surface modifications by adsorption of bacteria and their effect on aqueous lead adsorption. J Chem Technol Biot 76:1209–1215

Rivera-Utrilla J, Bautista-Toledo I, Ferro-García MA, Moreno-Castilla C (2003) Bioadsorption of Pb(II), Cd(II), and Cr(VI) on activated carbon from aqueous solutions. Carbon 41:323–330

Rivera-Utrilla J, Prados-Joya G, Sánchez-Polo M, Ferro-García MA, Bautista-Toledo I (2009) Removal of nitroimidazole antibiotics from aqueous solution by adsorption/bioadsorption on activated carbón. J Hazard Mater 170:298–305

Rivera-Utrilla J, Sánchez-Polo M, Gómez-Serrano V, Álvarez PM, Alvim-Ferraz MCM, Dias JM (2011) Activated carbon modifications to enhance its water treatment applications, an overview. J Hazard Mater 187:1–23

Rivera-Utrilla J, Gómez-Pacheco CV, Sánchez-Polo M, López-Peñalver JJ, Ocampo-Pérez R (2013a) Tetracycline removal from water by adsorption/bioadsorption on activated carbons and sludge-derived adsorbents. J Environ Manag 131:16–24

Rivera-Utrilla J, Sánchez-Polo M, Ferro-García MA, Prados-Joya G, Ocampo-Pérez R (2013b) Pharmaceuticals as emerging contaminants and their removal from water. A review. Chemosphere 93:1268–1287

Rodriguez-Reinoso F, Linares-Solano A (1989) Microporous structure of activated carbons as revealed by adsorption methods. Chem Phys Carbon 21:1–146

Ros A, Lillo-Ródenas MA, Fuente E, Montes-Morán MA, Martín MJ, Linares-Solano A (2006) High surface area materials prepared from sewage sludge-based precursors. Chemosphere 65:132–140

Sánchez-Polo M, Rivera-Utrilla J (2003) Effect of the ozone–carbon reaction on the catalytic activity of activated carbon during the degradation of 1, 3, 6-naphthalenetrisulphonic acid with ozone. Carbon 41(2):303–307

Sánchez-Polo M, Rivera-Utrilla J (2006) Ozonation of naphthalenetrisulphonic acid in the presence of activated carbons prepared from petroleum coke. Appl Catal B-Environ 67:113–120

Shao L, Ren Z, Zhang G, Chen L (2012) Facile synthesis, characterization of a $MnFe_2O_4$/activated carbon magnetic composite and its effectiveness in tetracycline removal. Mater Chem Phys 135:16–24

Sheng GP, Yu HQ, Yu Z (2005) Extraction of extracellular polymeric substances from the photosynthetic bacterium Rhodopseudomonas acidophila. Appl Microbiol Biot 67:125–130

Simazaki D, Fujiwara J, Manabe S, Matsuda M, Asami M, Kunikane S (2008) Removal of selected pharmaceuticals by chlorination, coagulation-sedimentation and powdered activated carbon treatment. Water Sci Technol 58:1129–1135

Snyder SA, Adham S, Redding AM, Cannon FS, DeCarolis J, Oppenheimer J (2007) Role of membranes and activated carbon in the removal of endocrine disruptors and pharmaceuticals. Desalination 202:156–181

Srivastava VC, Mall ID, Mishra IM (2006) Equilibrium modelling of single and binary adsorption of cadmium and nickel onto bagasse fly ash. Chem Eng J 117:79–91

Stackelberg PE, Gibs J, Furlong ET, Meyer MT, Zaugg SD, Lippincott RL (2007) Efficiency of conventional drinking-water-treatment processes in removal of pharmaceuticals and other organic compounds. Sci Total Environ 377:255–272

Tally FP, Goldin B, Sullivan NE (1981) Nitroimidazoles: in vitro activity and efficacy in anaerobic infections. Scand J Infect Dis 13:46–53

Ternes TA, Meisenheimer M, McDowell D, Sacher F, Brauch HJ, Haist-Gulde B, Zulei-Seibert N (2002) Removal of pharmaceuticals during drinking water treatment. Environ Sci Technol 36:3855–3863

Traegner UK, Suidan MT (1989) Parameter evaluation for carbon adsorption. J Env Eng Div-ASCE (J Environ Eng-ASCE) 115:109–128

Turku I, Sainio T, Paatero E (2007) Thermodynamics of tetracycline adsorption on silica. Environ Chem Lett 5:225–228

USEPA (1991) Granular activated carbon treatment report EPA-540/2–91/024. US Environmental Protection Agency, US Government Printing Office, Washington, DC

Van der Bijl P, Pitigoi-Aron G (1995) Tetracyclines and calcified tissues. Ann Dent 54:69–72

Vidal CB, Seredych M, Rodríguez-Castellón E, Nascimento RF, Bandosz TJ (2015) Effect of nanoporous carbon surface chemistry on the removal of endocrine disruptors from water phase. J Colloid Interf Sci 449:180–191

Wang R, Wei RC, Liu T, Wang T (2008a) Sorption characteristics of veterinary antibiotics chlortetracycline on manure. Huan Jing Ke Xue 29:1363–1368

Wang X, Zhu N, Yin B (2008b) Preparation of sludge-based activated carbon and its application in dye wastewater treatment. J Hazard Mater 153:22–27

Wang YJ, Sun RJ, Xiao AY, Wang SQ, Zhou DM (2010) Phosphate affects the adsorption of tetracycline on two soils with different characteristics. Geoderma 156:237–242

Wang H, Wu X, Liu X, Cong P (2011) Application study of a modified phenolic resin as binder for hybrid fibers reinforced brake pad for railroad passenger-coach braking. J Macromol Sci Chem 48:261–270

Werther J, Ogada T (1999) Sewage sludge combustion. Prog Energ Combust 25:55–116

West CC, Harwell JH (1992) Surfactants and subsurface remediation. Environ Sci Technol 26:2324–2330

Wilke CR, Chang P (1955) Correlation of diffusion coefficients in dilute solutions. AIChE J 1(2):264–270

Yang S, Cha J, Carlson K (2005) Simultaneous extraction and analysis of 11 tetracycline and sulfonamide antibiotics in influent and effluent domestic wastewater by solid-phase extraction

and liquid chromatography-electrospray ionization tandem mass spectrometry. J Chromatogr A 1097:40–53
Yu Z, Peldszus S, Huck P-M (2008) Adsorption characteristics of selected pharmaceuticals and an endocrine disrupting compound-naproxen, carbamazepine and nonylphenol-on activated carbon. Water Res 42:2873–2882
Zhang L, Song X, Liu X, Yang L, Pan F, Lv J (2011) Studies on the removal of tetracycline by multi-walled carbon nanotubes. Chem Eng J 178:26–33
Zhou Q, Li Z, Shuang C, Li A, Zhang M, Wang M (2012) Efficient removal of tetracycline by reusable magnetic microspheres with a high surface area. Chem Eng J 210:350–356
Zhu X, Tsang DC, Chen F, Li S, Yang X (2015) Ciprofloxacin adsorption on graphene and granular activated carbon: kinetics, isotherms, and effects of solution chemistry. Environ Technol 36:3094–3102
Zielke U, Hüttinger KJ, Hoffman WP (1996) Surface-oxidized carbon fibers: I Surface structure and chemistry. Carbon 34:983–998
Zogorski JS, Faust SD (1977) Operational parameters for optimum removal of phenolic compounds from polluted waters by columns of activated carbon. AIChe Symp Ser 73:54–65
Zou YL, Huang H, Chu M, Lin JW, Yin DQ, Li YN (2012) Adsorption research of tetracycline from water by HCl-modified zeolite. Adv Mater Res 573:43–47

Chapter 7
Biosorption of Copper by *Saccharomyces cerevisiae*: From Biomass Characterization to Process Development

Pietro Altimari, Fabrizio Di Caprio, and Francesca Pagnanelli

Abstract Biosorption offers a competitive technological solution to the removal of heavy metals from wastewaters. Nevertheless, large-scale application of biosorption is still hindered by the absence of systematic methodologies for product and process design. Crucial role is played by the development of models that can describe the effect of operating conditions on biosorption equilibrium and kinetics and thus drive the rational process design. In this contribution, biosorption of copper onto a wild-type strain of *Saccharomyces cerevisiae* is analyzed. The analysis is structured in a way that allows reviewing and discussing the main stages of process development. Type and concentration of the biomass active sites were determined by fitting mechanistic equilibrium models accounting for the distribution of the site protonation constants to potentiometric titration data. Equilibrium biosorption tests were performed to determine the dependence of the biosorption capacity on pH (3 and 5) and metal liquid concentration (0–120 ppm). The immobilization of biomass by calcium alginate was performed to produce composite sorbent beads with different biomass contents. An experimental analysis was performed to characterize the kinetics and the equilibrium of copper biosorption onto the produced beads. These data were exploited to identify kinetic and equilibrium models describing the competitive biosorption of protons and copper ions onto the beads. A mathematical model was derived to describe the transport of copper and protons in a biosorption column packed by the produced beads. The column mathematical model was validated by recourse to the experimental data derived by the operation of lab-scale packed column (length 15 cm, diameter 1.7 cm).

Keywords Biosorption • Copper • Biomass

P. Altimari (✉) • F. Di Caprio • F. Pagnanelli
Department of Chemistry, Sapienza University of Rome, P.le Aldo Moro 5, 00185, Rome, Italy
e-mail: pietro.altimari@uniroma1.it; fabrizio.dicaprio@uniroma1.it; francesca.pagnanelli@uniroma1.it

© Springer International Publishing AG 2017
A. Bonilla-Petriciolet et al. (eds.), *Adsorption Processes for Water Treatment and Purification*, DOI 10.1007/978-3-319-58136-1_7

Contents

7.1 Introduction

The release to the environment of heavy metals by anthropogenic activities poses a severe environmental concern to the health and development of living organisms. Numerous industrial activities are accompanied by the release to the environment of heavy metals that accumulate in groundwater and surface water. The methodologies commonly implemented to remove heavy metals from water include chemical precipitation, reverse osmosis, evaporation, and ion exchange by resins. The extended application of these methodologies is frequently hindered by economic and technical limits. Economic limits can be determined by elevated energy consumption and/or by the elevated costs of employed materials and equipment, while technical limits are frequently determined by the impossibility to lower the heavy metal concentration below the threshold imposed by current regulation. For example, most of the aforementioned methodologies cannot allow achieving heavy metal concentration values around 10–100 ppm, which is frequently required by environmental regulation.

In this framework, a competitive technological solution is represented by biosorption (Wang and Chen 2009). This technique relies on the property of certain biomolecules (or types of biomass) to bind and concentrate selected ions or other molecules from aqueous solutions. Examples of biomasses employed in the removal of heavy metals from water include bacteria, algae, fungi, and yeasts (Davis et al. 2003; Volesky 2007; Ahluwalia and Goyal 2007; Wang and Chen 2009). The main advantages of biosorption are the application of an inexpensive and biodegradable adsorbing material and the achievement of elevated removal efficiency at low metal concentrations (0–100 ppm) (Ahluwalia and Goyal 2007; Wang and Chen 2009).

Numerous reviews have been reported over the past two decades that have thoroughly discussed the potential of different biomasses to bind and thus remove heavy metals from wastewaters. Despite the considerable interest raised by the scientific community toward biosorption, the industrial application of this technology is still hindered by the absence of a systematic and unified framework for process design and development. Crucial role is played by the development of mathematical models that can describe the effect of operating conditions on biosorption equilibrium and kinetics and thus drive the rational process design. The challenging task is represented by the development of mathematical models that can effectively describe the biosorption mechanisms enabled by the complex biochemistry of employed matrices and, at the same time, retaining the simplicity required to smoothly driving process development from lab-scale research to plant design and operation. Nevertheless, the study of biosorption has frequently included the recourse to mathematical representations developed through the analysis of adsorption by ion exchange resins or activated carbons (Wang and Chen 2006; Goldberg et al. 2007; Ghorbani et al. 2008; Padmavathy 2008; Cojocaru et al. 2009). This approach cannot account for biosorption mechanisms that are made possible by the complex chemistry of biomasses and are absent in synthetic adsorbing materials. For example, the central role played by protons in biosorption is neglected by the most popular mathematical description of the biosorption equilibrium data, the Langmuir model (Schiewer and Volesky 1995). A more effective mathematical representation of biosorption can be derived through analysis of the interaction between biomass functional groups and ions in solution (Schiewer and Volesky 1995; Schiewer and Wong 2000; Pagnanelli 2011).

Additional obstacles hindering the large-scale application of biosorption are imposed by the small particle size and low mechanical strength of employed biomasses. The application of freely suspended biomasses may cause clogging and make impractical biomass-effluent separation following heavy metal biosorption. Immobilization of biomasses into polymeric beads with controlled size and enhanced mechanical stability can overcome this difficulty (Veglio and Beolchini 1997). The immobilization matrix does not only determine the mechanical stability of the produced biosorbent but can also participate into the biosorption process by binding targeted heavy metal ions. Further, the immobilization process must be performed in a way that can reduce the transport resistance to the immobilized biomass and minimize the amount of polymeric material employed to attain satisfactory stability. The latter requirement is imposed by the need to minimize the cost of polymeric material employed in the immobilization.

In this contribution, we analyze the biosorption of copper ion by a wild-type strain of *Saccharomyces cerevisiae*. The analysis is structured in a way that allows discussing the main stages of process development. Type and concentration of the biomass active sites were determined by fitting mechanistic equilibrium models accounting for the distribution of the site protonation constants to potentiometric titration data. Equilibrium biosorption tests were performed to determine the dependence of the biosorption capacity on *pH* (3 and 5) and metal liquid concentration (0–120 ppm). The immobilization of biomass by calcium alginate was performed to produce composite sorbent beads with different biomass contents. An experimental analysis was performed to characterize the kinetics and the equilibrium of copper biosorption onto the produced beads. These data were exploited to identify kinetic and equilibrium models describing the competitive biosorption of protons and copper ions onto the beads. A mathematical model was derived to describe the transport of copper and protons in a biosorption column packed by the produced beads. The column mathematical model was validated by recourse to the experimental data derived by the operation of lab-scale packed column (length 15 cm, diameter 1.7 cm). Numerical simulations were performed providing guidelines for the optimal column design and operation.

7.2 Materials and Methods

7.2.1 Yeast Strain

The following *S. cerevisiae* strains, referred in the following to as wild type, were used in this study: BY4741 (MATa, met15Δ0, his3Δ0, leu2Δ0, ura3Δ0). The cells were grown on YPD (1% peptone, 1% yeast extract, 2% glucose, DIFCO) at 28 °C until reaching late exponential phase. Following growth, cells were centrifuged and rinsed twice with distilled water to remove residual traces of the culture medium. The obtained pellet was incubated at 65 °C for drying.

7.2.2 Potentiometric Titration

In potentiometric titrations, 1 g of dry biomass was suspended in 50 mL of $NaNO_3$ solution. Experiments were performed at $NaNO_3$ concentrations equal to 0.1 M and 1 M. In any performed test, the initial pH of the suspension was corrected to a value around 2–2.5 by using HCl 0.1 N. pH was measured by a Eutech pH 700. NaOH 0.1 M was used as titrant solution and was added to the suspension through a burette. After the addition of any titrant solution volume, pH was measured. Titration was ended when reaching pH = 10.

7.2.3 Immobilization into Calcium Alginate

Beads of calcium alginate with immobilized biomass (hereafter referred to as composite beads) were obtained by dropping a biomass suspension (biomass concentration 0.1 g/mL) with sodium alginate (3% w/v) into a solution of $CaCl_2$ (20% w/w). The sodium alginate suspension with suspended biomass was withdrawn and pumped through a flexible plastic tube (0.3 mm diameter) at flow rate 1.56 mL/min. For this purpose, a peristaltic pump was employed. The same methodology was implemented to prepare the beads of calcium alginate without biomass. In this latter case, a sodium alginate solution (3% w/v) was dropped into $CaCl_2$ (20% w/w) solution.

7.2.4 Batch Biosorption

Any batch test was performed under magnetic stirring with 0.35 g of beads suspended in 80 mL of copper solution. To characterize the kinetics of biosorption, an initial copper concentration of 50 mg/L was selected. The initial pH was 4. In these experiments, suspension samples were withdrawn at prescribed time intervals and were analyzed to determine the liquid copper concentration. The copper concentration in the beads is, by virtue of mass balance, equal to $q_M = (C_0 - C_M)V/m$, where C_0 and C_M are the initial and the current liquid concentrations of copper, respectively, V is the solution volume, and m is the weight of suspended beads.

Biosorption isotherms were determined by performing batch tests with initial copper concentration ranging between 10 and 120 mg/L at pH 3 and 5. For this purpose, the pH was periodically corrected during the test by addition of NaOH or HCl. The evolution of liquid copper concentration during biosorption was monitored in preliminary experiments indicating that a time interval equal to 6 h was largely sufficient to reach equilibrium. Samples were therefore withdrawn at about 7 and 8 h and were analyzed to determine the equilibrium liquid copper concentration C_M and, thus, derive by mass balance the concentration of copper in the beads.

To determine the liquid copper concentration, withdrawn suspension samples were centrifuged, and the supernatant was analyzed by atomic adsorption (analytic Jena contra 300).

7.2.5 Fixed-Bed Biosorption

The produced beads were stacked into a glass column (17 mm diameter). The copper solution was pumped upward through the column to ensure the uniform filling of the void volume fraction. Copper concentration and *pH* of the pumped solution were 40 mg/L and 3, respectively. Before any experiment, beads were maintained about 1 h in a magnetically stirred water solution of HCl at pH = 4. This pH value was selected greater than the pH of the solution fed to the column to enhance the variation of the pH at the outlet of the column. Contacting beads previously suspended in a solution at pH = 4 with a solution at pH = 3 can enhance the transfer of protons from liquid to solid and thus enable the formation of a proton concentration front travelling through the column. This is essential to estimate reliably the parameters that govern the transport of protons through the column (interphase mass transport coefficient, dispersion coefficient).

Experiments were performed at flow rate ranging between 4 and 18 mL/min and bed height of 15 cm. The temperature of the pumped copper solution was 25 °C. The void fraction (ε) of the bed was estimated as $\varepsilon = V_F/V_B$, V_F, and V_B demoting the volume required to fill the bed and the apparent volume of the bed, respectively.

7.3 Results

7.3.1 Identification of the Biomass Active Sites

Electric charges can form at the cell wall by dissociation or protonation of weakly acidic and basic groups. This allows the biomass to bind heavy metal ions present in solution (Wang 2002; Kordialik-Bogacka 2011). Titration experiments were performed to determine the evolution of the net superficial charge Q_H with pH. The following expression, derived by imposing electroneutrality, was employed to determine Q_H from the evolution of pH observed during titration:

$$Q_H = -\frac{\left([\mathrm{Na}^+] + [\mathrm{H}^+] - [\mathrm{OH}^-] - [\mathrm{Cl}^-] - [\mathrm{NO}_3^-]\right)\mathrm{V}}{\mathrm{m}} \tag{7.1}$$

where m is the mass of suspended biomass and V is the suspension volume.

The evolution of the Q_H with pH is displayed in Fig. 7.1a. Q_H is positive at low pH values and monotonically decreases as the pH is increased. Particularly, negative Q_H values are found at pH values lower than 6.3. This evolution can be explained by the high concentration of protonated groups at low pH values and by the progressive deprotonation of such groups induced by the increase in pH.

Analysis of variance with $\alpha = 0.05$ was conducted to evaluate the influence of different $NaNO_3$ concentrations in the potentiometric titration tests. No statistically significant difference was found as the $NaNO_3$ concentration was varied in the examined range.

Titration data displayed in Fig. 7.1a can be analyzed to identify the biomass active sites that can bind the heavy metals and thus enable biosorption. This analysis relies on the formulation of mechanistic equilibrium models that can describe the evolution of Q_H with pH. Any of such models is completely determined by the protonation reactions of the active sites that contribute to biosorption. The identification of the active sites can thus be performed as follows:

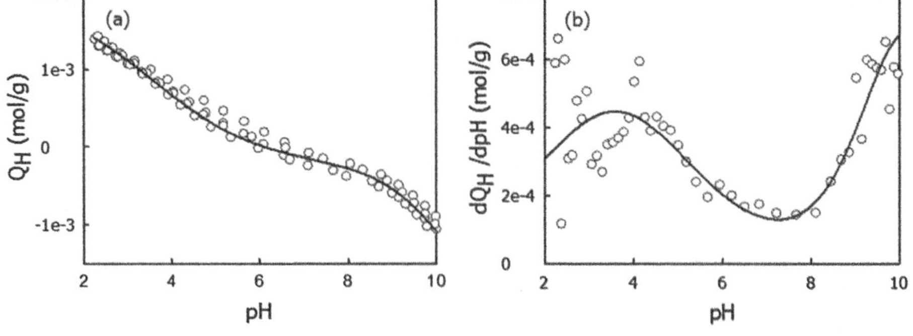

Fig. 7.1 Evolution of the net charge Q_H (**a**) and of the derivative dQ_H/dpH (**b**) with pH (*empty circles* and *solid lines* describe experimental values and model predictions, respectively)

- Identifying candidate reaction schemes that can justify the recorded evolution of Q_H with pH
- For any identified reaction scheme, formulating a mechanistic model to describe the evolution of Q_H with pH
- Comparing the abilities of the different formulated models to reproduce experimental titration data

The identification of the candidate reaction schemes requires preliminarily analyzing the biochemistry of the cell wall. The main groups present on the outer yeast surface are carboxylic, phosphoric, phosphodiesteric, hydroxyl, amine, and sulfide groups (Lesage and Bussey 2006; Di Caprio et al. 2014). Additional information about the candidate reaction schemes can be derived by analyzing the evolution of the derivative dQ_H/dpH with pH. Any peak appearing in the diagram dQ_H/dpH versus pH identifies a local maximum of buffering capacity and occurs when the pH equals the pK_a of an active site. Accordingly, the peaks allow identifying the protonation constants K_a of the active sites involved in biosorption (Pagnanelli 2011; Di Caprio et al. 2014, 2016). The evolution of dQ_H/dpH with pH is displayed in Fig. 7.1b. Two distinct peaks are found which correspond to pK_a values around 3.5 and 10. Reaction schemes including two protonation reactions can therefore be considered to reproduce titration data.

In accordance with the illustrated characterization of the cell wall biochemistry, carboxylic, phosphoric, phosphodiesteric, hydroxyl, and amine groups were considered as candidate functional groups determining acid-base properties. Two different reaction schemes were considered in our study, each of them including, in accordance with the analysis of dQ_H/dpH versus pH, two protonation reactions. The two reaction schemes are described in Table 7.1. For each scheme, the sites that can be involved in any protonation reaction are indicated. In reaction scheme A, variations in the net superficial charge Q_H with pH are attributed to the dissociation of carboxylic, phosphoric, and phosphodiesteric groups in the range pH < 4 and to the dissociation of amine groups in the range pH > 8. Dissociations of the hydroxyl

Table 7.1 Candidate reaction schemes and corresponding mathematical models describing the dependence of net superficial charge Q_H on pH. For each reaction, the expected range of pKa values derived from literature data is reported

Model	Reaction scheme	Site chemistry	pKa	Q_H(pH)
A	$S_1^- + H^+ \leftrightarrow S_1H$	S_1: carboxylic, phosphoric or phosphodiesteric	0.2–6	$-\frac{S_{1\,tot}}{1+(K_{H1}[H^+])^{m_1}} +$
	$S_2 + H^+ \leftrightarrow S_2H^+$	S_2: amine	8–11	$+\frac{S_{2\,tot}(K_{H2}[H^+])^{m_2}}{1+(K_{H2}[H^+])^{m_2}}$
B	$S_1 + H^+ \leftrightarrow S_1H^+$	S_1: hydroxyl[a]	<1	$\frac{S_{1\,tot}(K_{H1}[H^+])^{m_1}}{1+(K_{H1}[H^+])^{m_1}} +$
	$S_2^- + H^+ \leftrightarrow S_2H$	S_2: hydroxyl[b]	>11	$-\frac{S_{2\,tot}}{1+(K_{H2}[H^+])^{m_2}}$

[a]Reaction: R-OH + H + ↔ R-OH2+

[b]Reaction: R-O- + H + ↔ R-OH

group are considered responsible for the variation of Q_H at both pH < 4 and pH > 8 in the formulation of the reaction scheme B. This latter possibility was considered because of the large concentration of hydroxyl group found in cell wall (Lesage and Bussey 2006). A range of possible pK_a values can thus be associated to any reaction based on literature data (Ahluwalia and Goyal 2007; Di Caprio et al. 2014, 2016). We note that reactions are formulated in Table 7.1 as site protonation reactions and not dissociation of protonated sites. Accordingly, for any reaction, pK_a equals $logK_H$, K_H being the reaction equilibrium constant, also referred to as affinity constant for protons.

The second step of the illustrated procedure is to derive a mathematical model describing the evolution of Q_H with pH for any selected reaction scheme. This entails solving for any scheme the algebraic system including reaction equilibrium and site balance equations. The models derived by this approach were modified in the present study to describe surface heterogeneity. Surface heterogeneity imposes that identical functional groups can exhibit different affinity values for ionic species in solution depending on their chemical and physical surroundings. A range of affinity constants, rather than a fixed affinity constant value, is therefore found for any functional group. This was taken into account by implementation of the methodology illustrated in (Pagnanelli 2011). The expressions of the dependence of the net charge Q_H on $[H^+]$ obtained by this approach with the two reaction schemes are reported in Table 7.1. Heterogeneity is described by the introduction of the coefficients m_i ($0 < m_i < 1, i = 1, 2$). A value $m_i = 1$ corresponds to the case that all the functional groups of type i exhibit identical affinity constant and, hence, dissociate around the same pH value. As m_i is decreased, a broader range of affinity constants for the group is found, and acid dissociation takes place over a larger pH range.

The third step of the procedure is to compare the abilities of the formulated models to describe titration data. For this purpose, the two derived models were fitted to titration data displayed in Fig. 7.1. The estimated model parameters were the total concentrations $S_{i,tot}$, the affinity constants, and the heterogeneity factors m_i of the active sites. Satisfactory fitting of titration data was achieved with both models A and B. The residuals (squared differences between experimental and predicted Q_H values) attained with the optimal parameter values were, however, much greater in the application of model B than in the application of model A. Further, no reliable estimation could be attained for the protonation constants of the site S_2 in case of application of model B. This statistical analysis indicates that the model A should be implemented to describe proton binding by the biomass. Accordingly, it can be concluded that biosorption is determined by the dissociation/protonation of carboxylic, phosphoric, and phosphodiesteric groups in the range pH < 4 by the dissociation/protonation of amine groups in the range pH > 8.

The optimal parameters estimated by nonlinear regression of titration data with model A are reported in Table 7.2. The evolutions of Q_H and dQ_H/dpH with pH predicted by the model with the optimal parameters are displayed in Fig. 7.1a, b, respectively, along with experimental titration data.

Table 7.2 Optimal parameter values and relative confidence intervals determined for the model A (Table 7.1)

	Average	±Δ
$S_{1,tot}$ (mol/g)	0.0022	0.0001
$S_{2,tot}$ (mol/g)	0.0019	0.0001
$\log K_{H1}$	3.6	0.1
$\log K_{H2}$	10.2	0.1
m_1	0.35	0.03
m_2	0.59	0.03

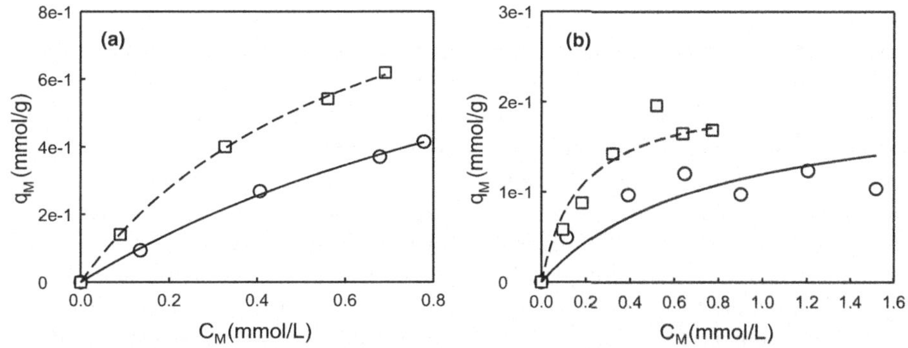

Fig. 7.2 Equilibrium isotherms found with calcium alginate (**a**) and composite beads (**b**) at different pH values. *Empty circles* and *squares* describe experimental values found at pH = 3 and pH = 5, respectively. *Solid* and *dashed lines* describe the predictions of Eqs. (7.4) and (7.5) with optimal parameter values reported in Table 7.3

7.3.2 Biosorption by Calcium Alginate Beads Under Batch Operation

7.3.2.1 Biosorption Isotherms

Biosorption experiments were performed with the produced beads under batch operation. Our analysis was aimed at developing a model that describes the uptake/release of protons and metal ions by the beads. For this purpose, equilibrium isotherms were derived at pH values 3 and 5.

Equilibrium data are displayed in Fig. 7.2a, b for calcium alginate beads and composite beads, respectively. The copper concentration in the composite beads is in Fig. 7.2b defined as the mass of copper divided by the mass of beads, which is the sum of the masses of calcium alginate and of biomass. In accordance with Fig. 7.2, the bead biosorption capacity increases with pH at any copper liquid concentration C_M. This can be explained by the increase in the concentration of negatively charged active sites with pH. Further, reported data indicate that the biosorption capacity of calcium alginate beads (Fig. 7.2a) is at any C_M and pH larger than the biosorption capacity of composite beads (Fig. 7.2b). This result, which is determined by the reduced biosorption capacity of the yeast compared to calcium alginate, may discourage the application of composite beads and suggest the application of beads composed of only calcium alginate. This analysis would neglect, however, that the application of calcium alginate alone increases processing costs. Unlike calcium alginate, the yeast can be obtained at zero or even at negative cost (in case it is would be wasted). This latter advantage becomes evident if the biosorption capacity of composite beads is computed by dividing the mass of copper contained in the beads by the amount of calcium alginate that composes the beads and not by the mass of the beads as in Fig. 7.1b. With this evaluation, the biosorption capacity of the composite beads becomes larger than the biosorption capacity of beads composed of only calcium alginate at any pH and concentration C_M. To verify it, it can be noted that beads with immobilized biomass contain about 18% w/w of calcium alginate (dry basis). This implies that, for composite beads, the mass of copper divided by the mass of calcium alginate is about five times larger than the mass of copper divided by the mass of beads. In industrial practice, a trade-off must be found between the reduction in the biosorbent cost and the increased biosorption capacity that can be attained by increasing biomass and calcium alginate fractions, respectively.

In accordance with the analysis illustrated in Sect. 7.3.1, two active sites S_1 and S_2 are responsible for biosorption by the selected yeast. The pK_a of the active site S_2, which is representative of amine group, is equal to 10.3, implying that the site S_2 cannot appreciably contribute to biosorption within the selected pH range 3–5. The only active site S_1 that can contribute to biosorption by the yeast within this latter pH range is representative of carboxylic, phosphoric, and phosphodiesteric groups. On the other hand, carboxylic groups also determine biosorption by calcium alginate. A one-site model can therefore be elaborated to describe the dependence of the equilibrium copper concentration q_M and proton concentration q_H in the beads on the copper liquid concentration C_M and solution pH, irrespective of whether beads composed of only calcium alginate or composite beads are employed. The model is completely defined by the reaction scheme that describes the competitive binding of metal ions and protons to the site S_1. Different reaction schemes were analyzed in the present study by modifying the stoichiometry of the site-metal-binding reaction. For any reaction scheme, analytical expressions were determined for q_M and q_H and fitted to equilibrium biosorption data displayed in Fig. 7.2 to determine equilibrium constant and biosorption capacity (equivalently, site concentration). The only reaction scheme that could ensure satisfactory fitting with physically admissible parameters was:

$$S_1^- + Cu^{2+} \leftrightarrow S_1Cu^+ \tag{7.2}$$

$$S_1^- + H^+ \leftrightarrow S_1H \tag{7.3}$$

By solving the algebraic system including equilibrium and site balance equations, the following expressions were found for q_M and q_H:

$$q_M = [S_1Cu^+] = \frac{q_{max}K_MC_M}{1 + K_MC_M + K_HC_H} \quad (7.4)$$

$$q_H = [S_1H] = \frac{q_{max}K_HC_H}{1 + K_MC_M + K_HC_H} \quad (7.5)$$

where K_M is the equilibrium constant of reaction Eq. (7.2) (affinity of the metal to site S_1) and $q_{max} = S_{1tot}$ is the maximum biosorption capacity. It must be noted that the overall concentration S_{1tot} and the affinity constant K_H refer to the beads and thus take, owing to the presence of calcium alginate, values different from those estimated for freely suspended yeast (Table 7.2). Further, it must be taken into account that concentration of sites available to metal binding may be different from the concentration of sites that can bind protons (Di Caprio et al. 2014). This may be caused by the different mobility of protons and copper ions, which can be enhanced by the formation of copper complexes, and may lead to a maximum biosorption capacity of the metal by the yeast lower than that site concentration determined by analysis of potentiometric titration data.

It is must be remarked that Eq. (7.2) does not fulfil electroneutrality. The biosorption of the bivalent copper ion by the monovalent group (carboxylic, phosphodiesteric) may be enabled by the interaction of the bivalent copper with different groups with delocalized negative charge. Alternatively, Eq. (7.2) could be explained by the formation of monovalent copper complexes, including, for example, $CuOH^+$, which can likely form around pH = 5.

It is important to note that Eq. (7.2) was introduced following the analysis of alternative metal-binding reactions fulfilling electroneutrality. However, these latter reactions led to expressions for q_M and q_H that could not ensure satisfactory fitting of equilibrium biosorption data.

Fitting Eqs. (7.2) and (7.3) to equilibrium biosorption data allows computing optimal values for K_M, K_H, and q_{max} reported in Table 7.3. Biosorption isotherms generated by the model with optimal parameter values are compared with experimental equilibrium data in Fig. 7.2. Satisfactory fitting was attained demonstrating the ability of the developed model to reproduce biosorption equilibrium data.

7.3.2.2 Biosorption Under Batch Operation

Batch experiments were performed with initial pH = 4 and liquid copper concentration $C_0 = 50$ ppm to characterize the kinetics of biosorption by calcium alginate

Table 7.3 Optimal parameter values estimated by fitting Eqs. (7.4) and (7.5) to biosorption equilibrium data displayed in Fig. 7.2

	Composite beads	±Δ	Calcium alginate beads	±Δ
q_{max} (mmol/g)	2.2669e−01	7.4035e−02	1.2015e+00	1.3040e−01
K_M (L/mmol)	4.7563e+00	4.6473e+00	1.5258e+00	2.9933e−01
K_H (L/mmol)	2.5624	2.1531	1.2627e+00	1.4141e−01

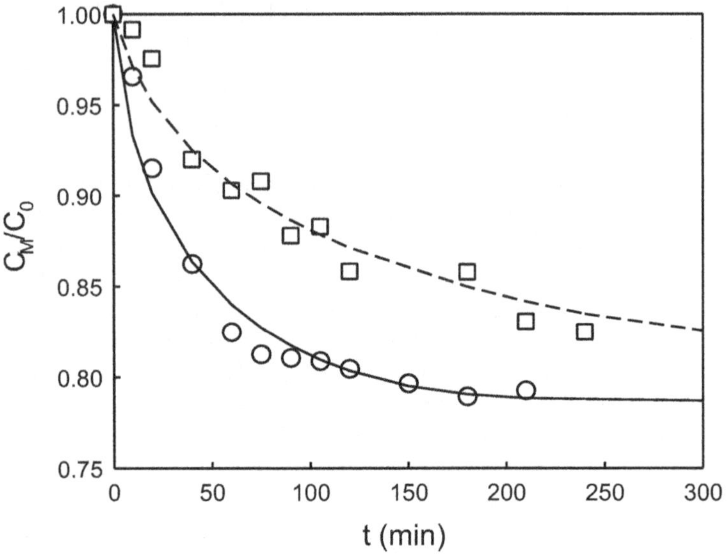

Fig. 7.3 Biosorption kinetics found under batch operation with calcium alginate and composite beads. *Empty circles* and *squares* describe experimental values found with calcium alginate and composite beads, respectively. *Solid* and *dashed lines* describe the predictions of Eqs. (7.6), (7.7), (7.8), (7.9), (7.10), (7.11), (7.12), and (7.13) with optimal parameter values reported in Table 7.4

and composite beads. Figure 7.3 displays the evolutions of the ratio C_M/C_0 recorded in the experiments performed. The characteristic times required to approach the equilibrium are around 100 and 300 min for calcium alginate beads and beads with immobilized biomass, respectively.

Differences between the evolutions of the liquid copper concentration are determined by differences between the microstructures of the produced beads. These ultimately influence the transport of copper ions through the beads. Crucial role is in this framework played by the identification of mathematical models that can describe the transport of copper ions through the beads. Fitting these models to biosorption kinetics data can be performed to identify the transport mechanisms governing metal uptake and thus derive estimates for the main transport parameters.

In the present study, a heterogeneous mathematical model was found to describe the evolution of metal uptake displayed in Fig. 7.3. Model formulation includes separate mass balances for copper and protons in the liquid and solid phases that compose the beads. The following assumptions are adopted in the formulation of the model:

Beads have spherical geometry, and the concentrations of copper and protons in the beads are uniform at any radial coordinate r.

The resistance to the transport of copper ions and protons from the bulk of the liquid to the bead surface can be neglected. The latter assumption can be fulfilled by ensuring sufficiently vigorous mixing of the bead suspension. In the present study, we verified that the evolution of the copper concentration does not vary within a wide range of the agitation rate around the value selected.

- At any time t and radial coordinate r, copper and proton concentrations in the beads are in equilibrium with the copper and proton concentrations in the liquid.
- The liquid phase is perfectly mixed.

In accordance with these assumptions, mass balance equations for metal ions and protons can be written as follows:

$$\varepsilon\frac{\partial C_M}{\partial t}+\rho_B\frac{\partial q_M(C_M)}{\partial t}=D_M\left[\frac{1}{r^2}\frac{\partial}{\partial r}\left(r^2\frac{\partial C_M}{\partial r}\right)\right] \quad (7.6)$$

$$\varepsilon\frac{\partial C_H}{\partial t}+\rho_B\frac{\partial q_H(C_H)}{\partial t}=D_M\left[\frac{1}{r^2}\frac{\partial}{\partial r}\left(r^2\frac{\partial C_H}{\partial r}\right)\right] \quad (7.7)$$

$$\frac{dC_{MB}}{dt}=-\frac{N4\pi R^2 D_M}{V}\frac{\partial C_M}{\partial r}\bigg|_{r=R} \quad (7.8)$$

$$\frac{dC_{HB}}{dt}=-\frac{N4\pi R^2 D_H}{V}\frac{\partial C_H}{\partial r}\bigg|_{r=R} \quad (7.9)$$

where R is the bead radius, ρ_B is the bead density, q_M and q_H are described by Eqs. (7.4) and (7.5) with $C_M = C_M(r, t)$, C_{MB} and C_{HB} are the copper and proton concentrations in the bulk of the liquid, and N is the number of beads in the solution. Equations (7.6), (7.7), (7.8), and (7.9) must be solved with the following boundary and initial conditions:

$$\frac{\partial C_M}{\partial r}\bigg|_{r=0}=\frac{\partial C_H}{\partial r}\bigg|_{r=0}=0, \quad t>0 \quad (7.10)$$

$$C_M(R,t)=C_{MB}(t), C_H(R,t)=C_{HB}(t), \quad t>0 \quad (7.11)$$

$$C_{MB}(0)=C_0, C_{H0}=10^{-pH0} \quad (7.12)$$

$$C_M(r,0)=0, C_H(r,0)=10^{-pH1} \quad (7.13)$$

where pH 0 is the initial pH of the liquid and pH 1 is the pH of the solution employed to store the beads prior starting the test. By application of finite difference techniques, the mathematical model (Eqs. 7.6, 7.7, 7.8, 7.9, 7.10, 7.11, 7.12, and 7.13) was reduced to a set of ordinary differential equations which was numerically integrated by the software Matlab. The numerical solution was fitted to transient concentration data displayed in Fig. 7.3 to estimate the diffusion coefficients D_M and D_H. The other model parameters were fixed to the following values: $N = 25$, $\rho_B = 1.7$ kg/L, $V = 0.08$ L, $R = 0.6$ mm, pH $0 = 3$, pH $1 = 4$, $C_0 = 0.8$ mmol/L, and $\varepsilon = 0.01$. Diffusion coefficient values determined by nonlinear fitting are reported in Table 7.4.

The evolution of the liquid copper concentration predicted by the model is included in Fig. 7.3 along with experimental concentration data. Comparison

Table 7.4 Diffusion coefficients estimated by fitting Eqs. (7.6), (7.7), (7.8), (7.9), and (7.10) to transient concentration data displayed in Fig. 7.3

	Composite beads	Calcium alginate beads
D_M (m_2/s)	1.7072e−09 +/− 5.3887e−10	6.3245e−09 +/− 1.9963e−09
D_H (m_2/s)	2.1229e−10 +/− 6.6201e−10	5.7226e−09 +/− 1.1983e−09

between predicted and experimental patterns confirms the ability of Eqs. (7.6), (7.7), (7.8), (7.9), and (7.10) to correctly describe the proton and metal uptake under batch operation.

7.3.3 Biosorption Under Fixed-Bed Operation

Fixed-bed experiments were performed with the produced beads at different feeding flow rates. The flow rates were selected to evaluate the effect of the liquid through the column. For this purpose, the column void fraction was preliminarily evaluated in any experiment by the procedure illustrated in Sect. 2.5, and the liquid flow rate was imposed equal to $v_z \varepsilon_b A$, with v_z, ε_b, and A denoting the selected velocity, the column void fraction, and the column section, respectively.

Void fraction and liquid velocity values imposed in any test are reported in Table 7.5. The velocity was varied from about 18 to 5 cm/min, while the column void fraction was fixed around 0.3. The evolutions of the dimensionless outlet metal concentration C_M/C_0 versus the dimensionless time t/τ, C_0 and τ denoting the metal concentration in the feed (~40 mg/L) and the column residence time, recorded during the different tests, are displayed in Fig. 7.4. The evolution of outlet metal concentration and pH recorded in test C was not included in Fig. 7.4 to avoid multiple overlapping of the displayed curves and thus ensure sufficient clarity.

Increasing the liquid velocity causes the outlet metal concentration to increase progressively faster. This effect can be attributed to the increase in the ratio between the characteristic time of interphase mass transport and the residence time with liquid velocity. In the liquid velocity range 8–18 cm/min, differences between the evolutions of the dimensionless outlet metal concentration are significant in the very early stage of the experiment and are progressively reduced as larger volume liquids are processed by the column. This can be verified by comparing the evolutions of outlet concentrations found for tests A and B (Fig. 7.4). The differences between the outlet metal concentrations recorded at different liquid velocity values become significantly larger when velocity values below 8 cm/min are selected.

Data displayed in Fig. 7.4 show that the column filled with composite bead exhibits lower performances compared with the column filled with calcium alginate beads. This can be verified by comparing the evolutions of the outlet metal concentration found in tests B and E, which correspond to the application of calcium alginate and composite beads, respectively, with liquid velocity around

Table 7.5 Column operating conditions and parameter value coefficients estimated by fitting Eqs. (7.6), (7.7), (7.8), (7.9), and (7.10) to transient concentration and pH data

	Test	v_z (cm/min)	ε_β	$k_M \cdot 10^5$ (m/s)	±Δ	$k_H \cdot 10^5$ (m/s)	±Δ	$D_{zM} \cdot 10^7$ (m^2/s)	±Δ	$D_{zH} \cdot 10^6$ (m^2/s)	±Δ
Calcium alginate beads	A	18.1	0.22	7.0	0.6	11.6	4.5	5.2	2.3	1.3	0.4
	B	7.8	0.26	5.0	1.0	5.2	1.3	39.4	12	2.7	–
	C	6.2	0.32	3.7	2.3	3.6	2.1	1.38	–	0.84	–
	D	5.3	0.25	–		–		–	–	–	–
Composite beads	E	6.7	0.3	4.7	0.1	17	3.7	2.7	–	1.21	–

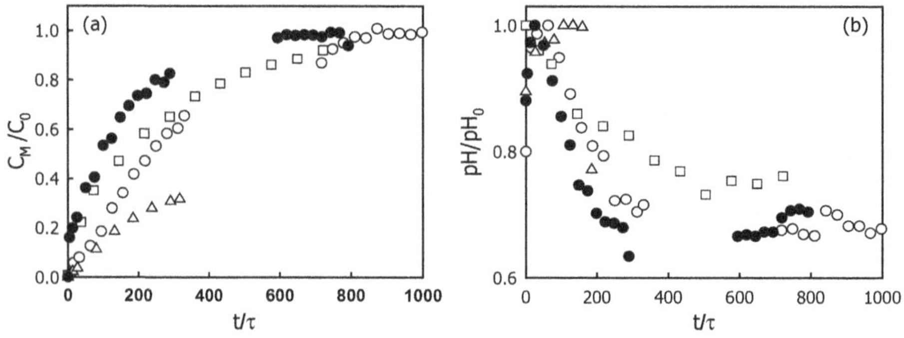

Fig. 7.4 Evolution of the outlet metal concentration (**a**) and pH (**b**) in fixed-bed experiments. *Empty squares*, *circles*, and *triangles* correspond to tests A, B, and D detailed in Table 7.5. *Filled circles* correspond to test E

7 cm/min. The dimensionless outlet concentration is at any t/τ (equivalently, at any processed volume of liquid) with calcium alginate greater than with composite beads.

In order to estimate how the characteristic times of the convection, interphase mass transport, and dispersion are affected by liquid velocity and by the selected beads, transient concentration and pH data were fitted to the numerical solution a heterogeneous column model. The main assumptions behind model formulation can described as follows:

1. Metal concentration and pH in the liquid and solid phases are at any prescribed time uniform over the column section and exclusively depend on the axial coordinate.
2. Concentration profiles within the beads are not described, meaning that an average concentration is adopted for protons and metal in the beads.

The mass balances that govern the temporal evolution of the spatial concentration profiles of metal and protons in the liquid and solid phases can thus be written as follows:

$$\frac{\partial C_M}{\partial t} + v_z \frac{\partial C_M}{\partial z} - D_{zM} \frac{\partial^2 C_M}{\partial z^2} + \frac{(1-\varepsilon_b)\rho_B}{\varepsilon_b} \frac{\partial q_M}{\partial t} = 0 \qquad (7.14)$$

$$\frac{\partial C_H}{\partial t} + v_z \frac{\partial C_H}{\partial z} - D_{zH} \frac{\partial^2 q_M}{\partial z^2} + \frac{(1-\varepsilon_b)\rho_B}{\varepsilon_b} \frac{\partial q_H}{\partial t} = 0 \qquad (7.15)$$

$$\frac{\partial q_M}{\partial t} = -k_M \left(q_M - q_{M,e}\right) \qquad (7.16)$$

$$\frac{\partial q_H}{\partial t} = -k_H \left(q_H - q_{H,e}\right) \qquad (7.17)$$

where z is the axial coordinate, D_{zM} and D_{zH} are the dispersion coefficients of metal and protons, respectively, and q_{Me} and q_{He} are the solid concentrations of metal and protons, respectively, in equilibrium with the liquid phase. Equations (7.4) and (7.5) are employed to compute the dependence of q_{Me} and q_{He} on C_M and C_H. Equations (7.14), (7.15), (7.16), and (7.17) were coupled to the following boundary and initial conditions:

$$v_z C_0 = v_z C_M(0,t) - D_{zM}\frac{\partial C_M}{\partial z}; \quad v_z C_0 = v_z C(0,t) - D_{zH}\frac{\partial C_H}{\partial z} \tag{7.18}$$

$$\frac{\partial C_M}{\partial z}(L,t) = 0; \quad \frac{\partial C_H}{\partial z}(L,t) = 0 \tag{7.19}$$

$$\frac{\partial q_M}{\partial z}(0,t) = 0; \quad \frac{\partial q_H}{\partial z}(0,t) = 0 \tag{7.20}$$

$$\frac{\partial q_M}{\partial z}(L,t) = 0; \quad \frac{\partial q_H}{\partial z}(L,t) = 0 \tag{7.21}$$

$$C_M(z,0) = 0; \quad C_H(z,0) = 10^{-pH1} \tag{7.22}$$

$$q_M(z,0) = 0; \quad q_H(z,0) = q_{H,e}\left(0, 10^{-pH1}\right) \tag{7.23}$$

The formulated mathematical model was reduced by finite difference techniques to a set of ordinary differential equations, which was successively integrated by Matlab. Fitting $C_M(L, t)$ and $C_H(L, t)$ numerical solution of the formulated model to the evolutions of outlet metal and proton concentrations allowed estimating k_M, k_H, D_{zM}, and D_{zH} for any experiment. The optimal parameter values estimated for any experiment in Table 7.4. For the sake of clarity, the evolutions of C_M/C_0 and pH/pH_0 recorded in the performed experiments and predicted with the estimated parameter values by the model are not included in Fig. 7.4 and separately displayed in Fig. 7.5.

Satisfactory fitting was achieved for any performed test with the exception of test E, which was performed with the lowest value of the selected liquid velocity range. In accordance with data reported in Table 7.4, comparable k_M and k_H values were estimated at any liquid velocity for calcium alginate beads. A k_H significantly larger than k_M was estimated in case of application of composite beads. Mass transport coefficients slightly increase with the liquid velocity. It is important to remark that no significant difference is found between the mass transfer coefficient values found with calcium alginate and composite beads. This implies that the difference observed in Fig. 7.4 between the evolutions of outlet metal concentrations for calcium alginate and composite beads cannot be imputed to different transport parameters but must rather be attributed to the differences in the equilibrium properties of two different beads (Table 7.3).

Dispersion values in agreement with literature values could be determined by fitting (Papageorgiou et al. 2008). However, no reliable confidence intervals could be derived for dispersion coefficients. To clarify it, the sensitivity of the model solution around the estimated parameter values was investigated. This analysis

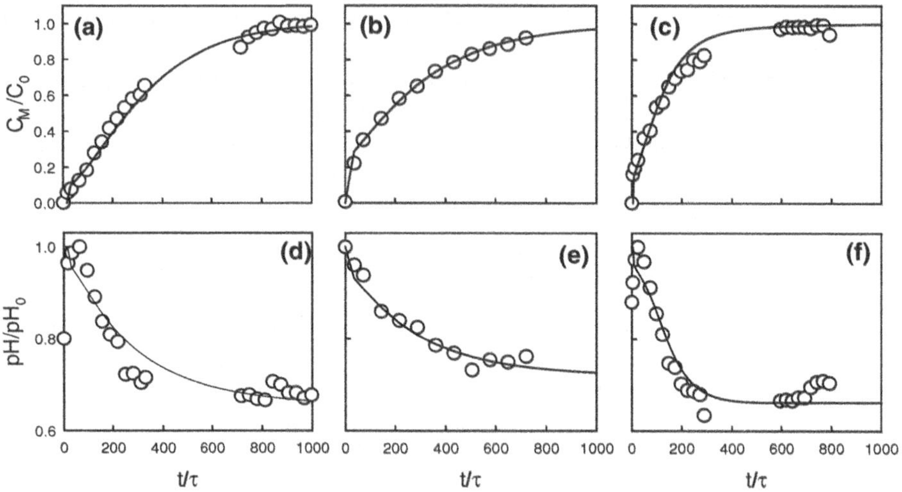

Fig. 7.5 Comparison between the evolutions of outlet concentration (**a**, **b**, **c**) and pH (**d**, **e**, **f**) predicted by the model (*solid line*) and recorded in the performed tests (*empty circles*); (**a**), (**d**) test A; (**b**), (**e**) test B; (**c**), (**f**) test E

revealed that model predictions are scarcely sensitive to variations in the dispersion coefficients. Particularly, varying the dispersion coefficients by about two orders of magnitude did not produce significant variations in the predicted outlet concentrations of metal and protons. This result can be explained by the large Peclet numbers $Pe_i = v_z L/D_{z,i}$ $(i = M, H)$ found with the estimated parameter values. Based on the estimated parameter values reported in Table 7.4, Pe_M values equal to about 1107 and 620 are found in tests C and E, respectively. Accordingly, the liquid velocity can be decreased below the selected range maintaining at the same time Peclet values $Pe_i >> 1$. This can enhance biosorption performance by increasing the ratio between the characteristic times of interphase mass transport and convection. Obviously, biosorption performances would deteriorate if Peclet values close to unity were attained. This would smooth the concentration fronts travelling through the front and would thus reduce the time needed for the outlet metal concentration to achieve any prescribed value.

7.4 Conclusions

Biosorption of copper by a wild-type strain of *Saccharomyces cerevisiae* was analyzed. The analysis was aimed at discussing the main stages of process development. A mechanistic equilibrium model accounting for the distribution of the site protonation constants was developed to describe proton binding and was identified by fitting to experimental titration data. Model development was driven by the preliminary analysis of the yeast wall biochemistry and allowed identifying type and concentration of the active sites responsible for biosorption. These results were successively exploited to develop a simplified equilibrium model describing the competitive binding of copper and protons to calcium alginate and composite beads. A one-site model could effectively reproduce the dependence of the biosorption capacity on pH (3 and 5) and metal liquid concentration (0–120 ppm).

A mathematical model describing the transport of metal ions and protons through the beads was numerically solved and fitted to transient concentration data recorded under batch operation. This allowed estimating the diffusion coefficients of metal and proton through the beads. The estimated metal diffusion coefficient was significantly greater with calcium alginate beads than with composite beads. This difference can be attributed to structural differences induced by the immobilization of the yeast in composite beads. Future studies should address the dependence of diffusion coefficient on the yeast fraction employed in the preparation of composite beads.

The analysis of biosorption under fixed-bed operation allowed evaluating the influence of liquid velocity on the mass transfer dispersion coefficients. For this purpose, the numerical solution of a one-dimensional heterogeneous model was fitted to the evolutions of the outlet metal concentration and pH recorded in the performed experiments. It was shown that the main factor limiting column performance in the selected liquid velocity range is represented by the ratio between the characteristic times of interphase mass transfer and convection. Accordingly, performances could be enhanced by decreasing the liquid velocity below the range selected in the present study.

References

Ahluwalia SS, Goyal D (2007) Microbial and plant derived biomass for removal of heavy metals from wastewater. Bioresour Technol 98:2243–2257

Cojocaru C, Diaconu M, Cretescu I, Savi J, Vasic V (2009) Biosorption of copper (II) ions from aqua solutions using dried yeast biomass. Colloid Surface A 335:181–188

Davis TA, Volesky B, Mucci A (2003) A review of the biochemistry of heavy metal biosorption by brown algae. Water Res 37:4311–4330

Di Caprio F, Altimari P, Uccelletti D, Pagnanelli F (2014) Mechanistic modelling of copper biosorption by wild type and engineered *Saccharomyces cerevisiae* biomasses. Chem Eng J 244:561–568

Di Caprio F, Altimari P, Zanni E, Uccelletti D, Toro L, Pagnanelli F (2016) Lanthanum biosorption by different *Saccharomyces cerevisiae* strains. Chem Eng Trans 49:415–420

Ghorbani F, Younesi H, Ghasempouri SM, Zinatizadeh AA, Aminia M, Daneshi A (2008) Application of response surface methodology for optimization of cadmium biosorption in an aqueous solution by *Saccharomyces cerevisiae*. Chem Eng J 145:267–275

Goldberg S, Louise JC, Turner DR, Davis JA, Cantrell KJ (2007) Adsorption–desorption processes in subsurface reactive transport modeling. Vadose Zone J 6:407–435

Kordialik-Bogacka E (2011) Surface properties of yeast cells during heavy metal biosorption. Cent Eur J Chem 9:348–351

Lesage G, Bussey H (2006) Cell wall assembly in *Saccharomyces cerevisiae*. Microbiol Mol Biol R 70:317–343

Padmavathy V (2008) Biosorption of nickel (II) ions by baker's yeast: kinetic, thermodynamic and desorption studies. Bioresour Technol 99:3100–3109

Pagnanelli F (2011) Equilibrium, kinetic and dynamic modelling of biosorption processes. In: Kotrba P, Mackova M, Macek T (eds) Microbial biosorption of metals. Springer, Dordrecht, pp 59–120

Papageorgiou SK, Kouvelo EP, Katsaros FK (2008) Calcium alginate beads from *Laminaria digitata* for the removal of Cu^{+2} and Cd^{+2} from dilute aqueous metal solutions. Desalination 224:293–306

Schiewer S, Volesky B (1995) Modeling of the proton–metal ion exchange in biosorption. Environ Sci Technol 29:3049–3058

Schiewer S, Wong MH (2000) Ionic strength effects in biosorption of metals by marine algae. Chemosphere 41:271–282

Veglio F, Beolchini F (1997) Removal of metals by biosorption: a review. Hydrometallurgy 44:301–316

Volesky B (2007) Biosorption and me. Water Res 41:4017–4029

Wang J (2002) Biosorption of copper (II) by chemically modified biomass of *Saccharomyces cerevisiae*. Process Biochem 37:847–850

Wang J, Chen C (2006) Biosorption of heavy metals by *Saccharomyces cerevisiae*: a review. Biotechnol Adv 24:427–451

Wang J, Chen C (2009) Biosorbents for heavy metals removal and their future. Biotechnol Adv 27:195–226

Chapter 8
Transition Metal-Substituted Magnetite as an Innovative Adsorbent and Heterogeneous Catalyst for Wastewater Treatment

Shima Rahim Pouran, Mohammad Saleh Shafeeyan, Abdul Aziz Abdul Raman, Wan Mohd Ashri Wan Daud, and Abolfazl Bayrami

Abstract Iron oxides are conventionally used as adsorbent and/or heterogeneous catalyst because of their abundance, easy magnetically separation, affordability, and applicability in broad pH range. This is especially reported for magnetite due to the presence of Fe^{2+} cations in its structure. However, the pure magnetite has lower adsorption capacity and degradation rate in Fenton reaction, which led to the introduction of transition metal-substituted magnetite (TMSM). This section gives an overview on the adsorption potential and Fenton catalysis performance of various transition metal-substituted magnetite samples. This recently introduced group is produced with incorporation of appropriately identified transition metal/metals into the naturally available magnetite with simple synthesis method. TMSM has showed a great capacity for treating polluted water bodies using physical and chemical processes. A combination of factors affects the activity: the increased

S. Rahim Pouran (✉)
Department of Chemical Engineering, Faculty of Engineering, University of Malaya, 50603, Kuala Lumpur, Malaysia

Research Laboratory of Advanced Water and Wastewater Treatment Processes, Department of Applied Chemistry, Faculty of Chemistry, University of Tabriz, 51666-16471, Tabriz, Iran
e-mail: rahimpooran@yahoo.com; sh.rahimpooran@tabrizu.ac.ir

M.S. Shafeeyan
School of Chemical Engineering, College of Engineering, University of Tehran, 11155/4563, Tehran, Iran
e-mail: ms.shafeeyan@gmail.com

A.A. Abdul Raman • W.M.A. Wan Daud
Department of Chemical Engineering, Faculty of Engineering, University of Malaya, 50603, Kuala Lumpur, Malaysia
e-mail: azizraman@um.edu.my; ashri@um.edu.my

A. Bayrami
Department of Biology, Faculty of Basic Sciences, University of Mohaghegh Ardabili, Ardabil, Iran
e-mail: a_bayrami@uma.ac.ir

© Springer International Publishing AG 2017

A. Bonilla-Petriciolet et al. (eds.), *Adsorption Processes for Water Treatment and Purification*, DOI 10.1007/978-3-319-58136-1_8

adsorption capacity of the samples evidenced by larger surface area, the participation of thermodynamically favorable redox pairs in regeneration of Fe^{2+} and $^{\bullet}OH$ radical generation, and the presence of oxygen vacancies serving as active sites on the surface of TMSM. Nevertheless, there is a need for further understanding and expansion of this class of adsorbents and heterogeneous catalysts.

Keywords Heterogeneous catalyst • Magnetite adsorbents • Oxidation processes • Transition metal-substituted magnetite

Contents

8.1 Introduction

Water is a key element on earth for survival of living beings, which plays a crucial role for the appropriate functioning of the terrain and aquatic ecosystems. However, water resources are contaminating continuously due to the discharge of various pollutants such as heavy metal ions, anions, dyes, organics, and microbes into the environment (Herney-Ramirez et al. 2010). Several factors including the growth in the world population, civilization, industrialization, agricultural functioning, and other geological and universal changes have contributed to the water crisis and environmental pollution (Ali and Gupta 2007). Literature reveals an increasing rate in the generation of wastewaters with refractory properties from the many of industrial activities (Shukla et al. 2010; Rahim Pouran et al. 2015b). The strategies for augmenting freshwater resources had better involved not only the prevention and minimization of water pollution but treating polluted water bodies to the degree that can be reused in another sector. In light of this, developing advanced systems for efficient water treatment and recycling have attracted considerable attention worldwide, especially in countries with a growing scarcity of water resources (Munoz et al. 2015).

Over the last decades, different approaches have been proposed and employed for water treatment, including physical methods (screening, filtration and centrifugal separation, micro- and ultrafiltration, reverse osmosis, crystallization, sedimentation and gravity separation, flotation, and adsorption), chemical methods (precipitation, coagulation, oxidation, ion exchange, and solvent extraction), electrical approaches (electrodialysis and electrolysis), thermal technologies (evaporation and distillation), and biological processes (aerobic and anaerobic processes) (Ali and Jain 2005; Diya'uddeen et al. 2015a). Out of these, adsorption is considered as one of the practical options because of its ease of operation, low cost, and applicability for the separation of soluble and insoluble organic, inorganic, and biological contaminants (Ali 2012). Adsorption process is especially promising at nanoscale where the specific surface area of the adsorbent is relatively high. Iron oxide nanoparticles have especially attracted a wide interest due to their great magnetic characteristics that make the separation process much easier. Literature is replete with studies signifying the efficiency of iron nanomaterials as adsorbent for decontamination of heavy metal polluted aqueous solutions (Hua et al. 2012).

Nevertheless, in the most industries, the treatment methods are not able to produce effluents that comply with the effluent discharge standards (Shestakova et al. 2015). In several cases, the purification strategies basically relocate the contaminants from one phase to another (Shukla et al. 2010; Nitoi et al. 2013). Therefore, the use of such approaches is often limited due to the development of secondary wastes. For example, adsorption processes generate spent adsorbents that can be either hardly regenerated – by environmentally incompatible ex situ operating conditions – or it becomes a solid waste, commonly for industrial wastewater, that needs to be disposed (Delmas et al. 2009). The disposal of the wastes, after the water treatment process, has become a serious environmental issue that should be addressed (Diya'uddeen et al. 2015b; Shestakova et al. 2015). Consequently, advanced treatment methods are being standardized and several processes for the recovery of the spent adsorbents are being developed.

Recently, advanced oxidation processes (AOPs) have attracted a great deal of attention due to their potential for degrading numerous organic pollutants and complete mineralization of them to CO_2, H_2O, and environmentally harmless inorganic compounds, without production of secondary wastes (Comninellis et al. 2008; Wang and Xu 2011; Nichela et al. 2013). Fenton chemistry has been extensively described in recently published reviews (Pignatello et al. 2006; Malato et al. 2009). The main Fenton equations are given as Eqs. (8.1) and (8.2):

$$Fe^{2+} + H_2O_2 \rightarrow Fe^{3+} + OH^{-1} + OH^{\bullet} \tag{8.1}$$

$$Fe^{3+} + H_2O_2 \rightarrow Fe^{2+} + HO_2^{\bullet} + H^{+} \tag{8.2}$$

This process presents some advantages over the conventional approaches including simple equipment, efficient removal within a short reaction time, and potential for complete oxidation and mineralization of contaminants to benign end products under appropriate operational conditions. The Fenton reaction initiated by heterogeneous Fe^{2+} or Fe^{3+} compounds or some other transition metals at low oxidation states such as Co^{2+} and Cu^{2+} is referred as Fenton-like reaction (Nichela et al. 2013). Fenton-like reaction (Eq. 8.2) has a lower rate compared to Fenton reaction (Eq. 8.1) (0.01–0.002 vs. 42–79 L/mol S) due to the unbound transfer of the reactants in the homogeneous reaction site. The relative abundance and low cost of iron minerals as well as their simple magnetic separation render them as suitable candidates as adsorbents and for heterogeneous Fenton treatment of recalcitrant wastewaters. Accordingly, several researchers have focused on improving the efficiency of iron oxides and enhancing the breakdown rate of contaminant molecules through structural modifications.

One of the recently studied alterations is to substitute the structural iron species of iron minerals with other active transition metals. The effectiveness of transition metal-substituted magnetite (TMSM) as an innovative adsorbent and heterogeneous catalyst for water treatment is presented in the following sections.

8.2 Transition Metal-Substituted Magnetite

Magnetite is the most dominant iron mineral that has been employed for TMSIO synthesis. Iron in the magnetite structure can be substituted isomorphically by other transition metals, wherein the integrated transition metal/metals should have similar ionic radius to Fe^{2+}/Fe^{3+} cations and the same or with one or two unit differences in the oxidation states to the exchanged iron species. For instance, magnetite octahedral Fe^{3+} is replaced by Cr^{3+} in $Fe_{3-x}Cr_xO_4$ with the similar ionic radii (64.5 vs. 61.5 pm) (Magalhães et al. 2007) and Fe^{3+} is replaced by Nb^{5+} with the same ionic radius (64 pm) (Oliveira et al. 2008; Rahim Pouran et al. 2015a). Concerning the replacements with differing charges, the same amount of Fe^{3+} is reduced to Fe^{2+} based on the electrovalence equilibrium (Pearce et al. 2015). On the other hand, the structural dislocations could be adjusted by prompting oxygen vacancies for the substitutions in the absence of reduction (Moura et al. 2006). These oxygen vacancies are believed that perform as active sites for generation of hydroxyl radicals in Fenton process.

The most widely used preparation approach is the coprecipitation of highly pure ferrous and ferric salts (Fe^{2+}/Fe^{3+} in the molar ratio of 1:2) plus a predetermined amount of the selected transition metal salt under an inert gas environment and a few drops of hydrazine to prevent the oxidation of ferrous cations (Fig. 8.1) (Yang et al. 2009a; Liang et al. 2012b). This process can be continued by thermal treatment at 400–430 °C (Costa et al. 2003, 2006; Lelis et al. 2004).

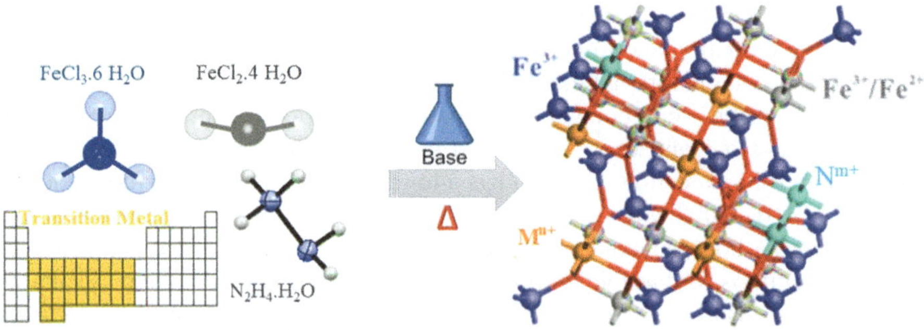

Fig. 8.1 Coprecipitation of Fe^{2+}, Fe^{3+}, and M^{n+} and/or N^{m+} as a TMSM

Yang et al. (2009a) represented the following set of reactions (Eqs. (8.4), (8.5), (8.6), and (8.7)) involved in synthesis of $Fe_{3-x}Ti_xO_4$ that were considered by Sugimoto and Matijević (1980):

$$(3-x)Fe^{2+} + xTi^{4+} + (6+2x)OH^- \rightarrow Fe_{3-x}Ti_x(OH)_{6+2x} \quad (8.3)$$

$$Fe_{3-x}Ti_x(OH)_{6+2x} + (1-x)NO_3^- \rightarrow Fe_{3-x}Ti_xO_4 + (1-x)NO_2^- + (3+x)H_2O \quad (8.4)$$

$$Fe_{3-x}Ti_x(OH)_{6+2x} + (2-2x)NO_2^- \rightarrow Fe_{3-x}Ti_xO_4 + (2-2x)NO + (2+2x)H2O + (2-2x)OH^{-1} \quad (8.5)$$

$$5Fe_{3-x}Ti_x(OH)_{6+2x} + (2-2x)NO \rightarrow 5Fe_{3-x}Ti_xO_4 + (2-2x)NH_3 + (8x+12)H_2O \quad (8.6)$$

TMSIOs of other iron oxides are often prepared under air atmosphere (dos Santos et al. 2001; Alvarez et al. 2006; Guimaraes et al. 2009) because they only contain Fe^{III} species. Meanwhile, the preparation procedure, type and quantity of the loaded transition metal, and the temperature range influence the properties of the developed TMSIO. The preparation of different catalysts through the impregnation of magnetite with transition metal/metals has been extensively reported in the literature. Most of the studies have explored the incorporation of the period 4 transition metals such as Ti (Yang et al. 2009a, b; Liang et al. 2012a, b; Zhong et al. 2012), V (Liang et al. 2010, 2012b), Cr (Magalhães et al. 2007), Mn (Oliveira et al. 2000; Costa et al. 2003, 2006; Coker et al. 2008), Co (Costa et al. 2003, 2006; Lelis et al. 2004; Coker et al. 2008), Ni (Costa et al. 2003, 2006; Coker et al. 2008), Cu (Lee and Joe 2010), Zn (Coker et al. 2008), and other metals like Al (Jentzsch et al. 2007) into the magnetite structure. The schematic presentation of the preparation set up is shown in Fig. 8.2.

The investigation on the recent studies indicates that this group of chemicals can be proposed as a novel promising adsorbent and heterogeneous Fenton catalyst in the degradation of organic pollutants.

8.3 Physicochemical Changes in Modified Magnetite

The incorporated transition metal may give rise to significant changes in magnetite physicochemical properties (Magalhães et al. 2007; Zhong et al. 2012). The main structural changes in magnetite structure through the incorporation of various transition metals are given in Table 8.1. The degree of advancement in physicochemical properties is mainly dependent on the synthesis method, type and percentage of the host metal/metals, and nature of the occupied site (Oliveira et al. 2000; Ramankutty and Sugunan 2002; Costa et al. 2003; Magalhães et al. 2007; Lee et al. 2008; Zhong et al. 2012; Liang et al. 2013). Nonetheless, the spinel structure of magnetite is often kept unchanged after the incorporation.

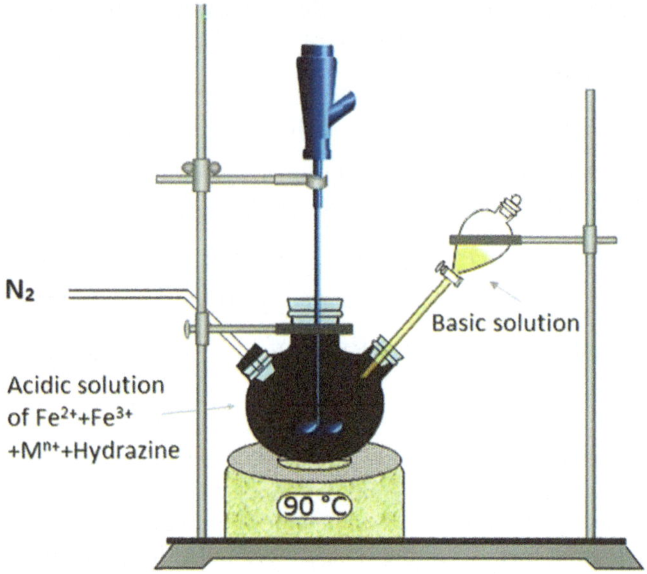

Fig. 8.2 The schematic presentation of the TMSM preparation set up (Rahim Pouran et al. 2015c)

Several literature on the characteristics of TMSM samples using Brunauer-Emmett-Teller (BET) surface area analysis reported a major growth in the surface area, primarily caused by a decrease in the particle size and/or pore diameter (Silva et al. 2009; de Souza et al. 2010; Liang et al. 2012b; Zhong et al. 2012). For example, in the $Fe_{2.93}Cr_{0.07}O_4$ sample, the pore diameter decreased from meso- to micro-size via the substitution of Fe^{3+} by Cr^{3+} in which the surface area was significantly increased (Magalhães et al. 2007). On the other hand, there was indistinct variation in the surface area and porosity of magnetite after the incorporation of Al, as reported by Jentzsch et al. (2007). It is worth mentioning that the magnetic property of magnetite should be preserved after the modification, as it is required for facile recovery of the sample from the treated water (Liang et al. 2012b). This characteristic can be affected by the cationic arrangement in the tetrahedral and octahedral sites, production condition, and the size of magnetite (Lelis et al. 2004). For instance, a decrease in the particle size to a few nanometers can intensify the magnetic order on the surface of the magnetite particles (Haneda and Morrish 1988).

8.4 Adsorption

Surface characteristics of a hetero-catalyst define its activity in a solution. The electrostatic interaction between the probe molecule and the catalyst surface is a major controlling parameter, so that the probe molecule removal from the target

Table 8.1 Preparation method and characteristics of transition metal-substituted iron oxide catalysts (Rahim Pouran et al. 2014)

Heterogeneous catalyst	Characteristics	Preparation method	References
$Fe_{3-x}Cr_xO_4$ (x = 0.00, 0.07, 0.26, 0.42, and 0.51) Fe_3O_4 $Fe_{2.93}Cr_{0.07}O_4$ $Fe_{2.74}Cr_{0.26}O_4$ $Fe_{2.58}Cr_{0.42}O_4$ $Fe_{2.49}Cr_{0.51}O_4$	Spinel crystalline, replacing Cr^{3+} by Fe^{3+}_{oct} and for higher Cr contents by Fe^{2+}_{oct} and Fe^{3+}_{tet}, substitution by Cr^{3+} decreased the pore diameter from meso- to micropore with a significant increase on the BET surface area	Conventional coprecipitation method	Magalhães et al. (2007)
$Fe_{3-x}Ti_xO_4$ (x = 0.00, 0.20, 0.46, 0.71, and 0.98) Fe_3O_4 $Fe_{2.80}Ti_{0.20}O_4$ $Fe_{2.54}Ti_{0.46}O_4$ $Fe_{2.29}Ti_{0.71}O_4$ $Fe_{2.02}Ti_{0.98}O_4$	Well-crystallized spinel structure, contains Ti^{4+} in octahedral sites, hydrophilic surface, decrease in particle size (≈82 nm) and pore diameter plus significant increase in surface area along with the increase in Ti content	Precipitation-oxidation method	Zhong et al. (2012)
$Fe_{3-x}Mn_xO_4$ (x = 0.21, 0.26, and 0.53) Fe_3O_4 $Fe_{2.79}Mn_{0.21}O_4$ $Fe_{2.74}Mn_{0.26}O_4$ $Fe_{2.47}Mn_{0.53}O_4$	Existence of the spinel phase, Mn^{2+} replacing mainly Fe^{2+} in the octahedral site, i.e., $[Fe^{3+}]_{tetrahedral}[Fe^{3+}Fe^{2+}_{1-x}M^{2+}_x]_{octahedral}O_4$, phase transformation of magnetite to maghemite and hematite due to the presence of Mn	Coprecipitation of the precursor ferric hydroxyl-acetate containing the metal Mn	Costa et al. (2003, 2006), Oliveira et al. (2000)
$Fe_{3-x}Co_xO_4$ (x = 0; 0.19; 0.38 and 0.75) Fe_3O_4 $Fe_{2.81}Co_{0.19}O_4$ $Fe_{2.62}Co_{0.38}O_4$ $Fe_{2.25}Co_{0.75}O_4$	Existence of the spinel phase, Co^{2+} replacing mainly Fe^{2+} in the octahedral site, i.e., $[Fe^{3+}]_{tetrahedral}[Fe^{3+}Fe^{2+}_{1-x}M^{2+}_x]_{octahedral}O_4$, the increase in hyperfine magnetic field for the octahedral iron with the increase in structural Co	Coprecipitation of the precursor ferric hydroxyl-acetate containing the metal Co	Costa et al. (2003, 2006), Lelis et al. (2004)

(continued)

Table 8.1 (continued)

Heterogeneous catalyst	Characteristics	Preparation method	References
$Fe_{3-x}Ni_xO_4$ (x = 0; 0.10; 0.28 and 0.54) $Fe_3\ O_4$ $Fe_{2.90}Ni_{0.10}O_4$ $Fe_{2.72}Ni_{0.28}O_4$ $Fe_{2.46}Ni_{0.54}O_4$	Existence of the spinel phase, Ni^{2+} replacing mainly Fe $^{2+}$ in the octahedral site, i.e., $[Fe^{3+}]_{tetrahedral}$ $[Fe^{3+}Fe^{2+}_{1-x}M^{2+}_x]_{octahedral}O_4$	Coprecipitation of the precursor ferric hydroxyl-acetate *containing* the metal Ni	Costa et al. (2003, 2006)
$Fe_{3-x-x'}Ti_xV_{x'}O_4$ (x = 0.00, 0.40, 0.42, 0.54, and 0.69) (x′ = 0.00, 0.03, 0.08, 0.13, and 0.32) $Fe_3\ O_4$ $Fe_{2.31}Ti_{0.69}O_4$ $Fe_{2.43}Ti_{0.54}V_{0.03}O_4$ $Fe_{2.50}Ti_{0.42}V_{0.08}O_4$ $Fe_{2.47}Ti_{0.40}V_{0.13}O_4$ $Fe_{2.68}V_{0.32}O_4$	Well-crystallized spinel structure, occupancy of mainly octahedral sites by Ti^{4+} and V^{3+}, no apparent effect of Ti-V on the magnetite structure, size: less than 100 nm, magnetic, higher adsorption activity of Ti-V magnetite catalysts than pure magnetite (more dependent on Ti than V), increase in specific surface area compared to Fe_3O_4	Coprecipitation method	Liang et al. (2012b)
$Fe_{3-x}Ti_xO_4$ (x = 0.00, 0.17, 0.23, 0.37, 0.50, 0.78) Fe_3O_4 $Fe_{2.83}Ti_{0.17}O_4$ $Fe_{2.77}Ti_{0.23}O_4$ $Fe_{2.63}Ti_{0.37}O_4$ $Fe_{2.50}Ti_{0.50}O_4$ $Fe_{2.22}Ti_{0.78}O_4$	Spinel structure, bigger lattice parameter than magnetite, average diameters of 120 nm, Ti^{4+} replacing mainly Fe^{3+} in the octahedral site. Simultaneous increase in oxidation and transition temperature by increase in Ti content, increase in surface area from 6.65 $m^2\ g^{-1}$ of pure magnetite to 20.7 $m^2\ g^{-1}$ in titanomagnetite	New soft chemical method	Yang et al. (2009a, b)

$Fe_{3-x}V_xO_4$ (x = 0.00, 0.16, 0.26, 0.34) $Fe^{2+}_{0.61}Fe^{3+}_{2.39}O_4$ $Fe^{2+}_{0.45}Fe^{3+}_{2.39}V_{0.16}O_4$ $Fe^{2+}_{0.70}Fe^{3+}_{2.04}V_{0.26}O_4$ $Fe^{2+}_{0.73}Fe^{3+}_{1.93}V_{0.34}O_4$	V^{3+} mainly occupies the octahedral site (chiefly replaced Fe^{3+}), the increase in vanadium content causes a decrease in the total Fe and Fe^{3+} content, no apparent change in the average crystal size and surface area, increase in the superficial hydroxyl groups, and a decrease in the temperature of maghemite-hematite phase transformation	Precipitation-oxidation method	Liang et al. (2010)
$Fe_{3-x}Al_xO_4$ (x = 0.00, 1.48, 2.14, 5.49, 8.07 mol %) Fe_3O_4 $Fe_{2.83}Al_{0.17}O_4$ $Fe_{2.77}Al_{0.23}O_4$ $Fe_{2.63}Al_{0.37}O_4$ $Fe_{2.50}Al_{0.50}O_4$	Al^{3+} replacing mainly Fe^{3+}, the increase in aluminum content causes a decrease in the particle size, no obvious change in porosity of the particles and their surface area, samples containing greater amounts of aluminum were overall less reactive than undoped samples	Coprecipitation method	Jentzsch et al. (2007)
$Cu_xFe_{3-x}O_4$ (x = 0.00, 0.2, 0.3) Fe_3O_4 $Cu_{0.2}Fe_{2.8}O_4$ $Cu_{0.3}Fe_{2.7}O_4$	A single-phase cubic spinel structure of $Cu_{0.2}Fe_{2.8}O_4$ in magnetite and $Cu_{0.3}Fe_{2.7}O_4$, consists of a cubic spinel structure in magnetite and a hexagonal structure in hematite	Mechanical alloying	Lee and Joe (2010)

solution is largely determined by its adsorption on the catalyst surface (Yang et al. 2009b). Several factors such as contact time, pH, chemical properties, and initial concentration of contaminant affect the adsorption capacity of the catalyst (Hanna et al. 2008; Yang et al. 2009b; Ai et al. 2011a; Yuan et al. 2011; Liang et al. 2012a). Among surface properties, basicity is an important factor that arises from the hydroxyl groups on the surface of the catalyst. The ligand shell accomplishment of the surface Fe atoms leads to the formation of Fe-OH groups on surface of the catalysts in which the surface adsorption is largely controlled by these groups (Sun et al. 1998). Accordingly, pH plays a dominant functional role in the catalytic action of the iron oxides. The pH of point of zero charge (PZC) is a key parameter that is defined as the pH in which the charge of the surface of the iron oxide is zero or the total number of the $FeOH^{2+}$ and FeO^- groups on the catalyst surface is the same. Conventionally, the determination of the pH_{pzc} is crucial for identifying the solution pH influence on the catalyst surface charge and consequent interaction with probe molecule.

In magnetite, protonation and deprotonation are the main reactions that occur on the surface, which are given by Eqs. (8.7), (8.8), and (8.9):

$$\equiv \mathrm{Fe(II-III)} \sim \mathrm{OH} + \mathrm{H}^+ \leftrightarrow \mathrm{Fe(II-III)} \sim \mathrm{OH}_2^+ \quad \mathrm{pH} < \mathrm{pH}_{\mathrm{pzc}} \tag{8.7}$$

$$\equiv \mathrm{Fe(II-III)} \sim \mathrm{OH} \leftrightarrow \mathrm{Fe(II-III)} \sim \mathrm{O}^- + \mathrm{H}^+ \quad \mathrm{pH} > \mathrm{pH}_{\mathrm{PZC}} \tag{8.8}$$

$$\equiv \mathrm{Fe} \sim \mathrm{OH} + \mathrm{OH}^- \rightarrow \mathrm{Fe} \sim \mathrm{O}^- + \mathrm{H_2O} \quad \mathrm{pH} > \mathrm{pH}_{\mathrm{pzc}} \tag{8.9}$$

At higher pH values than pH_{pzc}, the magnetite surface is negatively charged, and at lower pH values, it is positive (Petrova et al. 2011). The pH_{pzc} of magnetite at room temperature changes between 6.0 and 6.8 in an aqueous medium wherein the surface charge is of about neutral at this range (Sun et al. 1998; Cornell and Schwertmann 2003). Accordingly, the surface of the magnetite samples is negatively charged at pH higher than pH_{pzc}. Hence, it is favored for the adsorption of cationic probe molecules such as methylene blue (MB), based on the electrostatic interaction, and vice versa. For example, Liang et al. (2012a) observed that MB removal through Fenton reaction catalyzed by Cr-substituted magnetite was significantly influenced by its adsorption on the sample surface at neutral pH value, whereas the samples indicated no adsorption to acid orange II (anionic dye) and the degradation of the investigated dyes demonstrated different removal mechanisms. Table 8.2 gives a number of examples on the modified magnetite adsorbents for eliminating various contaminants from the aqueous medium. The data shows that the adsorption is highly affected by the pH of the solution.

In the heterogeneous catalysis, the iron catalyst and the organic pollutant are stirred together for a period of time to achieve the adsorption equilibrium (Hanna et al. 2008). The maximum adsorption is normally attained in the first hour and it continues at a decreased rate to reach the equilibrium state. It can be ascribed to the progressive filling of the most active adsorption sites on the catalyst surface. Then, the adsorption rate decreases as a result of the decreased vacant sites and subsequent repulsion force between the catalyst surface and adsorbed molecules.

Table 8.2 Iron oxide-based adsorbents for contaminant removal through adsorption

Compound/initial amount (mg/L)	Adsorbent	Operational condition					Optimal performance	References
		Contact time (min)	pH_{pzc}	Solution pH	T (°C)	Shaking speed (rpm)		
Methylene blue (MB)/20	(M-MWCNTs)[1] 0.02 g/50 ml	120	6.5	2–10	25 ± 1	150	The adsorption capacity of MB increases with increasing solution pH from 2.0 to 7.0 and changes slightly when solution pH is above 7.0. Adsorption kinetics follows the pseudo-second-order model. Maximum monolayer adsorption capacity: 48.06 mg g^{-1}	Ai et al. (2011a)
Methyl orange (MO)/4.8	Mg-Al LDH[2]	50		5.0	25	> 420	The adsorption capacity of the Mg-Al LDH toward MO was 0.453 mol/kg. The adsorption kinetics and equilibrium adsorption data were well-described by the pseudo second order model and fitted well to both the Langmuir and Freundlich models, respectively	Ai et al. (2011b)
Cd^{2+}/100	P (MB-IA)-g-MNCC[3] 2 g/L	240	6.1	8.0	30	200	The maximum adsorption capacity of P (MB-IA)-g-MNCC was found to be 262.27 mg/g. Kinetic and isotherm data were described using pseudo-second-order kinetic model and Sips isotherm model, respectively	Anirudhan and Shainy (2015)
P/40	Magnetic iron oxide (MIO) 1 g/ 50 ml	120	5.8	> 5.0	35	NA	The maximum adsorption capacity of MIO was 15.2 mg P/g MIO. The adsorbed phosphorous was effectively detached from MIO within 30 min using 20 wt% NaOH solution. The adsorption kinetic of the phosphate adsorption on MIO fitted well on the pseudo-second-order and Elovich models	Choi et al. (2016)

(continued)

Table 8.2 (continued)

		Operational condition					Optimal performance	References
Compound/initial amount (mg/L)	Adsorbent	Contact time (min)	pH_{pzc}	Solution pH	T (°C)	Shaking speed (rpm)		
Cd^{2+}/44.6	MT-MN 0.53 g	240	NA	3.5	20	120	The maximum adsorption capacity of the MN-MT for Cd^{2+} was 52.05 mg g^{-1}. Kinetic studies revealed that the adsorption process followed the pseudo-second-order model	Guyo et al. (2015)
Cu^{2+}, Pb^{2+}, Cd^{2+}, Cr^{6+}, Ni^{2+}, 20 each	NMag-CS 2 g/L	120	NA	5.5	25	200	96% Cu^{2+} and 94.6 Cr^{6+}, Pb^{2+}, Cd^{2+}, and Ni^{2+} removal percentage on film surface. The adsorption data were well fitted by both the Langmuir and Freundlich isotherms and the pseudo-second-order kinetics	Lasheen et al. (2016)
Pb^{2+}	Mn-magnetite 1 g/L	60	7.1	4.5	25	600	The adsorption capacity of magnetite samples toward Pb (II) gradually increased with the increase in Mn content. The adsorption capacity q_m for the samples with C_{Mn} 20 Wt % was 36.27 mg g^{-1}	Liang et al. (2014)
Cr^{6+} and Pb^{2+}/50 each	Fe_3O_4 NP	Up to 24 h	7.4	2 and 5	45	NA	The Sips and Langmuir models best described Cr^{6+} and Pb^{2+} adsorption on magnetite nanoparticles, respectively. The maximum Langmuir adsorption capacities were 34.87 (Cr^{6+}) and 53.11 (Pb^{2+}) mg/g. The pseudo-second-order model fits the adsorption kinetics	Rajput et al. (2016)

Cr^{6+}/50	MMT-Mag NP 0.5 g/100 ml	120	8.3	2–2.5	25 ± 2	160	Kinetics of the adsorption followed the pseudo-second-order model. Adsorption data fit well with the Langmuir and Freundlich isotherm equations. Adsorption capacity per unit mass of magnetite: 15.3 mg/g	Yuan et al. (2009)

M-MWCNTs are magnetite-loaded multi-walled carbon nanotubes. Mg-Al LDH is layered double hydroxide. P (MB-IA)-g-MNCC is 2-mercaptobenzamide modified itaconic acid-grafted-magnetite nanocellulose composite. MT-MN is maize tassel-magnetite nanohybrid. NMag-CS is nano-magnetite chitosan film. MMT-Mag NP is montmorillonite-supported magnetite nanoparticles

In a study conducted by Liang et al. (2012b), the substitution of Ti^{4+} and V^{3+} improved the adsorption activity of magnetite such that all the $Fe_{3-x-x'}Ti_xV_{x'}O_4$ samples had greater saturated adsorbed content than Fe_3O_4 with much higher dependence on the amount of Ti^{4+} than V^{3+}. Similarly, $Fe_{3-x-y}Nb_xMo_yO_4$ samples showed a significantly higher adsorption capacity of 80% more than the pure magnetite in which the effects of Nb incorporation were prominent (Rahim Pouran et al. 2015c). This clearly indicates that the incorporation of transition metals positively affected the magnetite adsorption capacity, primarily resulting from the enlarged specific surface area and, accordingly, the amount of magnetite surface hydroxyl (Liang et al. 2014).

On the other hand, the adsorption kinetics provides valuable understanding of the reaction pathways and the adsorption mechanism and describes the solute uptake rate. A number of models can be employed to express the mechanism of solute adsorption onto a sorbent. To explore the adsorption mechanism, a pseudo-first-order equation of Lagergren (1898) based on solid capacity, a first-order equation of Bhattacharya et al. (1984) based on solution concentration, and a pseudo-second-order equation based on solid phase adsorption rate are used to determine the characteristic constants of adsorption. Details of both models are provided in Chap. 3.

The pseudo-first-order model proposes that the experimental data is only well fitted to an initial period of the first reaction step. However, the pseudo-second-order model provides the best correlation of the experimental data over a long period in the studied systems (Ho and McKay 1999). Consequently, in the most adsorption studies using modified magnetite samples, the adsorption kinetics were well described by pseudo-second-order model in kinetics (Table 8.2). For instance, in a study on the MB adsorption on co-substituted Nb-Mo-magnetite samples, the pseudo-second-order model presented the best fit to the kinetic data at 25, 50, 100, and 200 mg L^{-1} MB concentrations (Rahim Pouran et al. 2015c). However, it should be borne in mind that the kinetic models are not adequate to describe the adsorption process. Indeed, adsorption is a complex multistep process, and the kinetic studies provide valuable insights of the adsorption mechanisms which involve mass transfer, diffusion, and surface reaction phenomenon. In addition to the kinetic studies, it is recommended to investigate the adsorption data using various isotherm models and thermodynamic evaluations. Lastly, the merits accompanied the adsorption process, such as easy operation, low cost, and huge sludge-handling processes could be completed with a more efficient method that helps for effective contaminant removal. Heterogeneous Fenton process is an excellent candidate for this purpose.

8.5 Oxidation Process

Transition metal-substituted magnetite (TMSM) has received growing interest for treatment of wastewaters using Fenton reaction, due to their higher adsorption capacity and reactivity in the degradation reaction compared to pure magnetite

(Rahim Pouran et al. 2014). The degradation process is started by adsorption of contaminant molecules on the catalyst surface before H_2O_2 addition and starting Fenton reaction.

From the reports, the enhancement in the catalytic activity of the modified magnetite samples has been resulted from the existence of the thermodynamically favorable redox pairs of the imported cations on the surface of the catalysts. These redox pairs enhance the Fenton degradation of probe molecule via (i) direct involvement in Fenton oxidation cycle and generation of $^{\bullet}OH$ radicals through Haber-Weiss mechanism, (ii) regeneration of Fe^{+2} cations, and (iii) acceleration of the electron transfer during the oxidation reaction in the magnetite structure (Costa et al. 2003).

Generation of oxygen vacancies from the adjustments for unequal charge replacements or cationic deficiency in the structure of modified iron oxide was proposed by Costa et al. (2006) as another possible reason for enhanced activities. These vacancies act as active sites in which they directly get involved in the degradation of probe molecules or indirectly in decomposition of H_2O_2 (Magalhães et al. 2007).

In photocatalysis process, the incorporated transition metals prevent the recombination of the photo-excited holes (h^+) and electrons (e^-) on the catalyst surface (Büchler et al. 1998) and extend the existence time of the charge carriers. For illustration, Fig. 8.3 shows the action of substituted Nb and Mo in magnetite samples for oxidation of MB (Rahim Pouran et al. 2015c). Other parameters including enlarged surface area and, accordingly, higher concentrations of OH groups on the surface of the catalysts are also reported in a number of studies (Liang et al. 2012a). However, the type and the quantity of probe molecule, Fenton reagent concentration, reaction duration and condition, and more importantly the elemental ratio of the imported transition metal play influential role in the

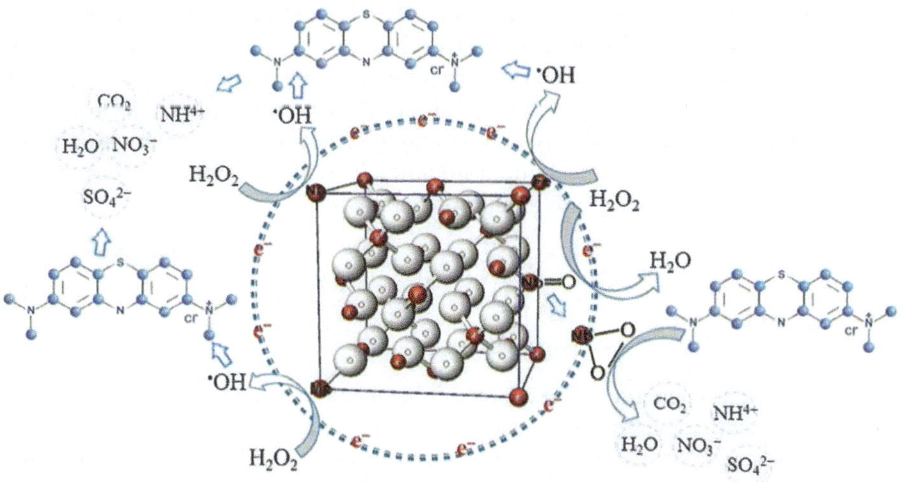

Fig. 8.3 Action of substituted Nb and Mo in magnetite samples for oxidation of MB through Fenton reaction (Rahim Pouran et al. 2015c)

degradation efficacy. For example, Costa et al. (2006) reported that although MB (50 ppm) removal was achieved within 10 min, the higher H_2O_2 concentrations (0.3 M) and Co (x = 0.75) and Mn (x = 0.53) loads were the major causes of the short reaction time. Liang et al. (2012a) observed that 59.3% of MB ($\approx$64 mg L^{-1}) was oxidized using the $Fe_{2.82}Cr_{0.18}O_4/H_2O_2$ (0.08 M) within 4 h, whereas $Fe_{2.33}Cr_{0.67}O_4/H_2O_2$ resulted in 95% color removal within the same reaction time. Furthermore, a long time of reaction (11 h) was utilized to degrade more than 90% of MB (70 mg g^{-1} of $Fe_{2.66}V_{0.34}O_4$ at pH 10 (Liang et al. 2013). On the contrary, the $Fe_{2.79}Nb_{0.171}Mo_{0.023}O_4$ catalyzed Fenton reaction could remove 100 mg/L of MB within 150 min (Rahim Pouran et al. 2015c), whereas the degradation was about 80% using $Fe_{2.73}Nb_{0.19}O_4$ sample (Rahim Pouran et al. 2015a).

The optimum portion of the integrated active cation to iron species drives higher activities and a concentration above this value may not improve the activity. For instance, Yuan et al. (2011) reported that the highest degradation percentage of dimethyl phthalate (DMP) by Si = FeOOH was detected at Si/Fe ratio of 0.2; however, this percentage decreased at lower and higher values than 0.2. It can be ascribed to the generation of suspended indigent catalyst at lower ratios and subsequent decrease in UV transmission into the solution. At higher values, the active sites are masked with high SiO_2 concentrations and lead to the formation of lower hydroxyl radical from H_2O_2 breakdown. Nevertheless, the increment in the content of the integrated Co and Mn leads to a remarkable increase in the catalyst activity where Fe_3O_4 demonstrated lower activity in comparison with the $Fe_{3-x}Co_xO_4$ and $Fe_{3-x}Mn_xO_4$ catalysts. In this study, the $Fe_{2.25}Co_{0.75}O_4$ and $Fe_{2.47}Mn_{0.53}O_4$ had the highest activities in the aforementioned reactions (Costa et al. 2003).

A combination of iron oxides and natural niobia (Nb_2O_5) led to the generation of a composite catalyst, of which maghemite (γFe_2O_3) and goethite (αFeOOH) were the chief constituents in its structure (Oliveira et al. 2007). The niobia load of the composite significantly influenced the discoloration rate, of which the niobia/iron oxide ratio of 1:5 only removed the half of the MB in solution, whereas in ratio of 1:1, the removal percentage was approximately 90%. Table 8.3 summarizes the data on the degradation of recalcitrant organic compounds using transition metal-substituted magnetite catalysts in Fenton reactions.

8.6 Conclusions

The research on magnetite as an adsorbent has been increasing due to its applicability in a wide range of pH, easy separation, and reusability. However, the adsorption capacity of magnetite can be improved via modification in its structure by enhancing its specific surface area and surface properties. One of the most promising methods that enhances its adsorption characteristic is the isomorphic substitution of the structural iron of magnetite with other transition metal/metals.

Table 8.3 Oxidation of various organic pollutants through Fenton reactions catalyzed by transition metal-substituted iron oxide (Rahim Pouran et al. 2014)

Compound	Catalyst	Operational condition				Optimal performance	References
		$[H_2O_2]$	pH	T (°C)	λ (nm)		
Methylene blue [MB], 50 mg/L	Cr-magnetite $Fe_{3-x}Cr_xO_4$ (15 mg)	0.3 mg/L	6.0	25	–	Higher degradation rate at lower Cr content, decrease in discoloration rate, and TOC removal by increase in Cr content mainly due to the decrease in Fe^{2+}	Magalhães et al. (2007)
Tetrabromobis-phenol A [TBBPA], 20 mg/L	Titanomagnetite $Fe_{2.02}Ti_{0.98}O_4$ 0.125 g L^{-1}	10 mmol/L	6.5	25	–	>97% TBBPA degradation in UV/$Fe_{2.02}Ti_{0.98}O_4$/H_2O_2 system, ≈75% in UV/H_2O_{2tr} system within 240 min of UV irradiation	Zhong et al. (2012)
Methylene blue [MB], 100 mg/L	$Fe_{2.46}Ni_{0.54}O_4$ $Fe_{2.47}Mn_{0.53}O_4$ $Fe_{2.25}Co_{0.75}O_4$ (30 mg)	2.5 mol/L	–	25	633	10% color removal within 50 min using $Fe_{2.46}Ni_{0.54}O_4$, complete discoloration of the solution in 5 and 10 min using $Fe_{2.47}Mn_{0.53}O_4$ and $Fe_{2.25}Co_{0.75}O_4$, respectively	Costa et al. (2003)
Chlorobenzene [CBZ], 30 mg/L	Mn-magnetite $Fe_{3-x}Mn_xO_4$ (30 mg)	2.5 mol/L	–	25	–	14, 7, 5, and <1% chlorobenzene degradation for the reactions using $Fe_{2.47}Mn_{0.53}O_4$, $Fe_{2.74}Mn_{0.26}O_4$, $Fe_{2.79}Mn_{0.21}O_4$, and Fe_3O_4, respectively	Oliveira et al. (2000), Costa et al. (2003)
Methylene blue [MB], 100 mg/L	γ-Fe_2O_3 α- Fe_2O_3 $Fe_{3-x}M_xO_4$ (M = Co and Mn) (30 mg)	0.3 mol/L (10 mL)	5–6.5	25	–	Not remarkable discoloration with Fe_2O_3 oxides, complete color removal, and higher oxidation by $Fe_{3-x}M_xO_4$ within 5–10 min	Costa et al. (2006)
Chlorobenzene [CBZ], 20 mg/L	Mn-magnetite $Fe_{3-x}Mn_xO_4$ (30 mg)	0.3 mol/L	–	25	–	1, 5, 7, and 14% CBZ removal using Fe_3O_4, $Fe_{2.79}Mn_{0.21}O_4$, $Fe_{2.74}Mn_{0.26}O_4$, and $Fe_{2.47}Mn_{0.53}O_4$, respectively, within 30 min	Costa et al. (2006)

(continued)

Table 8.3 (continued)

	Catalyst	Operational condition				Optimal performance	
Compound		$[H_2O_2]$	pH	T (°C)	λ (nm)		References
Methylene blue UV [MB], 0.2 mmol/L (500 mL)	Ti-V-magnetite $Fe_{3-x-x'}Ti_xV_{x'}O_4$ (1.0 g/L)	10 mmol/L	7.0	25	365	Increase in MB discoloration from 48% to 96% by increase in Ti content from x = 0.0 to x = 0.69 after 120 min	Liang et al. (2012b)
Methylene blue [MB], 100 mg/L (400 mL)	Titanomagnetite $Fe_{3-x}Ti_xO_4$ (1.0 g/L)	0.30 mol/L	6.8	30	–	Higher activity for Ti-magnetite than pure magnetite, decrease in residual MB with the increase in Ti content	Yang et al. (2009a, b)
Methylene blue [MB], 100 mg L^{-1} (150 mL)	Nb-magnetite Mo-magnetite Nb-Mo-magnetite (1.0 g/L)	0.2 mol/L	7.0	25	–	Higher activity for modified magnetite than pure magnetite; co-substituted Nb-Mo-magnetite had the highest activity compared to single metal-substituted samples. Decrease in residual MB with the increase in Nb content compared to Mo	Rahim Pouran et al. (2015a, c)
Methylene blue [MB], 0.2 mmol/L (200 mL)	V-magnetite $Fe_{3-x}V_xO_4$ $0 \leq x \leq 0.34$ (1.0 g/L)	100 mmol/L	10.0	25	–	Color removal of 41, 60, 81, and 93% of MB within 11 h using Fe_3O_4, $Fe_{2.84}V_{0.16}O_4$, $Fe_{2.74}V_{0.26}O_4$, and $Fe_{2.66}V_{0.34}O_4$, respectively	Liang et al. (2010)
Methylene blue [MB], 50 mg/L (10 mL)	Nb-hematite $Fe_{2-x}Nb_xO_3$ (10 g/L)	0.3 mol/L	6.0	25	–	Low discoloration with pure hematite and Hm-Nb_2, 70% color removal, and 25% TOC removal after 60 min with Hm-Nb10	Silva et al. (2009)
Methylene blue [MB], 50 mg/L (10 mL)	Nb-hematite-maghemite treated by H_2O_2 (1 g/L)	8.0 mM 30% v/v	6.0	25	–	73% MB removal using treated catalyst compared to 30% MB removal by non-treated samples within 60 min	Silva et al. (2011)

Bromophenol blue, Chicago sky blue, Evans blue and Naphthol blue black [dye] 50 mg/	FeO-Fe_2O_3 (25 mg/mL)	100 mmol/L	6.6	30	–	90% color removal within 24 h, the fast decomposition rate at first hour	Baldrian et al. (2006)
Naphthol blue black [NBB] 500 mg/L, COD_0 1.80	MO-Fe_2O_3 (M: Mn, Co, Cu, Fe) 25 mg/mL	100 mmol/L	6.0–6.6	30	–	COD removal of 97%, 92%, 88%, and 75% and color removal of 85%, 67%, 53%, and 58% using the catalysts of Cu, Co, Fe, and Mn, respectively	Baldrian et al. (2006)
Dimethyl phthalate [DMP] 7.7 mg/L (100 mL)	Si-FeOOH 0.5 g L^{-1}	2 mmol/L	5.0	25	365	97% DMP degradation within 30 min	Yuan et al. (2011)
Methylene blue [MB] 50 mg/L	Nb-FeOOH (11% Nb) 1 g/L	0.3 mol/L	6.0	25	–	15% discoloration after 120 min by pure goethite, ≈ 85% color removal using Nb11-FeOOH within 120 min	Oliveira et al. (2008)
Quinoline [Q] 10 mg/L	Ni-FeOOH 1 g/L	0.1 mL 5% v/v	6.0	25	–	28% Q removal after 5 h by pure goethite, 70% Q removal within 5 h	de Souza et al. (2010)

The optimum transition metal content generally decreases the crystal size significantly, with concomitant increased specific surface area, leading to the higher capacities for the adsorption in the samples. Despite the good adsorption efficiencies of the modified magnetite, it incapacitates in contaminant degradation. Consequently, hydrogen peroxide was introduced to the system that in turn hydroxyl radicals were generated through catalytic action of iron/imported transition metals in the magnetite. These generated hydroxyl radicals are highly energetic to attack the pollutant molecules and oxidize them to water and carbon dioxide.

Finally, for further discovery and understanding this class of catalysts, exploring the best combinations for higher degradation efficiencies and investigation of the effects of various factors such as wastewater composition on the stability, lixiviation, and aging of the catalytic sites for longer and efficient use in Fenton treatment of recalcitrant wastewaters are recommended.

References

Ai L, Zhang C, Liao F, Wang Y, Li M, Meng L, Jiang J (2011a) Removal of methylene blue from aqueous solution with magnetite loaded multi-wall carbon nanotube: kinetic, isotherm and mechanism analysis. J Hazard Mater 198:282–290

Ai L, Zhang C, Meng L (2011b) Adsorption of methyl orange from aqueous solution on hydrothermal synthesized Mg–Al layered double hydroxide. J Chem Eng Data 56:4217–4225

Ali I (2012) New generation adsorbents for water treatment. Chem Rev 112:5073–5091

Ali I, Jain CK (2005) Wastewater treatment and recycling technologies. Water encyclopedia. Wiley, New York

Ali I, Gupta VK (2007) Advances in water treatment by adsorption technology. Nat Protoc 1:2661–2667

Alvarez M, Rueda EH, Sileo EE (2006) Structural characterization and chemical reactivity of synthetic Mn-goethites and hematites. Chem Geol 23:288–299

Anirudhan TS, Shainy F (2015) Adsorption behaviour of 2-mercaptobenzamide modified itaconic acid-grafted-magnetite nanocellulose composite for cadmium(II) from aqueous solutions. J Ind Eng Chem 32:157–166

Baldrian P, Merhautová V, Gabriel J, Nerud F, Stopka P, Hrubý M, Beneš MJ (2006) Decolorization of synthetic dyes by hydrogen peroxide with heterogeneous catalysis by mixed iron oxides. Appl Catal B-Environ 66:258–264

Bhattacharya A, Arun K, Venkobachar C (1984) Removal of cadmium (II) by low cost adsorbents. J Environ Eng 110:110–122

Büchler M, Schmuki P, Böhni H, Stenberg T, Mäntilä T (1998) Comparison of the semiconductive properties of sputter-deposited iron oxides with the passive film on iron. J Electrochem Soc 145:378–385

Choi J, Chung J, Lee W, Kim JO (2016) Phosphorous adsorption on synthesized magnetite in wastewater. J Ind Eng Chem 34:198–203

Coker VS, Pearce CI, Pattrick RAD, van der Laan G, Telling ND, Charnock JM, Arenholz E, Lloyd JR (2008) Probing the site occupancies of Co-, Ni-, and Mn-substituted biogenic magnetite using XAS and XMCD. Am Mineral 93:1119–1132

Comninellis C, Kapalka A, Malato S, Parsons SA, Poulios I, Mantzavinos D (2008) Perspective, advanced oxidation processes for water treatment: advances and trends for R&D. J Chem Technol Biot 83:769–776

Cornell RM, Schwertmann U (2003) The iron oxides: structure, properties, reactions, occurrences and uses. Wiley, Weinheim

Costa RCC, Lelis MFF, Oliveira LCA, Fabris JD, Ardisson JD, Rios RRVA, Silva CN, Lago RM (2003) Remarkable effect of Co and Mn on the activity of $Fe_{3-x}M_xO_4$ promoted oxidation of organic contaminants in aqueous medium with H_2O_2. Catal Commun 4:525–529

Costa RC, Lelis MFF, Oliveira LCA, Fabris JD, Ardisson JD, Rios RRVA, Silva CN, Lago RM (2006) Novel active heterogeneous Fenton system based on $Fe_{3-x}M_xO_4$ (Fe, Co, Mn, Ni): the role of M^{2+} species on the reactivity towards H_2O_2 reactions. J Hazard Mater 129:171–178

de Souza WF, Guimarães IR, Oliveira LCA, Giroto AS, Guerreiro MC, Silva CLT (2010) Effect of Ni incorporation into goethite in the catalytic activity for the oxidation of nitrogen compounds in petroleum. Appl Catal A-Gen 381:36–41

Delmas H, Creanga C, Julcour-Lebigue C, Wilhelm AM (2009) AD–OX: a sequential oxidative process for water treatment – adsorption and batch CWAO regeneration of activated carbon. Chem Eng J 152:189–194

Diya'uddeen BH, Rahim Pouran S, Abdul Aziz AR, Nashwan SM, Wan Daud WMA, Shaaban MG (2015a) Hybrid of Fenton and sequencing batch reactor for petroleum refinery wastewater treatment. J Ind Eng Chem 25:186–191

Diya'uddeen BH, Rahim Pouran S, Abdul Aziz AR, Daud WMAW (2015b) Fenton oxidative treatment of petroleum refinery wastewater: process optimization and sludge characterization. RSC Adv 5:68159–68168

dos Santos CA, Horbe AMC, Barcellos CMO, Marimon da Cunha JB (2001) Some structure and magnetic effects of Ga incorporation on α-FeOOH. Solid State Commun 118:449–452

Guimaraes IR, Giroto A, Oliveira LC, Guerreiro MC, Lima DQ, Fabris JD (2009) Synthesis and thermal treatment of cu-doped goethite: oxidation of quinoline through heterogeneous fenton process. Appl Catal B-Environ 91:581–586

Guyo U, Makawa T, Moyo M, Nharingo T, Nyamunda BC, Mugadza T (2015) Application of response surface methodology for Cd (II) adsorption on maize tassel-magnetite nanohybrid adsorbent. J Environ Chem Eng 3:2472–2483

Haneda K, Morrish AH (1988) Noncollinear magnetic structure of CoFe2O4 small particles. J Appl Phys 63:4258–4260

Hanna K, Kone T, Medjahdi G (2008) Synthesis of the mixed oxides of iron and quartz and their catalytic activities for the Fenton-like oxidation. Catal Commun 9:955–959

Herney-Ramirez J, Vicente MA, Madeira LM (2010) Heterogeneous photo-Fenton oxidation with pillared clay-based catalysts for wastewater treatment: a review. Appl Catal B-Environ 98:10–26

Ho YS, McKay G (1999) Pseudo-second order model for sorption processes. Process Biochem 34:451–465

Hua M, Zhang S, Pan B, Zhang W, Lv L, Zhang Q (2012) Heavy metal removal from water/ wastewater by nanosized metal oxides: a review. J Hazard Mater 211:317–331

Jentzsch TL, Lan Chun C, Gabor RS, Lee PR (2007) Influence of aluminum substitution on the reactivity of magnetite nanoparticles. J Phys Chem C 111:10247–10253

Lagergren S (1898) About the theory of so-called adsorption of soluble substances. K Sven Vetenskapsakad Handl 24:1–39

Lasheen MR, El-Sherif IY, Tawfik ME, El-Wakeel ST, El-Shahat MF (2016) Preparation and adsorption properties of nano magnetite chitosan films for heavy metal ions from aqueous solution. Mater Res Bull 80:344–350

Lee CS, Joe YH (2010) Structural and magnetic properties of cu-substituted magnetite studied by using Mössbauer spectroscopy. J Korean Phys Soc 56:85–88

Lee HJ, Kim G, Kim DH, Kang JS, Zhang CL, Cheong SW, Shim JH, Lee S, Lee H, Kim JY, Kim BH, Min BI (2008) Valence states and occupation sites in (Fe, $Mn)_3O_4$ spinel oxides investigated by soft x-ray absorption spectroscopy and magnetic circular dichroism. J Phys Condens Matter 20:295–303

Lelis MFF, Porto AO, Gonçalves CM, Fabris JD (2004) Cation occupancy sites in synthetic Co-doped magnetites as determined with X-ray absorption (XAS) and Mössbauer spectroscopies. J Magn Magn Mater 278:263–269

Liang X, Zhu S, Zhong Y, Zhu J, Yuan P, He H, Zhang J (2010) The remarkable effect of vanadium doping on the adsorption and catalytic activity of magnetite in the decolorization of methylene blue. Appl Catal B-Environ 97:151–159

Liang X, Zhong Y, He H, Yuan P, Zhu J, Zhu S, Jiang Z (2012a) The application of chromium substituted magnetite as heterogeneous Fenton catalyst for the degradation of aqueous cationic and anionic dyes. Chem Eng J 191:177–184

Liang X, Zhong Y, Zhu S, Ma L, Yuan P, Zhu J, He H, Jiang Z (2012b) The contribution of vanadium and titanium on improving methylene blue decolorization through heterogeneous UV-Fenton reaction catalyzed by their co-doped magnetite. J Hazard Mater 199:247–254

Liang X, He Z, Zhong Y, Tan W, He H, Yuan P, Zhu J, Zhang J (2013) The effect of transition metal substitution on the catalytic activity of magnetite in heterogeneous Fenton reaction: in interfacial view. Colloid Surface A 435:28–35

Liang X, He Z, Wei G, Liu P, Zhong Y, Tan W, Du P, Zhu J, He H, Zhang J (2014) The distinct effects of Mn substitution on the reactivity of magnetite in heterogeneous Fenton reaction and Pb(II) adsorption. J Colloid Interface Sci 426:181–189

Magalhães F, Pereira MC, Botrel SEC, Fabris JD, Macedo WA, Mendonca R, Lago RM, Oliveira LCA (2007) Cr-containing magnetites $Fe_{3-x}Cr_xO_4$: the role of Cr^{3+} and Fe^{2+} on the stability and reactivity towards H_2O_2 reactions. Appl Catal A-Gen 332:115–123

Malato S, Fernández-Ibáñez P, Maldonado MI, Blanco J, Gernjak W (2009) Decontamination and disinfection of water by solar photocatalysis: recent overview and trends. Catal Today 147:1–59

Moura FCC, Oliveira GC, Araujo MH, Ardisson JD, Macedo WAA, Lago RM (2006) Highly reactive species formed by interface reaction between Fe0–iron oxides particles: an efficient electron transfer system for environmental applications. Appl Catal A-Gen 307:195–204

Munoz M, de Pedro ZM, Casas JA, Rodriguez JJ (2015) Preparation of magnetite-based catalysts and their application in heterogeneous Fenton oxidation – a review. Appl Catal B-Environ 176:249–265

Nichela DA, Berkovic AM, Costante MR, Juliarena MP, García FSE (2013) Nitrobenzene degradation in Fenton-like systems using Cu(II) as catalyst. Comparison between Cu(II) – and Fe(III)-based systems. Chem Eng J 228:1148–1157

Nitoi I, Oncescu T, Oancea P (2013) Mechanism and kinetic study for the degradation of lindane by photo-Fenton process. J Ind Eng Chem 19:305–309

Oliveira LCA, Lago RM, Rios RVRA, Augusti R, Sousa PP, Mussel WN, Fabris JD (2000) The effect of Mn substitution on the catalytic properties of ferrites. Stud Surf Sci Catal 130:2165–2170

Oliveira LCA, Gonçalves M, Guerreiro MC, Ramalho TC, Fabris JD, Pereira MC, Sapag K (2007) A new catalyst material based on niobia/iron oxide composite on the oxidation of organic contaminants in water via heterogeneous Fenton mechanisms. Appl Catal A-Gen 316:117–124

Oliveira LCA, Ramalho TC, Souza EF, Gonçalves M, Oliveira DQM, Pereira MC, Fabris JD (2008) Catalytic properties of goethite prepared in the presence of Nb on oxidation reactions in water: computational and experimental studies. Appl Catal B-Environ 83:169–176

Pearce CI, Henderson CMB, Telling ND, Pattrick RAD, Charnock JM, Coker VS, Arenholz E, Tuna F, Van der Laan G (2015) Fe site occupancy in magnetite-ulvöspinel solid solutions: a new approach using X-ray magnetic circular dichroism. Am Mineral 95:425–439

Petrova TM, Fachikov L, Hristov J (2011) The magnetite as adsorbent for some hazardous species from aqueous solutions: a review. Int Rev Chem Eng 3:134–152

Pignatello JJ, Oliveros E, MacKay A (2006) Advanced oxidation processes for organic contaminant destruction based on the Fenton reaction and related chemistry. Crit Rev Environ Sci Technol 36:1–84

Rahim Pouran S, Abdul Raman AA, Wan Daud WMA (2014) Review on the application of modified iron oxides as heterogeneous catalysts in Fenton reactions. J Clean Prod 64:24–35

Rahim Pouran S, Abdul Aziz AR, Wan Daud WMA, Embong Z (2015a) Niobium substituted magnetite as a strong heterogeneous Fenton catalyst for wastewater treatment. Appl Surf Sci 351:175–187

Rahim Pouran S, Abdul Aziz AR, Wan Daud WMA (2015b) Review on the main advances in photo-Fenton oxidation system for recalcitrant wastewaters. J Ind Eng Chem 21:53–69

Rahim Pouran S, Abdul Aziz AR, Wan Daud WMA, Shafeeyan MS (2015c) Effects of niobium and molybdenum impregnation on adsorption capacity and Fenton catalytic activity of magnetite. RSC Adv 5:87535–87549

Rajput S, Pittman CU Jr, Mohan D (2016) Magnetic magnetite (Fe_3O_4) nanoparticle synthesis and applications for lead (Pb^{2+}) and chromium (Cr^{6+}) removal from water. J Colloid Interface Sci 468:334–346

Ramankutty CG, Sugunan S (2002) Surface properties and catalytic activity of ferrospinels of nickel, cobalt and copper, prepared by soft chemical methods. Appl Catal A-Gen 218:39–51

Shestakova M, Vinatoru M, Mason TJ, Sillanpää M (2015) Sonoelectrocatalytic decomposition of methylene blue using Ti/Ta_2O_5–SnO_2 electrodes. Ultrason Sonochem 23:135–141

Shukla P, Wang S, Sun H, Ang HM, Tadé M (2010) Adsorption and heterogeneous advanced oxidation of phenolic contaminants using Fe loaded mesoporous SBA-15 and H_2O_2. Chem Eng J 164:255–260

Silva AC, Oliveira DQL, Oliveira LCA, Anastácio AS, Ramalho TC, Lopes JH, Carvalho HWP, Rodrigues Torres CE (2009) Nb-containing hematites $Fe_{2-x}Nb_xO_3$: the role of Nb^{5+} on the reactivity in presence of the H_2O_2 or ultraviolet light. Appl Catal A-Gen 357:79–84

Silva AC, Cepera RM, Pereira MC, Lima DQ, Fabris JD, Oliveira LCA (2011) Heterogeneous catalyst based on peroxo-niobium complexes immobilized over iron oxide for organic oxidation in water. Appl Catal B-Environ 107:237–244

Sugimoto T, Matijević E (1980) Formation of uniform spherical magnetite particles by crystallization from ferrous hydroxide gels. J Colloid Interface Sci 74:227–243

Sun ZX, Su FW, Forsling W, Samskog PO (1998) Surface characteristics of magnetite in aqueous suspension. J Colloid Interface Sci 197:151–159

Wang JL, Xu LJ (2011) Advanced oxidation processes for wastewater treatment: formation of hydroxyl radical and application. Crit Rev Environ Sci Technol 42:251–325

Yang S, He H, Wu D (2009a) Decolorization of methylene blue by heterogeneous Fenton reaction using $Fe_{3-x}Ti_xO_4$ ($0 \leq x \leq 0.78$) at neutral pH values. Appl Catal B-Environ 89:527–535

Yang S, He H, Wu D, Chen D, Ma Y, Li X, Zhu J, Yuan P (2009b) Degradation of methylene blue by heterogeneous Fenton reaction using titanomagnetite at neutral pH values: process and affecting factors. Ind Eng Chem Res 48:9915–9921

Yuan P, Fan M, Yang D, He H, Lui D, Yuan A, Zhu J, Chen T (2009) Montmorillonite-supported magnetite nanoparticles for the removal of hexavalent chromium [Cr(VI)] from aqueous solutions. J Hazard Mater 166:821–829

Yuan B, Li X, Li K, Chen W (2011) Degradation of dimethyl phthalate (DMP) in aqueous solution by UV/Si–FeOOH/H_2O_2. Colloid Surf A 379:157–162

Zhong Y, Liang X, Zhong Y, Zhu J, Zhu S, Yuan P, He H, Zhang J (2012) Heterogeneous UV/Fenton degradation of TBBPA catalyzed by titanomagnetite: catalyst characterization, performance and degradation products. Water Res 46:4633–4644

Index

© Springer International Publishing AG 2017

A. Bonilla-Petriciolet et al. (eds.), *Adsorption Processes for Water Treatment and Purification*, DOI 10.1007/978-3-319-58136-1

The manufacturer's authorised representative in the EU is Springer Nature Customer Service Centre GmbH, Europaplatz 3, 69115 Heidelberg, Germany. If you have any concerns regarding our products, please contact ProductSafety@springernature.com

Printed and bound by CPI Group (UK) Ltd, Croydon, CR0 4YY
30/09/2026
02225229-0007